Evolution and Speciation in Animals

T. J. Pandian

Valli Nivas, 9 Old Natham Road
Madurai-625014, TN, India

CRC Press
Taylor & Francis Group
Boca Raton   London   New York

CRC Press is an imprint of the
Taylor & Francis Group, an **informa** business

A SCIENCE PUBLISHERS BOOK

Cover page: Comparative anatomy of mammalian forelimbs, embryos of bird and mammal showing gill slits (free hand drawing from different sources). Human evolution (modified and added from *medium.com*), fossils of *Tyrannosaurus rex* (*pixabay.com*) and *Asterocerus* nautilaoid (*pxhere.com*). Amber containing spider and imprint of a fish (*freeimages.com*). All the sources are gratefully acknowledged.

First edition published 2021
by CRC Press
6000 Broken Sound Parkway NW, Suite 300, Boca Raton, FL 33487-2742

and by CRC Press
2 Park Square, Milton Park, Abingdon, Oxon, OX14 4RN

© 2021 Taylor & Francis Group, LLC

CRC Press is an imprint of Taylor & Francis Group, LLC

Library of Congress Cataloging-in-Publication Data
Names: Pandian, T. J., author.
Title: Evolution and speciation in animals / T.J. Pandian.
Description: First edition. | Boca Raton : CRC Press ; Taylor & Francis
    Group, 2021. | "CRC Press is an imprint of the Taylor & Francis Group,
    an Informa bsiness." | Includes bibliographical references and index.
Identifiers: LCCN 2021007107 | ISBN 9781032009193 (hardback)
Subjects: LCSH: Animal species. | Animal ecology. | Animals--Variation. |
    Species diversity. | Evolution (Biology)
Classification: LCC QH380 .P36 2021 | DDC 577--dc23
LC record available at https://lccn.loc.gov/2021007107

ISBN: 978-1-032-00919-3 (hbk)
ISBN: 978-1-032-00920-9 (pbk)
ISBN: 978-1-003-17638-1 (ebk)

Typeset in Palatino
by Radiant Productions

Preface

Having single handedly authored a dozen books in two series for posterity, I was longing for academic retirement at the seniority of my age. But I was literally bombarded with 'heaven sent ideas' that had to be tested for correctness. This book is an outcome of the 'tested ideas'.

For a long time, the idea of evolution existed among scientists and even with religions such as, Hinduism. With keywords "Variations, Struggle for existence and Survival of the fittest by Natural Selection", Charles Darwin established the theory of evolution and its by-product speciation. Subsequently, a large number of publications by microbiologists, botanists and zoologists have confirmed the correctness of Darwin's evolutionary theory. Presently, there are more concerns for species diversity than for evolution. The year 2010 marked the International Year of 'Species Diversity'. This book identifies some life history features and environmental factors that accelerate species diversity and others that decelerate it. Some of these features and factors are known but are not adequately recognized. This requires quantification of the identified factors and features.

In Shakespearean language, one may say, 'Oh, variation, thy name is evolution'. Hence, the idea of quantification of the identified factors and features may sound odd and not be possible at a time, when information on *per se* is not known for many species and when taxonomy itself is in a fluid but dynamic state. However, I was a little emboldened, as taxonomy itself represents quantification of species, genus and so on, despite variation(s) among individuals within a species. The onerous task of quantification required much of a computer search and a few compromises on the number of some taxa. Therefore, the quantifications may neither be exhaustive nor precise. But the proportions arrived and inferred generalizations shall remain valid. A separate chapter to highlight new findings is not included, as there are too many (shown in italics) of them. The Holy Bible states: "Let your light so shine that people may see your good work and praise the Lord". Being innovative and informative, I earnestly hope that this book stands up to the Biblical statement.

December 2020   T. J. Pandian
Madurai 625 014

Acknowledgements

This book is an outcome of my earlier book series, which was initiated by Prof. T. Balasubramanian and R. Gadagar, to whom I remain grateful. It is with pleasure, I wish to place on record my grateful appreciation to Drs. P. Murugesan, J. Muthukrishnan and E. Vivekanandan for patiently reviewing the manuscript of this book and offering valuable comments. Thanks are due to The American College, Madurai for lending me old classical zoology books. The manuscript was ably prepared by Mr. T.S. Surya, M.Sc. and I wish to thank him for his competence, patience and cooperation.

I wish to thank many authors/publishers, whose published figures are simplified/modified/compiled/redrawn for an easier understanding. To reproduce original figures from the published domain, I welcome and gratefully appreciate the open access policy of BioCell, Korean Journal of Parasitology and permission issued by Prof. O. Kah. For advancing our knowledge on this subject by the rich contributions, I thank my fellow scientists, whose publications are cited in this book.

December, 2020   **T. J. Pandian**
Madurai

Contents

Part B: Life History Traits

B1: Sexuality

B2: Gametogenesis and Fertilization

1

General Introduction

Introduction

Many religions believe that the earth with all its organisms was created in the year 4004 BC, it is at the center of the universe and all organisms living therein remain unchanged. In contrast, the Hindus created Gods, the Triumvirate, Brahma, the creator, Vishnu, the protector and Shiva, the destroyer. They strongly believe in recycling inclusive of rebirth and evolution (as depicted by the Ten Incarnations [*Avadhars*]). One principle of Isaac Newton (1642–1727) is that for every action, there will be an equal and opposite reaction. In fact, Hinduism is more a scientific religion. Hindus believe that if man commits *karma* (sin), he will be punished, but for his dharma (performing benevolent duty), he will be blessed. A Hindu Tamil poet Manickavasagar (9th century)* had gone to the extent of sequencing the major events in cosmic and biological evolution. The sequence commences with the formation of the universe – earth – atmosphere surrounding the earth – light – prokaryotes – eukaryotes, existing and extinct life forms – each claiming its own territory and so on. Unfortunately, neither the Hindus brought evidence for this hypothesis nor the western scientists had access to the Tamil poem. Earlier, however, science has been dominated by Cosmology, Mathematics, Physics and Chemistry, but Biology was not even considered as an independent discipline. In cosmic science, only a few discoveries have an implication to biology. For example, the Polish astronomer Copernicus (1473–1543) proposed that the earth is not at the center of solar system. Providing detailed evidences, the Italian Galileo (1564–1642) transformed the Copernicus's hypothesis into an acceptable theory. Accordingly, the earth revolves around itself in a day causing day in one part of earth and night in the opposite part of it; incidentally, this finding has a profound implication to

---

* "*Vaanagi, Mannagi, Valiyagi, Oliyagi, Oonagi, Uyiragi, Unmayumai inmayumai, yan, yenathendru ...*"

—*Manickavasagar, Thiruvasagam*, 9th century

diurnal and nocturnal rhythms in animals and plants. The earth also revolves around the sun in 365 days causing successive seasons. It has implication to organismic productivity and activity (as many organisms become dormant during winter), especially in temperate countries. It is only after the Swedish Naturalist Carl von Linne, who Latinized his name as Carolus Linnaeus (1707–1778), Biology, especially Taxonomy began to grow as an independent discipline of science. For example, the number of described species began to steeply increase only from the 19th century (Fig. 1.1).

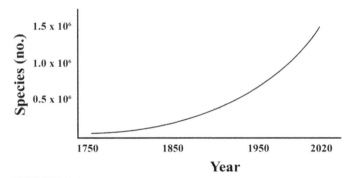

FIGURE 1.1

Number of described animal species as function of year (based on Mora et al., 2011).

1.1 Morphological and Molecular Taxonomy

Based on similarities of morphological and anatomical structures, animals are grouped into a sort of planned arrangement, which is named as Classification or Taxonomy or Systematics. The scheme of classification begins with the concept of species, which is defined differently. Perhaps the best of these definitions is that of Ernst Mayr, which defines that "a species is a group of actually or potentially interbreeding population that is reproductively isolated from all other groups". Some species like lions and tigers resemble more closely with each other than, say, dogs. Hence, lions and tigers are brought together under the genus *Felis* but dogs under another genus *Canis*. To distinguish lions from tigers, the former is named as *Felis leo* and the latter as *F. tigris* (see Hyman, 1940). Thanks to Linnaeus, this system is now established for all organisms as Binomial System of Nomenclature. Incidentally, a not well known historic anecdote should be brought to light. The Keralite Ayurvedic therapists of India always named medicinal herbs dually and the like. For example, they distinguished the Malayala (Kerala, receiving up to 600 cm rain over 150 d/y) tulasi *Oscimum sanctorum* as distinctly different from Madras (Tamil Nadu, receiving ~ 60 cm rain over

30 d/y) tulasi. Impressed by the dual naming system, the then Dutch Governor and a botanist Hendrick Van Rheede included the system in his monograph *Hortus malabaricus* (1678–1693). It was *Hortus malabaricus* that inspired Linnaeus to devise the Binomial System of Nomenclature. In fact, the word 'Nomen' (Name in English, Namen in German) originated from the Sanskrit word 'Namum' (see Pandian, 2002). The 2,000 year old Tamil Academy (Sangam) poems enumerate < 122 plant species and describe their flower color, fragrance and medicinal properties. In his poetic compendium named as *Malaipadukadam*, another poet lists > 100 flora of the montane zone. In ten thousand year old South Indian temples, there are sculptures depicting hundreds of animal species. Indeed, taxonomy was born in India.

Based on more and more studies recognizing different levels of structural similarities, the animal kingdom is divided into Phylum, Class, Order, Family, Genus and Species. The thus far recognized phyla are categorized into nine major phyla, Porifera, Cnidaria, Acnidaria, Platyhelminthes, Annelida, Arthropoda, Mollusca, Echinodermata and Vertebrata, and 26 minor phyla (see Pandian, 2021). The minor phyla comprise aberrant clades that are not usually considered in the mainstream of evolution. In the phylogenetic tree, most of them terminate as blind offshoots, albeit a few are regarded as a link between two or more major phyla. For example, the velvet worm Onychophora provides a vital link for arthropod evolution.

Enormous efforts have been made to construct phylogenetic cladograms and to track the history of animal evolution. Based on the structural and embryological features, zoologists have traced animal evolution from simple to complex animals through gradual steps and bridging links, although they have also been very conscious that parasitism reverses the course from complex to simple forms, as in parasitic rhizocephalic (e.g. *Sacculina carcini*) crustaceans (see Pandian, 2016). For classification, the conventional cladogram considers the absence of coelom, and presence of pseudocoelom, hemocoelom, lophophorate schizocoelom and (enteric) eucoelom (Fig. 1.2A). On the other hand, molecular biologists use ribosomal RNA genes and others (e.g. Littlewood et al., 2015) to construct a molecular cladogram (Fig. 1.2B). These two major classifications are more or less the same. Both of them classify the diploblastic radial and triploblastic bilateral symmetry in animals. They also recognize the distinct difference between Protostomia and Deuterostomia. In the former, the blastopore differentiates into a mouth but into an anus in the latter. The difference appears between these two cladograms, once the morphological cladogram classifies coelomates into Dueterostomia, Lophophorata and Protostomia, whereas the molecular cladogram classifies it into Lophotrochozoa and Ecdysozoa. The Ecdysozoa are characterized by the presence of a cuticle, which is periodically molted during the life cycle. Within it, Tardigrada, Onychophora and Arthropoda are grouped into Panarthropoda. The other Cycloneuralia includes Nematoida (with Nematoda, Nematomorpha) and Scalidophora (with Priapulida, Kinorhyncha, Loricifera) (see Blaxter and Koutsovoulos, 2015, Hejnol, 2015).

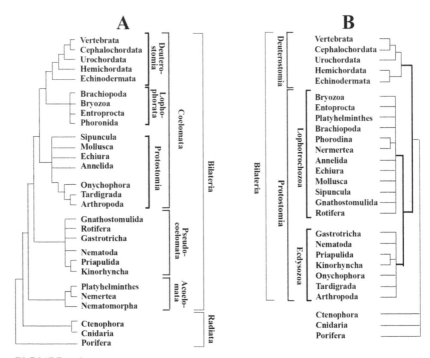

FIGURE 1.2

Metazoan phylogenies. A. The traditional phylogeny based on morphology and embryology. B. The new molecule (rRNA)-based phylogeny (compiled, modified and redrawn from Hyman, 1940, Adoutte et al., 2000).

1.2 Evolution and Species Diversity

Classification of plants and animals has increasingly convinced both morphological and molecular taxonomists that the organisms have been evolving from structurally simpler to more and more complex organization and evolution is an ongoing process. For example, with the same basic pattern, the skeletal organization is modified to suit for swimming (fish), walking (lizards), jumping (frogs), running (horse) and flying (birds). The same holds true for crustaceans and insects for swimming, leaping, walking and burrowing. On their part, embryologists came in support of evolution, and showed the presence of pharyngeal slits during the development of chicks and humans, recalling that they have evolved from a fish-like ancestor. These embryological events are succinctly stated as 'ontogeny recapitulates phylogeny'. In his book Principle of Geology, Charles Lyell (1797–1875) reported that the age of the earth is much older than the religionist's view that the earth was created in the year 4004 BC. Based on the so called 'carbon dating', the geological time scale indicates that the earth's age is about

4 billion years, the life has originated some ~ 3.2 billion years ago and invertebrates were abundant some –600 million years ago (see Table 25.1). A large number of species, that became extinct between 600 million years ago and the present, have left a part or the entire body as fossils and the like (e.g. amber). From these fossils, geologists have brought adequate and convincing evidence in support of evolution. Remarkably, a series of evidence comes from the sequence of fossils to show that horses have evolved from donkey-like ancestors. Thus, the hypothesis of evolution is now a tested and proven theory. Nevertheless, evidence for the origin of life and species are required to complete the compendium of evidences in support of the theory of evolution.

1.3 Life and Species: Their Origin

Other than water, living matter is composed of 71% proteins, 7% nucleotides and others like carbohydrates (5%), lipids (12%), and inorganic and organic molecules (5%). Proteins are made up of amino acids and the nucleotides with nitrogen containing base, sugar and phosphate (see Wallace, 1991). Hence, the formation of amino acids must have been the first step in origin of life. To trace the origin of life, many enthusiastic scientists have successfully carried out ingeniously designed experiments. Of them, only a few are mentioned. Miller (1953) subjected a mixture of methane, ammonia, water vapor and hydrogen to series electrical charges, approximately simulating conditions prevailed in the primitive earth. After a week, the inorganic molecules combined to form amino acids. Subsequently, Miller's experiment has been extended by many. These experiments have shown the production of substances like purines and pyrimidines and sugars, all of which are essentially important steps in origin of life. Fox (1965) found that complex molecules, the polypeptides are routinely formed under abiotic conditions. On dropping dilute solutions of simple molecules, believed to have existed in the primitive earth on hot sand, rock or clay, the solutions were vaporized and concentrated into complex ones, which Fox called proteinoids. Oparin (1938) reported that colloidal protein molecules, in which the particles are suspended in a gel-like state, tend to clump together into increasingly complex masses and form coacervative droplets to form 'protobionts', 'microspheres' and so on. To synthesize a specific protein, however, the Deoxyribo Nucleic Acid (DNA) and Ribo Nucleic Acid (RNA) are required. The double-stranded helical structure of the DNA was described by the Nobel Laureates Watson and Crick (1953). DNA is now known as the heredity material. How DNA replicates during cell division and its genetic codes go to synthesize a specific protein are all now taught to high school and college students. With formation of proteins, nucleic acids and others like sugars and lipids, life has originated as moneral organisms, which are considered to be the most primitive. All other organisms namely Protista, Plantae, Fungiae and Animaliae were evolved subsequently. Based on their characteristics (Table 1.1), they are classified into five kingdoms and the possible of course of evolution is depicted in Fig. 1.3.

With regard to origin of species, Charles Darwin (1809–1882) and Jean Baptiste de Lamarck (1744–1829) are the most towering figures. Unable to untie himself from the giraffe, Lamarck wrongly conceived the 'use and disuse hypothesis'. Based on the fact that the fossilized simpler organisms are found in the older layers of rock than those of complex ones in the recent rocks, he recognized that there was some 'force of life' that caused an organism to generate new structures and organs to meet its biological

TABLE 1.1

Characteristics of the five kingdoms (compiled and modified from Wallace, 1991)

Characteristics	Prokaryotes	Eukaryotes			
	Monera	Protista	Plantae	Fungiae	Animaliae
Nuclear membrane	Absent	Present	Present	Present	Present
Membrane bound organelles	Absent	Present	Present	Present	Present
Photosynthetic ability	Some do	Some have	Present	Absent	Absent
Chromosomes	Nucleic acids only	Include nucleic acids and proteins			
Motility	Some do it	Some do it	Non motile		Yes

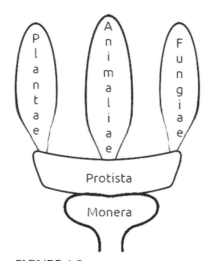

FIGURE 1.3

Origin and classification of the five kingdoms of organisms.

need; once formed, the organ continued to develop through use and its development was inherited. Unlike Lamarck, Darwin had two unique opportunities that led him to propose his hypothesis. 1. As a naturalist, he undertook a voyage on the HMS Beagle round the world (1831–1833). His stay and study on the Galapagos Islands is the most innovative event in conceiving variations within closely related species (e.g. Darwin's finches) as an important cause for evolution and origin of species. The Galapagos are located on the western side of the tropical America. The islands may have received the parental species from America. Having been isolated, some of these species underwent evolution and speciation much faster than those on the main land and picturized the long history in a short-time story. 2. Thomas Malthus (1766–1834) published his Essay on the Principle of Population in 1829. According to Malthus, if humans continue to reproduce at a rate prevailing then, the geometric progression of the population would deplete food supply and the world will be full of misery and vice. Impressed by the prediction, Darwin calculated that even a pair of elephants, known as 'notoriously slow breeders' could produce 19 million progenies within 750 years. That it was not happening led Darwin to recognize that some 'unknown factor' regulates the reproductive output of elephants (see Wallace, 1991). He named the 'unknown factor' as 'Natural Selection'. Finally, he comprehensively synthesized his ideas in the following sequence: 1. Organisms tend to produce far more numbers of progenies than that can be supported by resources. 2. Some of these progenies display one or more variation(s) between themselves and their parents. 3. Encountering intense competition for resource (and a mate), a few progenies displaying suitable variation(s) are found as the 'fittest' by natural selection and others with unsuitable variation(s) are weeded out immediately or eventually; the unfit individuals, species and up to order but not phylum of organisms have become extinct. Thus, 'variations', 'struggle for existence' and 'Natural Selection of the fittest' became the keywords in Darwin's proposed hypothesis of Origin of Species in 1859. Subsequent experiments on microbes, plants and animals by several authors have all brought adequate evidences in support of the Darwinean hypothesis and the hypothesis became an acceptable theory of 'Origin of Species'. Yet, the Darwinean Theory did not explain how the observed variations from parents are inherited by progenies. In the same year 1859, an Austrian monk Gregor Johann Mendel (1822–1884), speaking and writing only the German language, presented his findings on the Principles of Inheritance to the Brunn Natural Science Society, Vienna. Unaware of Mendel's contribution, Darwin was struggling up to his death over the enigma of heredity, as to how the variations/traits are actually transmitted from parents to progenies.

1.4 Inheritance and Sex

Inheritance: Darwin had missed not only Gregor Mendel but also Aristotle. In his *'Generation of Animals'*, Aristotle astutely noted that all children inherit features (traits) from their mothers and fathers (see Mukherjee, 2016). Prior to describing Mendel's contribution, Muller (1928) should be mentioned, who reported that the X-ray induced mutation(s) (equivalent to Darwin's variations) in fruit fly is inheritable. It is now known that a number of mutagenic chemicals can also induce mutations. Returning to Mendel, his choice of pea plant and simple but ingenious design of experiment are models for biology students. Luckily for Mendel, all the seven traits of the pea plant selected by him carried only dominant or recessive traits. From his extensive 8-year long experiments, Mendel concluded that a trait is inherited according to the: (1) Principle of dominance and recessive: When parents differ from a trait, their offspring may be a hybrid ('bastard' as it was then called by German botanists) of that particular trait; however, the offspring bear the trait of one parent instead of showing a blend of the traits of both parents. Hence, each trait is a discrete factor (borne in a gene). (2) Principle of segregation: When a hybrid (for a trait) reproduces, the gametes will be of two types; half will carry one (dominant or recessive) trait, provided by one parent and the other half will carry the other (subsequently named as) allele. (3) The principle of independent assortment: When the parents differ in two or more traits, the occurrence of any one trait in the progenies will be independent of the occurrence of the other. Thus, two or more traits are inherited independently. Briefly, Mendel had almost discovered that the factor (gene) is the material responsible for transmission of a trait from parents to progeny. Of course, the principles of dominance and independent assortment, for example, sex-linked traits, were subsequently modified. But his discovery of the factor (gene) as responsible for transmission of a trait from parents to progeny remains unquestionable.

Sex-linked traits: As a vast majority of animals are gonochores, it becomes necessary to know the mechanism of sex determination. In this regard, the contribution by the Nobel-laureate Thomas Hunt Morgan should be described. Again Morgan's choice of the fruit fly led to his success. Morgan and Bridges (1916) obtained puzzling results but Morgan ingeniously surmised that the determinants (traits) responsible for sex and eye color of the fly are not independently assorted; they are linked and harbored on the same sex chromosome. According to Morgan, of the four pairs of chromosomes of the fruit fly, three pairs are autosomes and the remaining one pair consists of sex chromosomes. In males, the sex chromosomes are heterologous and consist of an X chromosome and J-shaped Y chromosome. But in the female, the sex chromosomes consist of homologous XX chromosomes. At spermatogenesis, the heterologous X and Y chromosomes line up together at meiosis and produce half of the sperms containing X chromosome and the other half Y chromosome. As females have only XX chromosomes, every egg contains one X chromosome. If a Y-carrying sperm fertilizes an egg, the offspring will be XY male. But if an X-carrying sperm fertilizes an egg, the progeny will be XX female. Considering more popular evidence, the sex-linked inheritance of the bleeding disease hemophilia among European Royal families is described. The gene responsible for causing hemophilia is recessive and carried by one of the X chromosomes in the female. Queen Victoria suffered a mutation for hemophilia. She carried another X chromosome, in which the gene responsible for hemophilia was dominant. Hence, the hemophilia gene was not expressed in her but she was a "carrier" of the recessive gene in her other X chromosome. However, her son Leopold with a Y chromosome letting the expression of the hemophilia gene harbored on his X chromosome suffered the bleeding disease. His sister Alexandra, like her mother Queen Victoria, was also a carrier and passed it to her daughter Alice, who was married to the Czar Nicholas II of Russia. Their son Alexei with XY chromosome suffered hemophilia. Hence, the hemophilia history in the European Royal families confirmed the Morgan's discovery of the linkage between sex and eye color of the fly and chromosomal mechanism of sex determination.

Incidentally, scientific developments have greatly changed history. The development in chemistry has led the production of hand and cartage-aided bombs, against which, for example, many small territory-holding chieftains have surrendered not to British Country but to a small British commercial company. The developments in nuclear physics led to the production of the atom bomb, which after dropping brought an end to the Second World War. Genetics is the only branch of biology that has changed history at least a couple of times. Unaware of the sex-linked inheritance mechanism of hemophilia, Alice, intending to save her hemophilia-afflicted son Alexei, had a wrong association with the monk Grigory Rasputin, whose selfish interference with the government led to the ruthless massacre of him and Russian revolution. Another untold history is that of Adolf Hitler. Convinced by eugenics, Hitler legally prohibited all physically handicapped Germans from giving birth to even a single child.

1.5 Phylum vs Species Diversity

Based on the presence of an organ, the presence and nature of coelom, the phyla are categorized into (i) Aorganomorpha (a new term coined to indicate the lack of organs in line with Acoelomorpha), (ii) Acoelomorpha, (iii) Pseudocoelomata, (iv) Hemocoelomata, (v) Lophophorate Schizocoelomata and (vi) Eucoelomata in major and minor phyla. The number of phyla within each of these categories and number of species/ phylum in each of these categories are shown in Table 1.2. Among them, the pseudocoelomates with six phyla, each with the highest mean number of 5,251 species/minor phylum, have generated more often aberrant clades (Fig. 1.4A). Surprisingly, they are all eutelics, i.e. after hatching, mitosis is ceased in their somatic cells, a feature that has hitherto not been adequately recognized. Notably, major phyla have no representative to Pseudocoelomata and Schizocoelomata (Fig. 1.4B). In them, the Hemocoelomata comprising Arthropoda and Mollusca, with the mean number of 680,246 species/major phylum are the most speciose. Interestingly, this number may be compared with just 300 species/minor phylum among hemocoelomatic minor phyla. The hemocoelomatic arthropods and mollusca are relatively more motile than their counterparts in minor phyla. *Hence, motility, especially among gonochores has been a driving force to foster and facilitate species diversity* (see also Pandian, 2016).

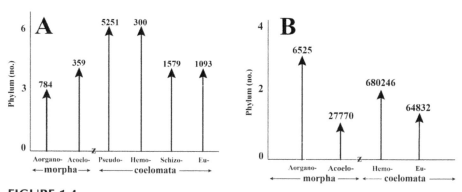

FIGURE 1.4

Number of phyla and species/phylum within major clades of A. minor and B. major phyla (Figure A in the left panel is from Pandian, 2021).

Whereas the number is limited to 9 in major phyla, it runs to as many as 26 in minor phyla (Table 1.2). The number of species ranges from just 2 in Cycliophora to 27,000 in Nematoda among minor phyla but from 166 in Acnidaria to 1,242,040 in Arthropoda among major phyla. At present, the cumulative species number is in the range of 46,687 and 1,496,509 for the minor and major phyla, respectively. Of 1,543,196 species included in the animal kingdom, 97% of them belong to the major phyla. Not surprisingly, the minor phyla are not as speciose (1,795 species/phylum) as major phyla (166,279 species/phylum). *Hence, the relatively more motile major phyletics have evolved more toward species diversity, while the relatively low motile or sessile minor phyletics have gone for more phylum diversity. Interestingly, sessility has limited the plant diversity to 374,000 species, i.e. 24% of all animal species. In animals,*

TABLE 1.2

Number of phyla and approximate species number in major and minor phyla (from Pandian, 2021 and other sources)

Major phylum	Species (no.)	Minor phylum	Species (no.)
Aorganomorpha			
Porifera	8,553	Placozoa	3
Cnidaria	10,856	Mesozoa	150
Acnidaria	166	Myxozoa	2,200
Acoelomorpha			
Platyhelminthes	27,700	Loricifera	34
Turbellaria	5,500	Cycliophora	2
Non-Turbellaria	16,310	Nemertea	1,300
		Gnathostomulida	100
Psuedocoelomata			
		Rotifera	2,031
		Gastrotricha	813
		Kinorhyncha	200
		Nematoda	27,000
		Nematomorpha	360
		Acanthocephala	1,100
Hemocoelomata			
Arthropoda	1,242,040	Priapulida	19
Crustacea	54,384	Sipuncula	160
Myriapoda	11,800	Echiura	230
Insecta	1,020,007	Tardigrada	1,047
Chelicera	155,620	Onychophora	200
Mollusca	118,451	Pentastomida	144
Lophophorata			
		Entoprocta	200
		Phoronida	23
		Brachiopoda	391
		Bryozoa	5,700
Eucoelomata			
Annelida	16,911	Chaetognatha	150
Echinodermata	7,000	Hemichordata	130
Vertebrata	64,832	Urochordata	3,000
Major phyla total	1,496,509	Minor phyla total	46,687
Grand total		1,543,196	
166,279 species/major phylum; 1,795 species/minor phylum			

motility has increased the diversity to > 4 times than that in sessile plants (see Pandian, 2022).

1.6 Biodiversity: Species Diversity

The most widely used definition of biodiversity comprises the concepts that include three levels of biodiversity: genome, species and habitat. Hence, biodiversity is defined as the collection of genomes, species and habitats occurring as interacting components. The importance of genes and genomes is being increasingly realized; creation of genetically engineered organisms may increase the world's stock of biodiversity. Recent developments in molecular biology make it possible to move genes across completely unrelated organisms—from mammals to yeasts and from bacteria to insects. Similarly, it is also possible to transplant the sperm (e.g. Kirankumar and Pandian, 2004) or Primordial Germ Cells (PGCs) of a fish species to another (see Pandian, 2011b, Chapter 8). These have added an entirely new dimension of the possibilities of allowing genetic resources to human use, so that organisms once thought as totally insignificant, may now be used to increase commercial gains (elaborated in Section 26.4). Understandably, collection, cataloging and conservation of genomes of agriculturally and pharmaceutically important biological resources are made by a number of organizations/institutions. For example, International Rice Research Institute, Manila has a collection of over 90,000 rice germplasm accessions from different parts of the world. However, the collection of such genomes and germplasms by a private organization and a biotechnologically rich but biodiversity-wise poor country may lead to 'Gene Imperialism', which should be avoided (see Pandian, 2002).

Nevertheless, genome diversity is a nascent subject. Hence, it is not considered in this book. Voluminous information is available on evolution and speciation. But, it is almost silent on the environmental factors and life history features that accelerate or decelerate species diversity in animals. According to Costello and Chaudhary (2017), higher temperatures, e.g. prevailing in tropics, cause faster metabolism and mutation rate, and shorter generation time (e.g. Crustacea, Pandian, 2016) and thereby accelerate the scope for speciation. Diversification of environmental and biological niches also accelerates it. On earth, 11% of all animal species are microscopic (< 1 cm), 86% macroscopic and the remaining 3% are large (> 10 cm). In the oceans alone, these proportions are 25, 74% and 1% (see Costello and Chaudhary, 2017). According to WoRMS (World Register of Marine Species), the inventory of 232, 625 species (see Table 2.1) may include 98% of all the described marine species; other estimates by Zhang (2011) and Integrated Taxonomy Information System (ITIS) may not agree with this number. Nevertheless, the 25% of microscopic marine animals are not readily amenable for identification at species level and recognition of their diversity. Not surprisingly, animal taxonomy is in a fluid but dynamic state. Due to the erection of new phyla and species, recognition of synonyms and reclassification, changes in species number are an ongoing process. Hence, any quantification of species number may only be an approximation, albeit generalizations based on such approximations can be valid.

Presently, there are more concerns for species diversity than for evolution. In recognition of its importance, countries like India have minted special coins. The year 2010 marked the International Year of 'Species Diversity'. Hence, the objective theme of this book is to identify life history traits and environmental factors that foster and promote species diversity in animals and those that deter and limit it. For a long time, some of these factors and traits were known but not adequately recognized. Relevant information on some of these traits or factors is not yet available, or when available, it is widely scattered and remains not quantifiable. Undertaking an onerous task, the book is a synthesis of collection, collation, sequencing and quantification of relevant information to discover many new findings and opens new avenues for research. It represents perhaps the first attempt in this direction and elaborates the role played by selected environmental and biological factors that govern and regulate the process of species diversity.

Part A
Environmental Factors

"Grow and expand; they are the signs of life"—said the Hindu monk Vivekananda. Space is required for existence and expansion, and input of matter and energy is obligatory to ensure growth. In Part A, environmental factors namely space and food are identified to accelerate or decelerate species diversity. Temperature and its annual oscillation range decreases from the tropics to the polar zone. The same holds true for altitudes in a montane terrain. Hence, temperature distribution profoundly alters the number of species and number of individuals in a species. On land, water availability is the second most-important factor that significantly affects abundance and distribution of flora and fauna. These dual factors namely temperature and availability of water (in liquid form) foster or deter species diversity. Increased packing of animal species in a terrestrial habitat has led to coevolution. Being heterotrophs, animals acquire food from the environment. Depending on the food type, animals invest different amounts of resources on feeding and digestive devices and duration. Acquisition of low molecular nutrients against osmotic gradient from the surrounding aquatic medium by free-living animals or the host gut by parasitic animals can be the costliest and may demand a relatively larger surface area. Microphagy, i.e. suspension feeding can be costlier, as it requires relatively more investments on the feeding device and a longer duration. Fluid feeders acquire rich food but the duration of suction can be longer than that of macrophagy. Not surprisingly, the time and energy costs of food acquisition foster or deter species diversity.

2

Spatial Distribution

Introduction

Life has existed in the sea longer than on land. Fossils reveal the existence of bacteria over 3.7 and 3.1 billion years ago (BYA) in the oceans and on land, respectively. On land, the greater variety of environmental niches provides a better scope for specialization, which leads to larger occupation of space and in turn, to more biological niches. Species richness in land seems to have overtaken that in the oceans about 125 million years ago (MYA). This coincided with diversification of flowering plants, which constitute 79% of all photosynthetic eukaryotic species (see Pandian, 2022). Flowering plants increased terrestrial productivity (52% of global productivity, Field et al., 1998), leading to herbivory three times more than that in the oceans (Costello and Chaudhary, 2017). As the speciation rate is the fastest in insects, they should be mentioned. In his book *Six Legged Science*, Hocking (1968) described several important roles played by insects in ecosystem functions like pollination, nutrient recycling, necrophagy and so forth. For example, the total valuation of pollination and recycling services rendered by insects in the USA alone is conservatively estimated as US$ 57 billion. A more recent estimate indicates that between US$ 235 billion and 577 billion worth annual global food production relies on the contribution by pollinators (Intergovernmental Science-Policy Platform on Biodiversity and Ecosystem Services, 2017). In admiration of insects, Wilson (1991) stated "If all mankinds were to disappear, the world would regenerate back to the rich state of equilibrium that existed ten thousand years ago. If insects were to vanish, the environment would collapse into chaos".

Returning to the ocean and land, it must, however, be noted that the oceans and land mass are not isolated entities. There are a lot of exchanges between them. On a large scale, for example, the Nile, prior to the construction of dams, exported 1,820 million m^3 of water along with alluvium into the Mediterranean Sea (see Pandian, 1980). On the other hand, ~ 80 million metric tons (mmt) of fish are imported into the land from the seas by capture fisheries (Pandian, 2015). With regard to species exchange, spawning migration from the sea to freshwater by salmon and that from freshwater to the sea by eels are classic examples (see Section 15.2). Estuaries provide another example for the extended distribution between the sea and rivers. In Parangipettai (Tamil Nadu, India), the Vellar Estuary extends 1,800 m in length from the west (fresh water zone) to the east at the river mouth. The number of marine species penetrating into the estuary decreases from 5,125/m^2 at the river mouth to 298/m^2 at the west end of the estuary (Murugesan et al., 2007). Ephemeropterans (3,000 species), plecopterans (2,000 species), odonates (5,500 species) and many dipteran insects oviposit in freshwaters and their terrestrial adults emerge from waters after completion of larval development. The dragonfly may serve as an example for mini scale exchange between the land and limnic habitats. A sum of 7.27 million eggs are imported into the Idumban Pond (Palani, Tamil Nadu, India) through oviposition by *Brachthemis contaminata*, but only 23,990 are exported into the land through emerging adults (Mathavan and Pandian, 1977).

2.1 Habitat Distribution

The oceans cover 70% of the earth's surface with 97% of its water. Freshwater systems, however, cover only ~ 1% and hold as small as 0.01% of its water (see Pandian, 2011). The remaining 29% of the earth is covered by land. With water mass amounting to 1.36 billion km^3 (*jbutler@uh.edu*), the oceans provide 900 times more livable volume of space than that on land. At greater depths, they also provide a more stabilized habitat with almost constant temperature (> 4°C), salinity (35‰) and oxygen (~ 3 ml/l) levels. The absence of light and photosynthetic activity at a greater depth eliminates the herbivory but lets the existence of filter and sediment (deposit)-feeding and carnivorous animals. With increasing depth, the number of species is reduced; the reduction is from 41,350 species at depths between 0 and 100 m depth to 11,592 and 8,459 species at depth ranges of 100–200 m and 200–500 m, respectively, i.e. the reductions are 72 and 80% at these depth ranges (see Costello and Chaudhary, 2017). Considering the bathymetric distribution pattern on the horizontal (continental shelf) slope of the Bahamas up to 500 m depth, Maldonado and Young (1996) found that the number of sponge species decreased from ~ 10 to 4 and the number of individuals from 35/species to 10/species. Plotting the number of individual per species against the species number, they showed a statistically significant correlation between them, i.e. the decreases were from 10 species and 35 individual (indi)/species at the surface to 5 species and 2 indi/species at 450 m depth. Contrastingly, the distribution was almost zero with increasing depth at an upright surface providing no livable substratum for the sponges. However, the crevices and rugosity of this upright surface may provide micro-substrata for the small, slow motile and sessile animals with adhesive pad/holdfast (see Pandian, 2011a).

Through computer searches and compromises, an attempt has been made to assess the approximate number of species distributed in marine, freshwater and terrestrial habitats and thereby to arrive at some generalizations. The availability of fragmentary data on horizontal distribution are limited to one or other sub-phyla (e.g. Turbellaria: Schockaert et al., 2008), and are also contradictory in some taxa. Hence, a few compromises had to be made to arrive at the approximate values listed in Tables 2.1 and 2.2. For example, 605 Tardigrada species are reported from limno-terrestrial habitats (Gross et al., 2015). For want of adequate details and as a compromise, 302 species were assigned each to freshwater and terrestrial habitats. Similarly, from values reported for freshwater and marine 489 myxozoan species by Vidal-Martinez et al. (2016), the proportion for entire Myxozoa was arrived at. Amin et al. (2019) reported 43 acanthocephalan species from marine mullets; from it, 4% for marine acanthocephalan species was estimated. The proportion of 67% of terrestrial nematodes was also assessed from the reports by Appeltans et al. (2011) and Abebe et al. (2008), who estimated the values for marine (6,900 species) and freshwater (7% of 27,000 nematode species), respectively.

Tables 2.1 and 2.2 list the approximate species number in 9 major and 26 minor phyla and their distribution in marine, freshwater and terrestrial habitats. From them, the following may be inferred: (1) Understandably, all of them have their marine routes. For example, the exclusively terrestrial Onychophora also resembles the fossil hemocoelomates—the putative stem group representatives of Panarthropoda (Onychophora + Tardigrada + Arthropoda) (Mayer et al., 2015). (2) Perhaps for the first time, comprehensive data on habitat distribution of all animal phyla and species have been assembled. For each taxon, the number may change but the proportions for the marine, freshwater and terrestrial inhabitants may remain valid. Accordingly, (2a) A sum of 1,543, 196 animal species are known to exist. *Of them, 97.0% and 3.0% belong to the nine major and 26 minor phyla, respectively.* (2b) Minor phyletic species are more 'aquatics' and are occupants of marine (41.3%) and freshwater (15.4%) habitats (Table 2.2), in comparison to the major phyla that are more 'terrestrial' (78.1%) inhabitants. Insects (77.1% of Arthropoda) among the major phyla and Nematoda (67.0%) among the minor phyla are taxa that have successfully ventured and colonized the *terra firma.* (2c) *On the whole, 15.1, 7.8 and 77.1% species are distributed in marine, freshwater and terrestrial habitats, respectively* (Table 2.1). *Despite the varied environmental and biological niches provided by the lentic and lotic habitats, the freshwaters have not provided the scope for species diversity for two reasons: (i) it covers < 1% of the earth's surface and holds as small as 0.01% of the earth's water. (ii) Most of its habitats are ephemeral or transient.* Many taxa require more perennial aquatic system for their existence. For example, ephemeropteran mayflies (Balachandran et al., 2012) and most dipteran chironomids (Sankarperumal and Pandian, 1991) can flourish only in more perennial aquatic system.

TABLE 2.1

Approximate species number in major phyla and their habitat distribution

Phylum	Species (no.)				Reference
	Total	Marine	Freshwater	Terrestrial	
Aorganomorpha					
1. Porifera	8553	8296	257	0	Van Soest et al. (2012)
2. Cnidaria	10856	10831	25	0	Mapstone (2014), Jankowski et al. (2008)
3. Acnidaria	166	166	0	0	*Anim Diver Web*
Acoelomorpha					
4. Platyhelminthes	27700	17422	7902	2376	
Turbellaria	5500	4096	1404	0	see Pandian (2020)
Non-Turbellaria	22200	13326	6498	2376	
Hemocoelomata					
5. Arthropoda	1242040	46800	79800	1115440	Kohler (2008)
Crustacea	54384	46800	5200	2384	Dumont and Negrea (2002)
Non-crustacea	1187656	0	74600	1113056	Kohler (2008), Sabatini et al. (2008)
6. Mollusca	118451	89595	4856	24000	see Pandian (2017)
Eucoelomata					
7. Annelida	16911	13776	1939	1202	see Pandian (2019)
8. Echinodermata	7000	7000	0	0	see Pandian (2018)
9. Vertebrata	64832	19463	18966	26403	
Cephalochordata	35	35	0	0	see Pandian (2018)
Pisces	32510	18317	14193	0	see Pandian (2011)
Amphibia	5228	0	4117	1111	Vences and Kohler (2008)
Reptilia	9545	327	25	9193	Pincheira-Donoso et al. (2013)
Aves	10038	646	560	8832	Dehorter and Guillemain (2008)
Mammalia	5513	138	71	5304	Veron et al. (2008)
Major phyla total	1496509	213343	113745	1169421	
	97.0%[†]	14.3%	7.6%	78.1%	% of 1,496,509 species
Minor phyla total	46687	19282	7197	20208	
	3.0%[†]	41.3%	15.4%	43.3%	% of 46,687 species
Grand Total	1543196	232625	120942	1189625	
		15.1%	7.8%	77.1%	% of 1,459,676 species

[†] = as % of 1496509 species; 166,279 species/major phylum, 1769 species/minor phylum

TABLE 2.2

Approximate species number in minor phyla and their habitat distribution

Minor phyla	Species (no.)	Habitat (species no.)			Reference
		Marine	Freshwater	Terrestrial	
Aorganomorpha					
Placozoa	3	3	-	-	Schierwater (2005)
Mesozoa	150	150	-	-	GUWS Medical (2019)
Myxozoa	2200	733	1467	-	Gruhl (2015), Vidal-Martinez et al. (2016)
Acoelomorpha					
Loricifera	34	34	-	-	Kristensen (2002)
Cycliophora	2	2	-	-	Kristensen (2002)
Nemertea	1300	1265	22	13	Goransson et al. (2019)
Gnathostomulida	100	100	-	-	Hejnol (2015)
Pseudocoelomata					
Rotifera	2031	50	1981	-	Segers (2008), Wallace and Snell (2010)
Gastrotricha	813	430	382	-	*Anim Div Web*
Kinorhyncha	200	200	-	-	Pandian (2021)
Nematoda	27000	6900	1890	18210	Abebe et al. (2008), Appeltans et al. (2011)
Nematomorpha	360	5	355	-	Pandian (2021)
Acanthocephala	1100	43	-	1057	Nickol (2006), Amin et al. (2019)
Schizocoelomata/Hemocoelomata					
Priapulida	19	19	-	-	Wennberg (2008)
Sipuncula	160	160	-	-	Adrianov et al. (2011)
Echiura	230	230	-	-	Zhang (2011)
Tardigrada	1047	160	303	584	Gross et al. (2015)
Onychophora	200	-	-	200	*Anim Div Web*
Pentastomida	144	-	-	144	Christoffersen and Assis (2015)
Schizocoelomata/Lophophorata					
Entoprocta	200	198	2	-	Santagata (2015b)
Phoronida	23	23	-	-	Santagata (2015a)
Brachiopoda	391	391	-	-	Santagata (2015c)
Bryozoa	5700	4906	794	-	Wanninger (2015)

TABLE 2.2 Contd. ...

...TABLE 2.2 Contd.

Minor phyla	Species (no.)	Habitat (species no.)			Reference
		Marine	Freshwater	Terrestrial	
Eucoelomata					
Chaetognatha	150	150	-	-	Harzsch et al. (2015)
Hemichordata	130	130	-	-	see Pandian (2018)
Urochordata	3000	3000	-	-	see Pandian (2018)
Total	46687	19282	7196	20208	
		41.3%	15.4%	43.3%	

(3) In the animal kingdom, Echinodermata are among major phyletics and as many as 15 minor phyletics are exclusively marine inhabitants. Similarly, Onychophora and Pentastomida are exclusively terrestrial. (4) Restriction to marine habitat alone limits the species number to 7,000 in Echinodermata, whereas the species number goes to 186,193/phylum among the remaining major phyla, which are distributed in marine, freshwater and terrestrial habitats. This is also true of minor phyla. In them, the mean species number is limited to 366/phylum for each of the exclusively marine 14 minor phyla, in comparison to 9,437 species/phylum for Nemertea, Nematoda and Tardigrada, which are distributed in marine, freshwater and terrestrial habitats. These values are 172 species/phylum for Onychophora and Pentastomida exclusively restricted to terrestrial habitat. There are other phyla with 1,100 species/phylum distributed in marine and freshwater and 2,220 species/phylum for the phyla distributed in marine and terrestrial habitats. Notably, no phylum exists exclusively distributed within freshwater alone. *Briefly, the ability of animal species to access and colonize freshwater and/or terrestrial habitat(s) is decisively important for horizontal distribution and thereby speciation.* Being subjected to widely varied and fluctuating environmental factors, the *terra firma* remains a difficult domain to access. Yet, the nemerteans are the first to venture into it.

2.2 Vertical Distribution

2.2.1 Reproduction and Residency

With decreasing atmospheric pressure above the land, man could climb by flying up to 3–5 km altitude and land on the moon. In 2012, the film director and deep-sea explorer James Cameron successfully descended to greater depths (*https://www.livescience.com/23387-mariana-trench.html*). Nevertheless, animals are known to inhabit greater depths (Fig. 2.1). Of nine

Major Phyla

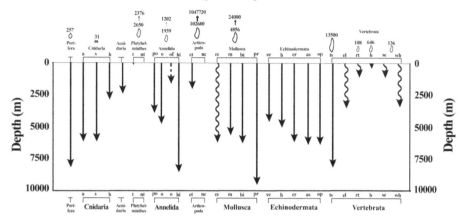

Minor Phyla

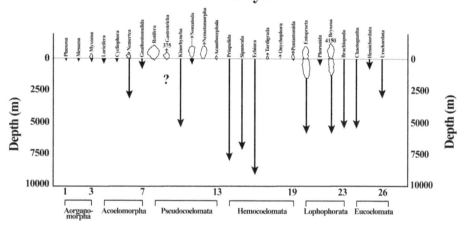

FIGURE 2.1

Vertical distribution of major and minor phyla. Thick inverted arrows indicate the penetrative depth of major clades and the wavy circles and thin arrows the distribution in freshwater and terrestrial habitats, respectively. Arabic numbers over the wavy circle indicate the altitudes, at which the clade occurs. In minor phyla, the initial bulges indicate the species rich zone in freshwater oligochaetes and bryophytes. The question mark represents the unknown depth (drawn from Hyman, 1951a, b, 1959, www.accessscience.com). For major phyla, wavy lines indicate nektonic cephalopods and higher vertebrates. a = Anthozoa, s = Scyphozoa, h = Hydrozoa, t = Turbellaria, nt = non-turbellarian platyhelminths, po = Polychaeta, o = marine Oligochaeta, of = freshwater Oligochaeta, hi = Hirudinea, ct = Crustacea, nc = non-crustacean arthropods, ce = Cephalopoda, m = Monoplacophora, bi = Bivalvia, pr = Prosobranchia, ec = Echinoidea, h = Holothuroidea, cr = Cranoidea, as = Astroidea, op = Ophiuroidea, ts = Teleosts, el = Elasmobranchs, rt = marine reptiles, b = birds, se = marine Carnivora, wh = Cetacea.

major phyla, eight of them have representative species penetrating to greater depths. Nemertea (*Ovicides jasoni*, 2719 m, Shields and Segonzac, 2007) and Crustacea (*Kiwa hirsuta*, 2,200 m, *Wikipedia*), Kinorhyncha (3,000 m, Meadow et al., 1994), and Chaetognatha (5,500 m, *Sagitta gazellae*, Hyman, 1959) are able to penetrate to greater depths. Notably, the structural complexity increases from the acoelomate Nemertea through pseudocoelomate Kinorhyncha and hemocoelomate Crustacea to eucoelomate Chaetognatha. Among them, Nemertea, Crustacea and Chaetognatha are carnivores but Kinorhyncha are sediment feeders. *S. gazellae* was first collected from 2,500 to 3,000 m depth by the German South Polar Expedition in the Antarctic. Subsequently, it was dredged from 5,500 m depth by the Scottish Antarctic Expedition. *Heterokrohnia mirabilis* was collected from 2,000, 3,000 and 5,000 m depths by the German South Polar Expedition. Apparently, *S. gazellae* and *H. mirabilis* are distributed from 2,500 to 5,500 m depth. The others like the polychaetes *Rarricirrus variabilis* and *Eusyllis blomstrandi* (4,000 m), tubificid *Abyssidrilus stilus* (see Pandian, 2019), sipunculid *Golfingia flagrifera* (3,000–5,000 m, Hyman, 1959) and Echiura (9,000 m, Dawydoof, 1959) are all burrowing sediment-feeders. The snailfish has been recorded from > 8,100 m depth in Mariana Trench and its genome has been sequenced (Wang et al., 2019).

The turbellarian flukes are not known to occur below 2 m depth. Among the relatively structurally simpler 26 minor phyla, only 8 have some representatives at the level (at 5,000 m depth) comparable to eight out of nine major phyla (Fig. 2.1). The herbivorous/microphagic Gnathostomulida are not found below the level of 400 m depth (Schmidt-Rhaesa, 2014). So are the nematodes that are not recorded below the depth of 800 m (see Heip et al., 1982). The deep sea inhabitants are divided into two major groups each with three sub-groups. To be a true resident, the listed (Table 2.3) species must be capable of reproduction at the respective depth. Hermaphroditism eliminates the need for mate searching, whereas gonochores have to depend on (i) gregarious residency and (ii) chemical cues to meet the mate. As > 3% (1,560 species) of prosobranchs and 9% (810 species) of bivalves are hermaphrodites, it is reasonable to assume that the deep sea molluscs are hermaphrodites (Heller, 1993, see Pandian, 2017). So are the deep sea crabs, as > 54 decapod species are hermaphrodites (see Pandian, 2016). The second hurdle is the indirect life cycle involving one or more larval stage(s), which may facilitate dispersal either locally by lecithotrophic larva or over a long distance by planktotrophic larva. The latter encounters two problems: (i) climbing, for example, a vertical distance of 6,000 m to reach the pelagic realm and to return to the depth by sinking. Incidentally, the sinking rate decreases to 240 m per day even at 1,500 m depth (Fig. 2.2 left panel). At the rate of 240 m per day, a pelagic larva may require a minimum of 25 days to descend to the depth of ~ 6,000. A common strategy is to climb vertical distance by reducing specific gravity through accumulation of water or lipid globules.

TABLE 2.3

Maximum depth of distribution of deep sea fauna

Phylum/Species	Depth (m)	Reference
1a. Hermaphoroditism, Indirect life cycle, Clonal potency		
Porifera, Cladorhizidae	8,840	Van Soest et al. (2012)
Bryzoa, *Arachnopusia monoceros*	5,719	Hyman (1959)
Urochordata, *Bathystyeloides mexicanus*	2,850	Darnell (2015)
1b. Hermaphroditism, Direct life cycle		
Oligochaeta, *Abyssidrilus stilus*	4,900	see Pandian (2019)
Nemertea, *Ovicides jasoni*	2,719	Shields and Sogonzac (2007)
Chaetognatha, *Sagitta gazellae*	5,500	Hyman (1959)
1c. Hemaphroditism, Indirect life cycle		
Mollusca, *Calyptogena phaseoliformis*	6,370	see Pandian (2017)
Buccinid	9,050	Brunn (1957)
Crabs	> 4,500	Brunn (1957)
2a. Gonochorism, Indirect life cycle, Clonal potency		
Cnidaria, *Marous orthocanna*	3,000	Mapstone (2014)
Acnidaria	3,000	*Anim Div Web*
Echinodermata, *Albatrossaster richardi*	6,035	see Pandian (2018)
Ophiacantha opercularis	6,035	see Pandian (2018)
2b. Gonochorism, Direct life cycle		
Kinorhyncha		
Isopoda, *Eurycope galatheae*	7,000	Brunn (1957)
2c. Gonochorism, Indirect life cycle		
Polychaeta, *Raricirrus variabilis*	4,000	see Pandian (2019)
Priapulida, *Priapulus caudatus*	8,000	see Wennberg (2008)
Sipuncula, *Golfingia flagrifera*	5,000	see Hyman (1959)
Echiura	9,000	Dawydoof (1959)
Brachiopoda, *Abyssothyris wyvillei*	5,500	see Hyman (1959)
Enteropneusta, *Spengelia sibogae*	275	see Hyman (1959)
Pisces, *Coryphaenoides*	> 6,000	Gaither et al. (2016)

With an obvious simpler structure and a consequent inability to climb to the pelagic, the hermaphroditic Porifera and Acnidaria, and gonochoric Cnidaria have opted for clonal multiplication. The others, which utilize the clonal strategy, are the gonochoric echinoderms with an indirect life cycle. Not surprisingly, the burrowing (thereby escaping from the pressure problem) polychaetes, priapulids, sipunculids and echiurans with trochophore or its equivalent larvae are buoyant to rapidly climb to the pelagic and sink to settle at a respective depth in ~ 25 days. Among epitokous polychaetes, adults of *R. variabilis* and *E. blomstrandi* may reduce the specific gravity to climb 4,000 m vertical distance, whereas the epitokous nereidid adults may use muscular strategy to climb vertical distance of ~ 50 m (Pandian, 2019). It is likely that the cyphonautes larva of the phoronids and tornaria of enteropneustus hemichordates, despite planktotrophism, seem to adopt the muscular strategy to reach the pelagic. This may be a reason for their vertical distribution being limited to < 100 m depth. The third strategy is to adopt a direct life cycle, as in hermaphroditic chaetognaths and oligochaetes, as well as gonochoric amphipods, isopods and tanaidaceans. A direct life cycle is also not uncommon among bivalves. Many brachiopod species are hermaphroditic brooders (Pandian, 2021, see Table 24.1). The indirect life cycle of decapod crabs commences at the advanced lecithotrophic zoeal stage.

Nekton: Cephalopods and aquatic amniotic vertebrates constitute the nektons. Of them, the amniotic vertebrates obligately require emerging to the surface to breathe and then diving to the depths. The nektonic cephalopods *Cirrothauma murrayi* and *Grimpoteuthis* sp inhabit at depths ranging from 2,430 to 4,850 m and from 4,802 to 4,848 m, respectively (Collins et al., 2001). The elasmobranch *Centrophorus squamosus* descends to maximum depth of 3,280 m (Priede et al., 2006). Among the amniotic vertebrates, there seems to be a relation between body weight and maximum descending depth, as detailed below (see also Schreer and Kovacs, 1997):

Species	Weight	Depth (m)	Reference
Gullemot bird *Uriaaalge*	990 g	180	*ecomare.nl*
Elapsid snake *Hydrophis* sp	115 g	200	Crowe-Riddeli et al. (2016)
Turtle *Dermochelys coriacea*	500 kg	1200	Bennet (2018)
Walrus *Odobennus rosamarus*	1250 kg	90	*oceanwide-expedition.com*
Seal *Mirounga*	3600 kg	100	ukfound.com

from Vivekanandan and Jeyabaskaran (2012)					
Species	Weight	Depth (m)	Species	Weight	Depth (m)
Sea cow *Dugong dugon*	300 kg	10	Dolphin *Ziphius carvirostris*	2700 kg	2000
Porpoise *Neophocaena phocaenoides*	32 kg	200	Whale *Balaenoptera edeni*	16000 kg	50
Dolphin *Sousa chinensis*	280 kg	5–30	Whale *B. musculus*	100000 kg	2000

Estimating the distance travelled for air-breathing by starved and fed *Channa striatus* fingerlings, Pandian and Vivekanandan (1976) showed that surfacing to breathe air in this obligately air-breathing fish costs considerable energy drain. This sort of wasteful energy drain may have limited the species number to 644, 138 and 108 in marine birds, mammals and reptiles, respectively, which may have to sink to a maximum depth of 2,500 m, in comparison to a large number of gill-breathing fish species at the depth of 8,000 m (Table 2.1). *Obviously, adoption of a strategy causing drain of wasteful energy does not facilitate speciation.*

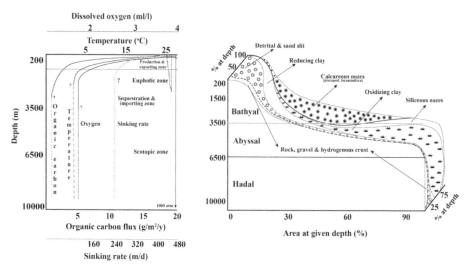

FIGURE 2.2

Left panel: Bathymetric distribution of temperature (values available up to 2000 m), light (Arizona State University, *skip.derra@asu.edu*) dissolved oxygen (3500 m), organic carbon (3000 m) and sinking rate of matter (up to 1500 m). Right panel: distribution of sediment types in the ocean. Distribution plotted on hypsographic curve to indicate approximate percentage at given depth and approximate area occupied by the sediment types (modified and redrawn from Brunn, 1957).

2.2.2 Distribution in the Oceans

Thanks to the establishment of new field stations and research vessels, the discovery of new equipment like acoustics and underwater video, our understanding of the oceans has largely increased. Yet, most sensors and others can measure temperature, dissolved oxygen and nutrients only up to 1,500–3,500 m depths. Using more sophisticated sensors, Gamo and Shitashima (2018) demonstrated that the hadal waters at 9,768 m depth have a slightly higher temperature and nitrate, and lower dissolved oxygen than waters at depths (9,600 m) outside the Izu-Ogasawara (Bonin) Trench, probably due to the effective accumulation of geothermal heat and active biological processes inside the trench. Firstly, the atmospheric pressure increases at the rate of 1 atm/10 mm with increasing depth (Fig. 2.2 left panel). Secondly, temperature declines to 4°C between 1,500 and 2,800 m depths and may remain around that level, as water density begins to decrease below

4°C. Thirdly, light penetration diminishes with increasing depth. In the total absence of light below 2,400 m depth, no photosynthesis occurs. Thereby, it reduces not only the nutrient availability but also dissolved oxygen. At a greater depth, oxygen levels are measured in μmol/l, i.e. 100 μmol/l is equal to 2.2 mg O_2/l. The sinking rate of dead and decaying organisms also decreases with increasing depth from about 400 m/day (d) at the pelagic to 240 m/d at 1,500 depths. On the ocean floor, 20–80% of sediment is distributed as calcareous oozes up to 2,500 m depth and ~ 10% at the hadal zone, which may be a reason for the restriction of hadals to a few species. The presence and proportions of detrital sand and silt up to > 3,500 m depth, oxidizing clay (up to the hadals) and reducing clay (up to 4,500 m depth) are shown in Fig. 2.2 (right panel). Based on the distribution of these factors, the ocean depths have been grouped into (i) shelf/pelagic (0–200 m), (ii) bathyal/ benthic (201–3,500 m), (iii) abyssal (3,501–6,500 m) and (iv) hadal (> 6,501 m) zones. Using area estimates in 12 geographical regions, Stohr et al. (2012) showed that the available area in the oceans is 30.4, 93.8, 252.3 and 2.2 million km^2 for the shelf, bathyal, abyssal and hadal zones, constituting 8.0, 24.8, 66.6 and 0.6% of the total available 378.7 million km^2 area, respectively. Hence, the largest area available is the abyssal and the least is the hadal. All organisms below the photosynthetic zone must depend upon the sinking dead/decaying organisms and thereby imported from the upper zone. At the scotopic zones, nutrients may appear from (ia) bacteria (ib) fungi, (ii) organic nutrients adsorbed on the silt/sediment and (iii) Dissolved Organic Matter (DOM). According to Zobell (1954), bacteria oxidize ~ 65% organic carbon to carbon dioxide and convert the remaining ~ 35% of organic carbon into their body substance. In general, bacteria are considered as the most important food source for deep sea fauna. However, no deep sea bacteria have so far been identified to species level. Several filamentous fungi and yeasts, recovered from deep-sea sediments of the Central Indian Basin from 5,000 m depth, grow under hydrostatic pressures of 20–40 MPa at 5°C (Raghukumar et al., 2010). The DOM is known as an important or the only source of nutrients for free-living (e.g. gutless oligochaetes, see Pandian, 2019) animals. Dissolved Organic Carbon (DOC) concentrations in the deep ocean are 5 to 10 times lower than surface values; DOC occurs in an extraordinary variety of forms (*Brittanica.com*).

The distribution of some taxa can readily be brought under these four recognized zones. For example, Gaither et al. (2016) provided quantitative data on depth-wise distribution of ~ 166 gadiformid fish species of the genus *Coryphaenoides*. Grouping them into these zones, a clear picture emerges, i.e. 8.9, 77.4, 13.1 and 0.6% species are distributed in the shelf, bathyal, abyssal and hadal zone, respectively. A similar analysis of data provided by Stohr et al. (2012) for ophiuroids revealed that of 3,298 species, 2,129 (64.6%), 2,136 (64.8%), 253 (7.7%) and 15 (0.45%) are distributed in the four zones. Clearly, the ophiuroid distribution overlaps between the shelf and bathyal zones. Blake (1996) reported taxonomic distribution of the Californian polychaetes.

From his informative description, Pandian (2019) traced their distribution to overlap between the shelf and bathyal for Orbiniidae, Paraonidae and Spionidae as well as even up to abyssal for Cirrutulidae. This type of overlapping distribution is also not uncommon among minor phyla; for example, the distribution of brachiopod *Pelagodiscus atlanticus* ranges from 1,500 m to 5,000 m depth (Hyman, 1959). Hence, the zone-wise quantitative assessment on distribution of the number of species is not possible for the shelf and bathyal. But, it may be possible to a certain extent for the abyssal and certainly for the hadal.

Abyssal zone: Brunn (1957) assembled the then available information for abyssal fauna, to which some are added in Table 2.4. While species names for some animals are indicated, many others are named at phylum or class level. At the phylum level, seven of nine major phyla (except Acnidaria and Platyhelminthes) and seven of 26 minor phyla are represented in the abyssal zone (see also Fig. 2.1). With the presence of an external shell and internal skeleton, almost all the major classes in Mollusca and Echinodermata are

TABLE 2.4

Approximate number of phyla and species existing in abyssal zone (compiled from Brunn, 1957, Hyman, 1951b,* 1959, Pandian, 2017, 2018, Gaither et al., 2016)[†]

Phylum		Species (no.)	Phylum	Species (no.)
Major phyla			Minor phyla	
Porifera		7	Kinorhyncha	1
Cnidaria		8	Priapulida	1
Polychaeta		7 + 5*	Sipuncula	5
Crustacea		22	Echiura	3
Mollusca		20	Bryozoa	2
Gastropoda	3		Brachiopoda	4
Bivalvia	10		Chaetognatha	3
Scaphophoda	5		Subtotal	19
Cephalopoda	2		Grand total	422
Echinodermata		187		
Holothuroidea	109			
Ephiuroidea	55			
Asterodea	20			
Crinoidea	2		*families	
Echinoidea	1			
Pisces	42 + 22[†]	66		
Subtotal (317 + 86)		= 403		

represented. For five polychaete families, species number is not provided. As the total number of abyssal species is double of that (165) hadal species, it is reasonable to double (86 species) the number of species for these five polychaete families. In all, 422 species (or 0.18% of 232,629 marine species, Table 2.1) were distributed in the abyssal zone. Of them, the most dominant group (212 species, 50%) was composed of burrowing sediment feeding holothuroids, spatangoid echinoid, polychaetes, priapulid, sipunculid and echiurans. Carnivores were the second largest group with 167 species, i.e. crustaceans, cephalopods, asteroids, ophiuroids, chaetognaths, kinorhynch and fish. Surprisingly, the number of filter-feeding species was 28 only (Porifera, Bivalvia, Bryozoa, Brachiopoda, *Lepas* and *Scalpellum* among crustaceans). However, Brunn (1957) emphatically stated that the number of individuals in sediment- and filter-feeding species outnumbered those of carnivores. The trochophore or its equivalent larvae of polychaetes, echiurans, and brachiopods may be able to reach the pelagic realm by reducing specific gravity and return to the abyssal zone by sinking. But the tadpole larvae of ascidians could do it only up to 3,000 m depth. And the structurally simple larvae of Porifera and Cnidaria may not do it at all. Obviously, they have to opt for clonal multiplication, as stated earlier. It is likely that echinoderms (see Pandian, 2018) and polychaetes, especially the oweniids (see Pandian, 2019) and bryozoans may also opt for clonal multiplication. All others may have to adopt an indirect life cycle with lecithotrophic larvae or a direct life cycle by brooding and/or releasing larvae at advanced developmental stage like zoea in crabs. In fact, Brunn (1957) recorded a female crab *Ethusa* carrying orange-yellow eggs.

That the larvae of some abyssal species rise to the surface by reducing specific gravity and return *per se* is supported by a few publications. *Echiurus abyssalis* larvae rise to the pelagic from the depth as 1,900 m (Pilger, 1978, 1987). More importantly, using very informative data reported by Schoener (1972) and Villalobos (2005), Pandian (2018) estimated the vertical distribution of eggs of some ophiuroids and asteroids. From Fig. 2.3, the following may be inferred: (1) Abyssal asteroids are capable of producing eggs hatching into planktotrophic (PLK) larvae from the depth of 4,000 m and eggs hatching into lecithotrophic (LEC) larvae from a depth as deep as 5,000 m (Fig. 2.3A). But the capacity of ophiuroids to produce LEC or PLK eggs is limited to the depth of 2,000–2,500 m only. (2) With increasing depth, fecundity of asteroids is reduced from $> 10 \times 10^5$ PLK eggs at 1,000 m depth to $> 10 \times 10^5$ eggs at 4,800 m depth. For LEC eggs, these values are from 10×10^4 to 10×10^3 eggs. However, the rate of reduction in fecundity of PLK ophiuroids is steep from 20×10^3 to 5×10^2 eggs at 2,500 m depth. (3) Interestingly, the PLK egg size of asteroids remains almost constant at 200 µm throughout the depth range up to 4,000 m. But the LEC egg size increases from ~ 500 µm at 1,000 m depth to 1,200 µm at 5,000 m depth for the asteroids and from 420 µm at 1,000 m depth to 600 µm at 2,000 m depth for the ophiuroids. Obviously, the non-feeding larvae appearing from the larger LEC eggs have to entirely

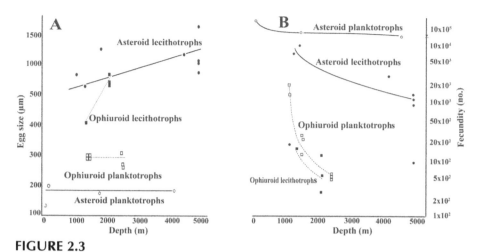

FIGURE 2.3

A. Egg size and B. fecundity as function of depth in some asteroids and ophiuroids (modified from Pandian, 2018).

depend on the yolk even for a short duration of dispersive stay at the pelagic realm. Hence, more number of abyssal ophiuroid species may opt for clonal reproduction at greater depths, while the asteroids have chosen the route, in which sexual reproduction is accomplished by producing a larger number of PLK or LEC eggs that are capable of rising to surface and the larvae returning to the abyssal.

Hadal: For hadal species diversity, Brunn's (1957) was perhaps the most informative contribution. Table 2.5 lists the names of species discovered for each of the representative phyla from seven trenches with depths ranging from 6,490 to 10,210 m. It also provides relevant information updated as far as possible. This list provides an excellent opportunity for further analysis and to arrive at the following inferences: (1) Of 162 + 1 hadal species, 27, 47 and 57 were discovered from Banta (Californian Pacific: 37.7° N, 121.4° W), Kuril (Russian Island: 32.6° N, 76.3° E) and Kermadec (South Pacific: 29.3° S, 177.9° W) Trench, respectively. Despite investigation up to 6,490–10,210 m depth, Sunda (Indonesia: 23.1° N, 73.2° E) with 15 species, the Philippines (10.1° N, 126.7° E) with 8 species, Puerto-Rica (18.2° N, 66.6° W) with 5 species and New Britain (41.7° N, 72.8° W) with 2 species remain environmentally more difficult to access and colonize (Table 2.5, lower panel). (2) Within them, the number of species decreases from 57 at depth ranges between 6,140 and 8,300 m in Kermadec to 2 between 8,810 and 8,940 m in New Britain. (3a) Among them, the burrowing, sediment-feeders are the dominant group, constituting 85 species or 52% of 162 hadal species, i.e. holothuroids: 25 species, polychaetes: 43, isopods + tanaidaceans: 16 (see Wilson, 2008) and priapulid: (3b) The next dominant group is constituted by filter feeders with 28 species, i.e. 17% (Acnidaria: 1 species,

TABLE 2.5

Upper Panel: Number of phyla and species existing in hadal zones (compiled from Brunn, 1957 and others) Lower panel: Number of species discovered at different depths in seven trenches (compiled from Brunn, 1957)

Phylum	Species (no.)		Phylum	Species (no.)	
	Major phyla			Minor phyla	
Porifera	3		Nemertea	1	
Cnidaria	16		Priapulida	3	
Acnidaria	1		Sipuncula	1	
Annelida (Polychaeta)	43+1(leech)		Echiura	2	
Mollusca	18		Entoprocta	1	
Arthropoda (Crustacea)	26		Bryozoa	1	
Echinodermata	36		Brachiopoda	1	
Vertebrata (Teleost)	3		Ascidiacea	3	
Subtotal	149 + 1		Subtotal	13	
Grand total			162 + 1		
Trench	Depth (m)	Species (no.)	Trench	Depth (m)	Species (no.)
New Britain	8810–8940	2	Kuril	6860–7230	47
Puerto-Rica	7625–7900	5		8380–9050	
Philippine	9790–10210	8		9700–9950	
Sunda	6740–7160	15	Kermadec	6140–7000	57
Banta	6490–6650	27		8210–8300	
	7250–7290				

Bryozoa: 1, Brachiopoda: 1, Entoprocta: 1, Cirripedia: 3, Bivalvia: 14, Ascidiacea: 3, Crinoidea: 4). (3c) The third dominant group is the carnivores with 29 species or 18%. The feeding habits of the remaining 20 species (12%) are not traceable but is likely to be sediment- or filter-feeders. (4) *More importantly and surprisingly, none of the hadal species existing in a trench is found in any other trench except for the two polychaetes Macellicecephala abyssicola found in Sunda and Kermadec Trenches and M. hadalis found in Banta, Kermadec and New Britain Trenches.* Brunn also listed the existence of Amphipoda with a direct life cycle in all the investigated seven trenches. In them also, species level identification may confirm the restricted existence of each amphipod species to a single trench alone. This may also hold true for the two scyphozoan species found in Kermadec and Kuril Trenches. Arguably, none of the hadal species seems to have a larva including trochophore that is capable of climbing a vertical distance of

6,490–10,240 m to reach the pelagic realms for feeding and/or dispersal (see Fig. 2.2). Briefly, all the identified hadal species remain completely isolated from each other of all the investigated trenches. Barring molluscs, crustaceans and teleosts, all other phyla have the potency for clonal multiplication. In the majority of crustaceans, the life cycle is direct in 36 isopod, amphipod and mysid species; they all brood and release young ones as manga. Brooding and viviparity are not uncommon among molluscs (see Pandian, 2017). It is likely that the hadal molluscan and crustacean species with no potency for clonal multiplication may either opt for brooding or producing lecithotrophic larvae at the advanced epimorphic stage. The chances for producing genetic diversity are minimal in brooding/viviparous species with considerably reduced progeny number; among clonal species, they are even more limited. *Though stabilized, hadal fauna may evolve and speciate at an extremely slow rate in this unusual habitat with limited or minimal scope for genetic diversity. Not surprisingly, they constitute 0.07% (162 of 232,629 marine species) of all marine animal species.*

2.2.3 Distribution in Freshwaters

Covering ~ 1% of the earth's surface and holding < 0.01% of its water, freshwaters affords habitats for only ~ 7.8% of all animal species (Table 2.1). Freshwater systems occur as perennial rivers and lakes, semi-permanent rivers and ponds, transient streams, pools, puddles, ditches, swamps and so forth. Man-made dams, reservoirs, canals and others abstract waters for domestic, agricultural, industrial and transport purposes, and thereby effectively minimize or prevent the rivers from flowing into the seas (e.g. Pandian, 1980). Paddy fields holding water up to 5–10 cm height offer an excellent habitat for oligochaetes (see Pandian, 2019), dipterans like mosquitoes (see Devi and Jauhari, 2004), snails, crabs and others. Much is known about the temperate aquatic system with continuing biological activity at the hypolimnion, while the epilimnion is covered by a sheet of floating ice. Relatively less is known about species diversity and productivity for the tropical freshwater system. Those interested in it may consult the special volume in Hydrobiologia (Pandian, 2000). Due to page limitation, this account, however, is limited to the deepest Lake Baikal on earth, ponds and others located at the higher elevated altitudes alone. Surprisingly, (a) the Lake Baikal is a well oxygenated aquatic system. Not surprisingly, many sediment-feeding oligochaetes flourish at 1,600–1,680 m depth: the enchytraeid: *Propappus glandulosus*, lumbriculid: *Stylodrilus asiaticus* and tubificid: *Balkaiodrilus maievici* (Martin et al., 1999), and (b) the existence of *Bucholzia appendiculata* at 2,000 m asl in the montane regions of South America is an another example (Schmelz et al., 2013). With increasing altitude, waters can be acidic (5.4 to 6.0), and may hold reduced dissolved oxygen, as a consequence of decreasing atmospheric pressure at the rate of 11%/km elevation; for example, the pressure is decreased to 89,

69 and 54% at altitudes of 1,000, 3,000 and 5,000, respectively (Jacobsen and Dangles, 2017).

In the Tibetan plateau alone, there are as many as 787 high altitudinal ponds and lakes with a cumulative area of 24,566 km² (*topchinatravel.com*). The Rigdonlabo Lake spreading over 3 hectare (ha) area is located at the world's highest altitude of 5,801 m asl. From the altitudes of 5,100–5,200 m, the Tibetan streams emerge (Jacobsen and Dangles, 2017). Not surprisingly, Tibet remains an attraction for altitudinal biologists. Unfortunately, authors reporting species diversity in relation to altitudinal distribution of animals have not provided temperature details except Ya'cob et al. (2016). However, vertical temperature decline for a Tibetan lake is available, indicating that the decline is from 15°C at surface water to 8°C at 36 m depth (Wang et al., 2014). This was used as a yardstick to fix the approximate temperature of some aquatic systems, for which relevant information is not reported. A couple of parthenogenic daphnids *Daphnia tibetana* and *D. dolichocephala* thrive in Tibetan ponds at 5,460 m asl, the 'roof of the world'. At the surface level of these ponds, the daphnids are known to produce dormant eggs encased in an ephippium, which can hatch after a diapausing period of 14–125 years (Pandian, 2016). In Tibetan ponds, the existence of pulmonates is reported for lymnaeid *Radix*, planorbid *Gyraulus* at 5,000 m asl (Oheimb et al., 2013), sphaeriid bivalves *Pisidium* and *Musculium* at 5,000 m asl (Clewing et al., 2013), and for the non-pulmonate snail *Valvata* sp at 4,300 m asl (Clewing et al., 2014). Incidentally, *Valvata*, *Pisidium* and *Musculium* are hermaphrodites (see Heller, 1993). To the pulmonate molluscs, characterized by hermaphroditism and a direct life cycle, the ability to breathe air and hibernation, when the water is frozen, is decisively important. The surface water pulmonates like *Pila globosa* are known to breathe air and survive by aestivation, when the pond water is dried for over 5 months (Haniffa and Pandian, 1978). Hence, it may not be surprising that a couple of pulmonates thrive in the Tibetan ponds at 5,000 m asl. However the presence of a couple of non-pulmonate species at 4,500 m asl and 4,300 m asl is surprising, respectively.

Among others, the vertical distribution is shown in Fig. 2.4A for vectors like mosquitoes in lentic waters and blackflies in streams, ecologically important chironomids and behaviorally interesting beetles. With the need to surface for breathing, the aquatic beetles are unable to penetrate below 750 m depth. Hanging to the surface with a respiratory cone to acquire atmospheric air, mosquito larvae and pupae thrive in rice fields (16 species up to 500 m asl) and exist up to 1,500–2,000 m asl mostly in holes containing limited water on trees and rocks. The hemoglobin containing chironomid larvae are found in sediment of aquatic bodies up to the altitude of 4,300 m. Remarkably, the number of species in all these insects decreases with increasing elevation (Fig. 2.4A). This observation is also substantiated by Jacobsen (2003), who reported that the number of insect families in the Ecquadorian streams decreases from 26 at 27°C (100–600 m asl) to 18 at 7°C (3,500–4,000 m asl). Notably, the odonates (present up to 2,600–3,000 m asl), hemipterans and

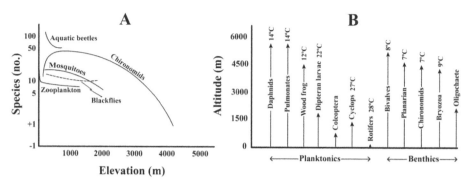

FIGURE 2.4

A. Vertical distribution of aquatic insects and zooplankton: the obligately air-breathing Tunisian beetles (Touaylia et al., 2013), the Peruvian chironomid larvae (Acosta and Prat, 2010), larval/pupal mosquitoes from Garhwal, India (Devi and Jauhari, 2004), blackfly larvae from Peninsular Malaysian streams (Ya'cob et al., 2016) and zooplankton from Kodayar River (Murugavel and Pandian, 2000). B. The planarian *Mesostomum arctica* (Hyman, 1951a), wood frog, *Rana kikinoris* (Zhao et al., 2009), cyclops: e.g. *Mesocyclops* sp, rotifer: *Brachionis* spp (Murugavel and Pandian, 2000), Bryozoa: *Fredericella australensis*, *Stolella agelis*, Lake Titicaca, Bolivia (Hyman, 1959), Oligochaeta: *Bucholzia appendiculata* (Schmelz et al., 2013).

megalopterans disappear at 3,500 m altitudes (Jacobsen, 2003). For all animal families, the decrease is from 55 to < 30 at 3,500 m asl. Therefore, *temperature seems to be the single most dominant factor that limits species diversity and speciation at altitudinal aquatic systems.* Further, it appears that the 'benthics' like bivalves, bryozoa, chironomid larvae, planarians and oligochaetes may occur at constant temperature of 7–9°C, irrespective of the altitudes at which they occur. However, the 'planktonics' like daphnids, pulmonates, wood frog, dipteran larvae, coleopterans, cyclopoids and rotifers are found at temperature rangings from 12 to 28°C, depending on the altitude, at which they are found (Fig. 2.4B).

In view of the fact that not much is known on aquatic systems in tropical elevated altitudes, this account has chosen to summarize the limited information available on changes of some abiotic and biotic factors in a few South Indian aquatic systems on the rain-capturing areas of the Western Ghats (Table 2.6). Notably, precipitation increases with elevation but temperature and pH drop. Comprehensive information is available for the Kodayar River, as it descends from 1,312 m from the upper dam down to the plains at Azhakia Pandia Puram (Table 2.7). With decreasing altitude, species diversity, density in each species and their productivity (both phyto- and zoo-plankton) were increased. Whereas temperature and pH increased, precipitation diminished. Flowing over the hilly gradient, dissolved oxygen level was higher at both lower and upper Kodayar dams. The incidence of *Macrocyclops pachysiphinosus* and *Odogonium* was restricted between the pH range of 6.75 and 7.0, but most other phyto- and zoo-plankton species were tolerant to a wide range of pH from 6.5 to 8.0.

TABLE 2.6

Changes in abiotic factors with increasing altitude in some South Indian aquatic systems (compiled from Murugavel and Pandian, 2000)

Aquatic system	Location	Altitude (m asl)	Temperature (°C)	Precipitation (cm/y)	pH
Idumban Pond	10°2'N	80	30	79	-
Bhavanisagar dam	11°3'N	280	23	-	7.8
Idukki dam	9°4'N	693	22	-	-
Kodayar dam	8°3'N	1312	24.5	366	6.7
Kodaikanal Lake	10°1'N	2285	20	300	6.6
Ooty Lake	11°3'N	2500	17	350	7.9

TABLE 2.7

Changes in abiotic and biotic factors during ascend of the Kodayar River from the upper dam down to the plain (compiled from Murugavel and Pandian, 2000). APP = Azhakia Pandia Puram

Factor	Upper dam	Lower dam	APP
Altitude (m asl)	1312	92	0
Temperature (°C)	24.5	27.9	30.0
Precipitation (cm/y)	366	175	127
Dissolved O_2 (mg/l)	7.4	7.4	7.0
pH	6.7	6.9	7.5
Phytoplankton			
Species (no./l)	5	14	> 14
Density (no./l)	159	203	412
Productivity (g C/m³/d)	0.11	0.79	1.97
Zooplankton			
Species (no./l)	8	9	17
Density (no./l)	30	44	585

2.4 Terrestrial Distribution

As much as 76% of all animals inhabit the widely varied niches on land, covering only 29% surface of the planet earth. On land, temperature and precipitation are decisively important factors that provide the highest scope for speciation. The earliest classification of the terrestrial habitats was reported by the Tamil poet Kappian some two thousand years ago. Based on water availability, he classified the habitats into (i) *Marutham*, the fertile land, in which paddy and the like are cultivated, (ii) *Kurinji*, the dry land, where the livestock are reared, (iii) *Mullai*, the montane area, where animal hunting is prevalent and (iv) *Neithal*, the coastal zone area, in which fishes are harvested and salt is manufactured. Amazingly, he had gone to the extent of describing the characteristics of these four terrestrial habitats, the flora, human professions and their deities.

In productive land, agriculture has largely increased productivity but enormously reduced species diversity. However, this account is limited to show how the harsh environments reduce species diversity and possibly speciation in elevated montane territories and deserts. Figure 2.5A depicts linear, negative but complicated (by latitudes) relations between temperature and altitudinal elevation in the montane areas. The relation is at a higher level for low latitudes than that at for the higher latitudes; for example, at 1,500 m altitude, prevailing temperatures are in the range of 20°C for lower latitudes but at 10°C for higher latitudes. A positive relation is also apparent for precipitation; however, the differences are wider between the lower and higher latitudes. Incidentally, the precipitation range is far higher on the rain-capturing areas than that on rain-shadow areas; for example, it is > 300 cm for the former but 100 cm for the latter in Tamil Nadu. Consequently, these two factors exert profound influence on vertical distribution of homeothermic birds and mammals (Fig. 2.5B) and poikilothermic insects (Fig. 2.5C). Notably, the limits for vertical distribution of birds, nesting on tall trees and rocks, are higher than those for the tree- or cave-dwelling bats with limited flying capacity and burrowing rodents. With different but lower (than vertebrates) ability for flight, the distribution of the insect is limited to 1,500 meters for the wasps and 2,800 meters for butterflies: their limit decreases in the following descending order: butterflies > mosquitoes > beetles > wasps. Notably, the orthropterans with a limited ability to fly are widely distributed up to 4,800 m. Strikingly, the species diversity in all of them decreases with increasing altitude except for honey bees, whose vertical

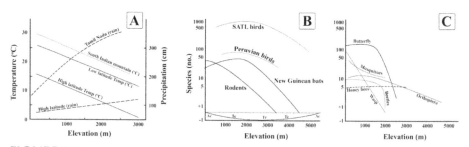

FIGURE 2.5

A. Changes in temperature (modified and redrawn from McCain and Gytnes, 2010) and precipitation as a function of altitude (data drawn from Tables 2.6, 2.7). B. Species number as a function of altitude in South American Tropical Land Birds (SATL), the Amazonian slope of the Andes, Peruvian birds, New Guinean birds and rodents (modified and redrawn from Gallen et al., 2016). C. Altitudinal distribution of butterflies, adult mosquitoes in Garhwal, India (Devi and Jauhari, 2004), African honeybee (Hepburn et al., 2000), Bulgarian beetles (Lobo et al., 2007), New Zealand wasp (Damken et al., 2012) and orthopterans of Colorado and Turkey (Alexander, 1951, Ciplak et al., 2008).

distribution seems to be correlated with those of flowering plants (Ramalho, 2004). Incidentally, birds can fly, soar and glide but insects like dragonfly can only fly and glide. Investigations on their aerodynamics have enormously enhanced the design of our aircrafts. For example, research on the ability of bats for echolocation have led to the discovery of the radar system. This is why we are often told about bird-hit but not bat-hit of the aircrafts. The former cannot echolocate but the latter escapes by echolocation from the climbing aircraft. Interestingly, dragonflies are the only creatures that have 'honeymoon' rather copulation, while flying in tandem.

The average annual precipitation over the globe is $5.77 \times 10^{14} \, m^3$, of which 79% falls over the oceans and 21% on the land (Agarwal, 2013). Understandably, tropical rain forests harbor 3 to 50 million species. Trees provide many more niches for animals. The redwood tree *Sequoia sempervirens* grows to the height of 91–116 m. With prevailing higher temperatures, tropical forests are more speciose than their temperate counterparts; for example, there can be as many as 480 tree species/ha in a tropical forest, in comparison to a dozen in temperate forests. A single tropical Amazonian bush may have more number of ant species than that in the entire British Isles. There are 667 mammal, 1,604 bird and 740 reptile species in tropical Indonesian forests, in comparison to 468, 888 and 365 species in the temperate forests of the USA (Butler, 2019). In the Arctic and Polar zones, only 67 (*caff@caff.is*) and 4 species (*worldwildlife.org*) mammals have been recorded. Briefly, *species diversity for a taxonomic clade progressively decreases with increasing latitudes in either direction from the equator*. The deepest depth, up to which animal species can burrow is 80 cm for the earthworm *Glossodrilus* (Jimenez and Decaens, 2000), 230 cm for mammals and 270 cm for the harvest ant *Pogonomyrmex* spp (Sample et al., 2014). The stigobiontoligocheates belonging to 42 genera in 17 families are reported from the subterranean aquatic habitats (see Pandian, 2019). The catfish *Kryptoglanis shajii* is reported to exist as stigobiont in the subterranean waters of Kerala (Vincent and Thomas, 2011).

Animals depend on plants for food and niche. Hence, plant diversity is considered in these harsh altitudes. The highest altitude, at which plants and animal are known to inhabit, is 6,480 m for the sandwort *Arenaria bryophylla*, 4,572 m for the flowering plant *Abies squamata* and 3,300–5,000 m for the Tibetan yak *Bos grunniens* (*Wikipedia*). The presence and abundance of plants on the montane territories depend on two critically important factors (i) photosynthesis and (ii) soil depth. The enzymes facilitating photosynthesis do not work efficiently at temperatures below 10°C (*Sciencing.com*). In the absence of soil, the montane rocks above 3,000 m asl may remain barren or covered with snow. As living sponges, soils retain moisture and supply it to plants with required minerals during non-precipitation times. The soil depth required for sustenance is 250–300 cm for the shrubs (e.g. the antelope bitterbush *Purshia tridendata*), 200 cm for the forbs (Sample et al., 2014) and less than these for trees. The Himalayan treeline of the pencil cedar *Juniperus tibetica* exist at the highest altitudes of 4,900 m; cool and moist conditions favor tree growth but higher temperature hampers its growth (Misra et al., 2020). Fixing the limits for global altitudinal treelines, Korner (2012) set the limit for tree lines to 10 cm soil depth with temperatures of 5–7°C. Above this limit, about 4,900 m asl, plants may not exist. *Above the altitudes of 3,500 m, species diversity and the speciation process of plants is limited by a combination of temperature and soil moisture.* Despite the existence of plants, reports are not available on the species diversity of insects and other fauna above 3,500 m altitude. Perhaps, reduction of atmospheric pressure to < 54% of that on the surface (see Jacobsen and Dangles, 2017) limits their existence.

Deserts: In deserts, water is scarce in liquid form. It is available in a solid frozen form in the Arctic (14.0 million km^2) and Antarctic (14.5 million km^2), in which biological activity is at the least. In the terrestrial arid deserts, evaporation exceeds precipitation (< 25 cm). Ward (2009) named as many as 23 deserts, which span over ~ 28.5 million km^2 (*edu.seattlepi.com*) and covers ~ 9% of land area, leaving only 20% of land area with relatively more productivity and species diversity. In some of these deserts, day temperature may shoot up to 57°C (e.g. in Libya) but night temperature may drop to freezing level (e.g. Gobi Desert, Ward, 2009). In deserts, precipitation does occur in pulses but more often as flash floods, as in the Arabian deserts. Utilizing these transient events, as many as 70 and odd species emerge and complete the life cycle after leaving dormant reproductive products the cysts and the like. To sustain their existence, for example, the cycle may be completed within 2 days after the flash flood on rock pool (e.g. rotifer: *Hexarthra* sp, Schroder et al., 2007). Emerging from the slumber, the American toad *Scaphiopus couchii* spawn once in a year within 12 hours following the flash flood and the life span of its tadpole lasts only for ~ 7 days (see Pandian, 2020). In deserts, there are reports claiming the existence of 1,200 species of plants, 200 vertebrate species, numerous insects and other invertebrate species (Tews et al., 2004); for example, Tews et al. (2004) described ~ 10 different niches on these plants and soil surfaces. They also estimated that 85% of all investigated desert flora and fauna hold a positive correlation between species diversity and niches as well as the adaptive strategies to sustain themselves in these harsh habitats. However, no one has yet provided a list for the faunal species diversity in deserts. In the Sonoran Desert of Arizona, temperature climbs to 40°C and precipitation averages ~ 24.5 cm/y. In it, the existence of 2,000, 60, 350, 20, 100 and 30 species have been reported for plants, mammals, birds, amphibians, reptiles and fish, respectively (National Park Service). On the other extreme, the Saudi Arabian Desert experiences 55°C and receives < 10 cm precipitation. Limited information available on the fauna and flora of the Arabian Desert indicates the existence of 85 plant species (Newton and Newton, 1997) and 476 vertebrate species (Mallon, 2011), in comparison to 2,000 plant species and 560 vertebrate species in the relatively cooler Sonoran Desert. However, no information is available on invertebrate species diversity. In South America, the Atacama-Sechura

Desert spreads from 10° S to 30° S with an area of 10,500 km², experiences 17°C temperature and receives the lowest precipitation of 1.5 (Guerrero et al., 2013) –4.5 cm/y (*en.climate-data.org*). The total number of vertebrate species thus far described is 52 for amphibians, 87 for reptiles (Vidal et al., 2009), four for birds and three for mammals—in all 146 vertebrate species exist in this desert. Hence, the number of vertebrate species decreases from 560 in the Sonoran Desert (40°C) receiving 24.5 cm precipitation/y to 476 in the Saudi Arabian Desert (55°C) with < 10 cm precipitation/y and to 146 in the Atacama-Sechura Desert (17°C) receiving 1.5–4.5 cm precipitation/y. In these deserts, *it is water availability rather than temperature that limits the vertebrate species diversity.*

An attempt is made to compare the approximate species number in the harsh habitats (Table 2.8). For deep sea fauna, the hadals make up to ~ 162 species (Table 2.5). The species number of 5,231 for the altitudinal freshwaters was estimated from those reported for insects by Jacobsen et al. (1997). The lowest number of 62 species for the montane altitudes was assessed from Peruvian birds (Fig. 2.5B). With availability of adequate information, the terrestrial vertebrates alone are considered among desert fauna; their average number is 381 species and is related to the total species number of only terrestrial vertebrates. The level of harshness of the spatial habitat decreases in the following descending order: altitudinal freshwater > montane altitude > hadal > desert. *A combination of pressure, temperature, nutrient availability and oxygen limits the species number to the least in the hadal. In the altitudinal waters, temperature is the dominant determinant factor that limits species number to < 4.32% of all freshwater faunal species. Temperature and possibly atmospheric pressure limit species diversity at altitudes. In deserts, it is water availability in a liquid (or moisture) form that critically limits species number. Briefly, the zones between 3,500 m depth in waters and 3,500 m asl on land represent the active biosphere on the planet earth, in which species diversity and scope for speciation occurs at a relatively faster rate.*

TABLE 2.8

Approximate number of species existing in harsh habitats

Habitat	Depth/Altitude (m)	Species (no.)		Species (%)
		Total	Habitat	
Marine	↓ > 6,500	232,677	162	0.07
Freshwater	↑ 3,500	120,896	5,231	4.32
Terrestrial	↑ 3,500	9,000	62	0.68
Desert	-	1,189,625	381	0.05

3

Coevolution and Diversity

Introduction

As stated earlier, the oceans cover 70% of the earth's surface, holding 97% of its water and providing 900 times more livable volume of space than that of land. On the other hand, land covers only 29% of the earth's surface, of which ~ 9% is covered by hostile deserts, leaving only 20% of the land area with relatively more productivity and species diversity. Yet, it provides adequate habitats, i.e. ecological niches to sustain > 77% of all animal species, whereas the larger oceans provide niches to sustain only 15% of all species (Table 2.1). Consequently, flowering plants and animals are so densely packed in terrestrial niches, resulting in their intimate interactions that have led to either a positive symbiotic relation, as observed between flowering plants and animals for pollination and seed dispersal, or a negative one, as between hosts and parasites. Briefly, the kind of mutual dependence among the densely packed flowering plants and animals on land has increased biological niches and coevolution. A large volume of literature is available on coevolution and many hypotheses are proposed. However, this account is limited to symbiotic and parasitic relations in the context of species diversity.*

3.1 Symbiosis and Diversity

The term coevolution has been defined differently by many authors. Perhaps, the best may be that it is an ongoing evolutionary process, in which closely interacting organisms respond to reciprocal selective pressures (Ragusa, 2020) and the number of very different types of biotic interactions leads to different diversities (Hembry et al., 2014). The concept of coevolution was first developed by Darwin (1859), who wondered "how

---

* This chapter was added at the end on the suggestion of a reviewer.

a flower and a bee might slowly become .... modified and adapted in the most perfect manner to each other .... by presenting mutual and slightly favorable deviations". Tracing the evolutionary history of coevolution between angiosperms and pollinators, Hu et al. (2009) reported that 86% of 29 extant basal angiosperm families were pollinated by insects during the early Cretaceous (see Table 25.1) with specialization increasing by the mid Cretaceous. However, Ricklefs (2010) noted that adaptive radiation to fill niche space by evolutionary diversification is limited to an initial period of rapid diversification and subsequent saturation by competing and parasitic species. For example, the numbers of species of mammals and tropical forest trees have diversified very little over the 60 million years since the early Tertiary recovery following the end of the Cretaceous extinction. A second reason for the saturation may be extreme specialization. For example, the Malagasy orchid *Angraecum sesquipedale* can be pollinated only by the hawkmoth with its exceptionally long tongue. Conversely, apid, dipteran and lepidopteran species can pollinate 79, 65 and 22% of all flowering plant species, respectively (see Table 27.3). *Hence, flexibility and non-specificity has facilitated species diversity in flowering plants and pollinators but specificity and extreme specialization deter it.* Though, it is widely accepted that angiosperm flowers and their insect pollinators have passed through a coevolutionary process, there is considerable controversy on the level of specialization in them. For more information, Pandian (2022) may be consulted for more details. Incidentally, there are examples for flowers, rewarding and cheating the insect pollinators. For example, the long tongued fly *Prosoeca ganglbaueri* pollinates the orchid *Zaluzianskya microsiphon* and *Disa nivea*. Whereas *Z. microsiphon* rewards *P. ganglbaueri* with nectar, *D. nivea*, a Batesian mimic, superficially resembles *Z. microsiphon* and so doing deceives *P. ganglbaueri* visiting the plant through a mistaken identity (Johnson and Anderson, 2010).

3.2 Escape and Radiate

During the checkered history of evolution, plants have developed defence against predation by structuring their cell wall with cellulose, lignin and others, for which animals produce the corresponding digestive enzymes cellulase, ligninase and so forth. Raguso (2020) has brought a new dimension in the race of prey-predator coevolution. Accordingly, escape occurs, when an escalating arms race between an herbivore and its host plant results in a novel adaptation (e.g. toxin) or counter-adaptation (e.g. detoxication enzyme). Radiation may follow, if the adaptive escape accelerates speciation in that lineage, as new niches become available. This process may generate species-rich lineages of interacting organisms. Apart from cellulose and lignin, plants have also repeatedly evolved gums, resins and latex as mobile chemical defenses against insect herbivores. Resin- or latex-producing plant lineages are significantly species-richer than their closest sister lineages, which suggest that the resin-latex defenses have a key innovation that increases species diversity in host plants. Sequential co-adaptation in resinous *Bursera* plant has led to diversification of 100 species but limited the *Blepharida* beetles to 45 species. For more examples, Raguso (2020) may be consulted.

3.3 Hosts and Parasites

Vertebrates provide an array of niches for parasites. For parasites, the host is the habitat. Considering the close interaction between the parasite and host, two alternative views have been proposed. According to Paterson et al. (1993), parasites tend to be host specific or stenoxenics, infecting one or two closely related host species, as in 71% of monogeneans. Hence, it is likely that the phylogenesis of coevolving parasites and hosts are mirror images of one another. The second view is that the colonization of a host species by the oioxenic parasite has played a major role in the diversification of parasitic lineages (Hoberg et al., 1997). An analysis of monogeneans, digeneans and copepods led Poulin (1992) to draw the following inferences: (A) The monogeneans are highly stenoxenics; their specificity is limited to one, two and three host species in 65, 25 and ~ 8%, respectively; they have more site-restricted attachment and their adhesive organs are highly specialized to specific microhabitats on the host's gills. This sort of highly specialized host specific coevolution has deterred species diversity. (B) Contrastingly, the ectoparasitic copepods, that have also an indirect life cycle involving no intermediate host, as monogeneans, are oioxenics and each of the parasitic copepod species can successively infect as many as 20 host species. So are the digeneans; at least 5% of them can colonize a dozen host species; the digeneans are not only oioxenics to their definitive host but also to their first and second intermediary hosts. *Not surprisingly, oioxenicity and colonization of new hosts have facilitated more species diversity in the ectoparasitic copepods and endoparasitic digeneans.*

4

Food and Feeding Modes

Introduction

The Hindus believe that food contains Lord Shiva,* who with 50% of his body (matter), bequeathed the remaining to the Goddess Shakti (energy). To be alive and active, organisms require input of matter and energy. Plants synthesize glucose from carbon dioxide, and water using solar energy and other substances required for their sustenance. Contrastingly, animals feed microbes/plants and/or other animals to acquire the required energy. Thus, there is a basic difference between autotrophic, non-motile plants and heterotrophic, motile animals (see Pandian, 1975). Elaboration of these basic differences has led to more species diversity in animals but more of racial diversity within the limited species diversity in plants. The ability of plants to synthesize amino acids and proteins is limited. As a result, they contain a maximum ~ 5% protein and in unit of nitrogen up to 7% of their body (see Fig. 4.1). Thanks to symbiotic microbes, the leguminous plants concentrate proteins up to 68% in their seeds (Anonymous, 2013). The blue-green algae are capable of directly fixing atmospheric nitrogen and accumulate proteins up to 24% (e.g. *Spirogyra*, Saragih et al., 2019).

Plants are also capable of synthesizing thousands of Secondary Plant Metabolites (SPMs). Some of them are useful in the kitchen (e.g. citric acid of lemon), industry (e.g. rubber) and medicine (e.g. azadirachtin of margosa). However, the synthesis of some SPS is induced by specific soil microbes. For example, jasmine cultivated in and around Madurai (Tamil Nadu, India) is unique for its fragrance but not in those cultivated elsewhere in Tamil Nadu. Similarly, an unidentified drug in *Phyllanthus amarus* can cure jaundice, a viral disease, for which no medicine is yet developed. However, the synthesis of the drug is induced by soil microbes in Tamil Nadu, India but not in Thailand (Thyagarajan et al., 2002). In defense of predation, plants do

---

* Southukulairukida Chockalingam.

synthesize an array of SPM like cellulose, lignin, toxins, enzyme inhibitors (e.g. trypsin inhibitor), as well as may also accumulate minerals over the body. Lacking cellulase and/or ligninase, most herbivorous animals may not feed on the plants that contain a high concentration of cellulose and/ or lignin. Nevertheless, a few fishes can tolerate and eat plants containing 28% cellulose (e.g. *Tilapia zilli* eats *Naja guadalupensis*), 12% lignin (e.g. *Hyporhampus melanochir* feeds *Heterozostrea tasmonica*) and 36% mineral content (e.g. *Ctenopharyngodon idella* consumes *Spirogyra maxima*). However, fishes are unable to utilize *Sargassum*, as it releases, on being digested, mannuronic acid rather than glucose (see Pandian, 1987c).

4.1 Food and Reproduction

For animals, food and its availability is a prime factor that determines survival, growth and reproductive output. Increase in the number of progeny enhances the scope for new gene combinations—the raw material for evolution and speciation. Availability of food in adequate frequency, quantity and duration at the right time are critically important determinants for fecundity. For example, Lifetime Fecundity (LF) of the detrivorous chironomid *Kiefferulus barbitarsis* is reduced from 1,080 eggs to 800 eggs with decreasing nitrogen content from 7% in *Chlorella* to 5% in pond silt (see Muthukrishnan, 1994). On feeding pollen-supplemented feed, the butterfly *Heliconius chartonius* lays 1,000 eggs, but on pollen deprivation, only 300 eggs (Dunlap-Pianka, 1977). Egg production efficiency (egg no ÷ food consumed × 100) of *Bombyx mori* is reduced from 3.5 to 2.8%, when feeding duration is restricted from 8 hour (h)/day (d) to 2 h/day (Haniffa et al., 1988). On reducing the ration below 70% of *ad libitum* level, fishes may not spawn (see Pandian, 1987c). Table 4.1 provides an idea about how ration and feeding frequency and consequent rationing reduce size and maturation of the fish *Gasterosteus aculeatus* as well as egg size and number of the prawn *Macrobrachium nobilii*. With reduction of 10% egg size, progeny survival may also be reduced.

4.2 Food Acquisition Cost

For animals, acquisition of food costs time and energy. Only limited information is available on this aspect and that too in different units and formats, making a comparison between species difficult. The passive ambush foraging iguanid lizard *Callisaurus draconoides* predates and eats

TABLE 4.1

Effect of ration and feeding frequency on maturation and fecundity of *Gasterosteus aculeatus* (modified and condensed from Wootton, 1973) and on egg size and fecundity of *Macrobrachium nobilii* (from Kumari and Pandian, 1991)

Parameters	Feeding frequency [time/week (w)]		
	7 times (*ad libitum*)	3 times (43% of *ad libitum*)	2 times (28% of *ad libitum*)
Gasterosteus aculeatus			
Mature fish (%)	63	63	38
Size at maturation (mg)	813	666	574
Gonado somatic index	6.5	7.4	3.7
Spawning frequency (time)	30	22	12
Lifetime fecundity (no)	2610	1212	727
Egg size (mg dry weight)	0.3	0.29	0.27
Ration (% *ad libitum*)	**Egg size (j)**	**Fecundity (egg no.)**	
		Batch	Lifetime
Macrobrachium nobilii			
0–45	-	0	0
60	414	677	2033
80	426	1245	3755
100	431	1160	3481

5.8 mg food/g live body weight/d, in comparison to 13.3 mg/g/d consumed by an active predator the tepid lizard *Cnemidophorus tigris* (Waldschmidt et al., 1987). In the Atlantic menhaden *Brevoortia tyrannus*, filter feeding involves swimming activity, an energetically costlier strategy. Its foraging duration is increased from 4 h/d at the swimming speed of 50 cm/second(s) to 24 h/d at the speed of 20 cm/s (see Pandian, 1987c). The foraging duration of the salamander *Plethodon jordani* increases with increasing Relative Humidity (RH) from 8 h/d at 80% RH to 16 h/d at 100% RH. At the low prey density, the salamander uses a high energy costing pursuit tactic (2.4×10^{-2} J/g/minute [min]) but switches to a low energy costing ambush tactic (1.9×10^{-2} J/g/min) at high prey density (see Seale, 1987). Combing relevant information reported by Tucker (1973), Dade (1977) and Schaffer et al. (1979), Muthukrishnan and Pandian (1987) provided the most complete estimate in the energy cost of food acquisition for the honey bee *Apis mellifera*. A colony consisting of 50,000 honey bees collects 259 kg nectar (1,590, 680 kJ) and 24 kg pollen (339,066 kJ) annually. To forage at an energy cost of 4.6 J/km, a bee travels over 3 km to collect 370.7 kJ worth nectar and pollen, i.e. for one unit energy spent on foraging, the bee gathers 27 units of energy from pollen and nectar. Energy cost of foraging of the bee is 3.7% of the food energy acquired.

Some wasps forage caterpillars or spiders and transport them to the nest to provide food for their larvae. Flying a distance of 68.4 km and transporting 190 spiders worth 24.1 kJ during a period of 11 hours and 36 minutes, the sphecid *Trypoxylon rejector* provides food for the larvae developing in nine-celled nest (Muthukrishnan and Senthamizhselvan, pers. comm). Investing 1.04 kJ on its flight lasting for 2 hours and 30 minutes, the wasp *Sceliphron violaceum* transports a number of spiders equivalent to 13.32 kJ to provide food to its larva developing in an unused hole of an electrical socket. Thus, the energy cost of foraging and provisioning the relatively more motile spider amounts to 5.2% of the food provided (Pandian, 1985). However, it is only 2.7% for foraging and providing the relatively less motile caterpillar for the wasp *Delta conoideus* larva (see Muthukrishnan and Pandian, 1987). To subdue and swallow the cricket prey by *Chaloides ocellatus*, the cost is 5 hours for foraging and feeding, but the whole process costs only 0.3% of the food energy gained (see Waldschmidt et al., 1987). On the whole, time cost of foraging and food acquisition ranges from 2.5 hours in *S. violaceum* to 5.0 hours in *C. ocellatus* and the energy cost from 0.3% of the food energy gained for the latter to 3.7% for the former. The high energy cost of foraging to acquire food by bees may be attributed to the energy diversity of nectar (6.1 kJ/g) and pollen (14.11 kJ/g) (Southwick and Pimental, 1981), in comparison to the energy-rich prey caterpillar (24.3 kJ/g) and spider (22.2 kJ/g) (Muthukrishnan and Senthamizhselvan, pers. comm). For more information on energy cost of food acquisition in insects, Southwick and Pimental (1981) may be consulted.

With regard to temporal cost of feeding, the hemimetabolous insects allocate less time for feeding than the time-maximizing holometabolous insects like Lepidoptera, whose larvae feed incessantly to maximize energy stored to be used during non-feeding pupal and adult stages. Fluid-sucking hemipterans (e.g. gerrid, *Podicus*) feed for a longer duration than chewing insects like the mantids. Satiation time for *Gerris remigis* (150 minutes) is longer than that of *Mantis religiosa* (40 minutes). Grazing on the periphyton and detritus adhering to the rock, the caddis fly *Discosmoecus gilvipes* spends nearly 18 h/d on food acquisition (see Muthukrishnan and Pandian, 1987). For satiation, the dragonfly nymphs of *Mesogomphus lineatus* weighing 50, 100 and 160 mg require 12, 17 and 39 min, respectively (Pandian et al., 1979). Time budget estimates indicate that the shell-drilling snail *Acanthina punctulata* spends 5% of its time on searching for a suitable prey, 48–70% of time on drilling into the shell of prey and the remaining time on eating. Anchor drilling snail *Nucella lapillus* requires 2.5 days to completely drill the oyster shell to obtain the meat of mussel. *Mirex virgioneus* requires 3 days to drill the shell of *Turbinella pyrum* (see Pandian, 2017).

4.3 Feeding Groups

4.3.1 Fluid Feeders

Based on the food type and feeding modes, animals may broadly be grouped into (1) Fluid feeders, (2) Microphagy or suspension feeders and (3) Macrophagy or parcel feeders. The first group may be divided into (i) osmotrophs and (ii) fluid suckers. For many groups, the exact number of species could not be obtained; hence, the numbers in sub totals are rounded. The incidence of osmotrophism is limited to free-living aquatic gutless oligochaetes and parasites. One liter of surface water of the ocean contains 1 mg total Dissolved Organic Matter (DOM)/l. Free amino acids, comprising 5% of DOM, occur at concentrations of 5×10^{-7} M/l in free water and 1.1×10^{-4} M/l in interstitial water of sediments (see Pandian, 1975), in which most of the soft-bodied annelids thrive. Indeed, sea water is an organic 'soup'. As early as in 1909, Putter rightly claimed that DOM may also be absorbed across the body surface and used as an energy source by animals. An argument against Putter's claim is that if DOM can be absorbed *per se*, the DOM from the body fluids can also be leaked through the body surface into the ambient sea water. Ferguson (1971, 1972) measured both influx and efflux of free amino acids in many invertebrates and found that the net influx of amino acids is overwhelmingly inward. Thus, the amino acid uptake alone contributes up to 25% of the nutrient requirements of *Nais eliguis* and possibly up to 20% in other polychaete species (see Pandian, 2019). However, there are 100 and odd free-living gutless tubificids that solely depend on osmotrophic uptake to satisfy their nutrient requirements (see Pandian, 2019, Section 1.4). For more information on DOM partially fulfilling the nutrient requirements of Porifera, Frost (1987), Cnidaria, Pandian (1975) and Sebens (1987), Nemertea and Turbellaria, Calow (1987), Polychaeta, Cammen (1987) and Southward and Southward (1987), and Bryozoa, Gordon et al. (1987) may be consulted. In fact, representatives belonging to 11 marine phyla are reported to absorb amino acids from the ambient water (see Pandian, 1975). Not surprisingly, it is difficult to starve aquatic animals.

The highly specialized Cestoda, Nematomorpha and Acanthocephala, and a few parasitic Turbellaria, bereft of a mouth and gut, rely solely on the body surface to acquire the entire spectrum of nutrients (Table 4.2). The external surface of cestodes, for example, is evaginated into finger-like structures, which amplify the functional surface area by two-three folds. The cestodes derive most of their energy from glucose, which is actively absorbed across the body surface by Na^+-mediated transport system. Their amino acid transport system is characterized by high affinity for D-amino acids, as

TABLE 4.2

Estimate on species number of fluid feeders and microphagous suspension feeders (Pandian, 2011a, 2016, 2017, 2018, 2019, 2020, see Table 22.4)

Taxa	Species (no.)	Taxa	Species (no.)
1. Osmotrophs		**1. Filter-feeders**	
A. Free-living		Porifera	8,553
Gutless tubificids	100	Acnidaria	166
B. Parasites		Rotifera	2,031
Mesozoa	150	Lophophorata	6,314
Myxozoa	2,200	Crustacea	12,484
Nematomorpha	360	Aquatic insects & others	?
Acanthocephala	1,100	Bivalvia, Scaphopoda	10,428
Platyhelminthes		Crinoidea	700
Monogenea		Polychaeta	2,000
Fecampiidae	10	Hemichordata	25
Acholadidae	2	Urochordata	3,000
Cestoda	4,647		
Crustacea: Rhizocephala	3	Cephalochordata	35
Gastropoda: Entocanchida	~ 12	Subtotal = 45,736 or 50,000	
Subtotal = 8,565 or 9,500		**2a. Aquatic sediment-feeders**	
2. Fluid suckers		Holothuroidea	1,000
C. Free-living		Living spatangoida	?
Nematoda	11,070	Acoelomata	~ 1,000
D. Pests		Pseudocoelomata	~ 1,000
Hirudinea	684	Hemocoelomata	409
Insecta: Psocoptera, Phthiraptera, Hemiptera	98,000	Polychaeta	8,335
		Gastropoda	10,600
Acari	54,617	Pisces	975
Subtotal	164,371	Subtotal	~ 23,500
E. Parasites		**2b. Terrestrial sediment-feeders**	
Nematoda	15,930	Lumbricina (earthworms & others)	~ 500
Pentastomida	144		
Trematoda	21,183	Arthropod fragmenters	~ 2,500
Crustacea	6,000	Subtotal	3,000
Subtotal	43,257	Total	26,500
Total	217,130	**Grand total**	76,500

against L-amino acids in mammals. The cestodes cannot synthesize lipids and cholesterol but rely on the host (see Pandian, 2020). Understandably, the absorption of low molecular substances from the surrounding ambient water by free-living osmotrophs and from the host by the gutless parasites may require a longer duration to acquire adequate nutrients at the highest cost of food acquisition. However, no data are yet available on this aspect.

4.3.2 Fluid Suckers

Fluid suckers comprise free-living nematodes, pests like leeches, hemipteran bugs, ticks, mites and others, as well as parasitic trematodes and 59% nematodes, nematomorphs and pentastomids (Table 4.2). They suck the body fluid or blood from the host, as much as required or as long as it is available. Feeding exclusively on the blood of the host toad *Scaphiopus couchii*, the monogenean gill parasite *Pseudodiplorchis americanus* sucks 5 µl blood/d (Tocque and Tinsley, 1994). Conversely, a leech may gorge sucking the blood 4- (e.g. *Limnatus*), 5- (e.g. *Hirudo*) or 10-times (e.g. *Haemolopsis*) of its own normal body weight in a single satiated meal. In them, the interval between successive meals may last from 20 days to 8 months (mo) (see Pandian, 2019). Acquisition of food by sucking body fluid may also cost time and energy but they may be less than those of osmotrophs and suspension feeders. Sanguivores make use of the pharyngeal pump to suck blood from the host. The pump is operated by contraction and relaxation of the pharyngeal dilator muscles. Depending on the viscosity of the blood, the animal changes the feeding duration as well as pumping frequency. Depending on the viscosity of the blood, the bug operates the pump at different frequencies; for example, the Vth instar *Rhodnius prolixus* (Table 4.3) sucks less (2.4 nl/s) of the more viscous (6.5 cP) blood per stroke of the pump and prolongs the feeding duration up to 25.6 minutes. However, the feeding duration is limited to 12.4 minutes with less viscous blood sucked at faster pumping frequency.

TABLE 4.3

Effect of the meal on the pharyngeal pump of *Rhodnius prolixus* (recalculated from Smith, 1979)

Viscosity (cP)	Pumping frequency (stroke/s)	Stroke volume (nl)	Feeding	
			Rate (nl/s)	Duration (min)
0.8	7.3	60	7.3	12.4
2.2	6.2	58	6.0	15.0
3.6	4.7	53	4.2	17.9
4.5	4.6	47	3.6	21.0
6.5	3.8	38	2.4	25.6

4.3.3 Suspension Feeders

Prior to the return of suspension feeders (Table 4.2), the following should be stated: 1. It is difficult to assign some groups like the dipteran mosquitoes to any specific feeding group, as their larvae are filter feeders, but their adult females are blood suckers, while their males are plant juice feeders. The Australian garfish *Hyporhampus tyrannus* is a diurnal herbivore and nocturnal carnivore. In fact, many fishes commence as planktivores/herbivores and switch to carnivory, as size increases or age advances. So are the amphibians, in which the tadpoles are herbivores and the adults switch to insectivory (Seale, 1987). Rarely, the green sea turtle *Chelonia mydas* switches from carnivory to herbivory at the size of 30–300 g (see Waldschmidt et al., 1987). 2. For a large number of animal species, relevant information is not available. When it is available, it has been difficult to assign some taxa to one or other group. For example, (a) planktotrophic larvae of most aquatic animal species are filter feeders. However, their adults continue as filter feeders in lophophorates but sediment feeders in holothurians and carnivores in asteroids. In lepidopterans, the larvae are herbivores but their adults are fluid feeders. On the basis that maximal feeding occurs during larval stages, they are considered as macrophagic herbivores. 3a. Cammen (1987) classified 50 of the 78 listed polychaete families as sediment feeders, 12 families as filter feeders, 5 as carnivores and ~ 11 as herbivores. Considering polychaetes with 13,002 species (Pandian, 2019), approximate values were assessed for sediment feeders (8,335 species), filter feeders (2,000 species), carnivores (833 species) and herbivores (1,833 species). 3b. Carefoot (1987) listed feeding rates of 29 gastropod species. Of them, 20+ (69%) species are herbivores, 6 (21%) species carnivores and 3 (10%) sediment feeders. Of 106,000 gastropod species (see Pandian, 2017), 73,140, 22,260 and 10,600 species were considered as herbivores, carnivores and sediment feeders, respectively. 3c. A detailed analysis of Love's (1980) report on food of 600 fish species by Pandian and Vivekanandan (1985) revealed that 85 + 2% are carnivores and scavengers, 4% omnivores, 6% herbivores and 3% detritivores. Considering 32,510 fish species (see Table 2.1), 975 species can be assigned as sediment feeders (e.g. mullets). Thus, it has been possible to assign some species more or less precisely to fluid feeders and microphagic suspension feeders. All the other species are assigned to macrophagic parcel-feeders (Table 4.5). Being the very first attempt, these estimates are by no means exhaustive and precise but the proportions arrived from the analysis may show how the food and feeding mode limits or facilitates species diversity.

   With regard to the cost of food acquisition among filter feeders, they may be grouped into (1) active but sessile-, (2) mobile-filter feeders and (3) passive filter feeders; their respective species number is listed below:

1. Active sessile filter feeders			2. Motile filter feeders	
Porifera		8,553	Rotifera	2,030
Mollusca		10,428	Branchiopoda	662
Bivalvia	9,853		Cephalochordata	35
Scaphopoda	575		Pelagic Ascidia	150
Lophophorata		6,314	Pisces (*Brevoortia, Engraulis*)	25?
Entoprocta	200		Subtotal	2,902
Phoronida	23		**3. Passive filter feeders**	
Bryozoa	5700		Anthozoa	10
Brachiopoda	391		Aquatic insects	4,000?
Benthic ascidians		2,850	(mayfly, caddisfly, blackfly)	
Arthropoda		13,824	Subtotal	4,010
Cirripedia	1,000			
Ostracoda	10,204			
Aquatic insects	2,000?			
Loricifera		34		
Subtotal		41,403		

Of 10,856 cnidarian species, some 411 species (Scyphozoa = 200 species + Cubozoa = 36 species + Siphonophora = 175 species) are motiles. The remaining 10,455 cnidarian species are sessile but carnivores. They belong to the sit and watch predator group. From this list and Table 4.4, the following may be inferred: (1a) The passively filter feeding aquatic insect larvae spend substantial resources and energy on construction of filtering devices, and reconstruct them, when they are torn and after each molt (see Muthukrishnan, 1994). (b) The motile active filter feeders have to invest more energy to propel a current of water against the filtering device by ceaselessly swimming fishes, or rhythmically fanning branchiopod crustaceans and larvae of some aquatic insects. (c) For each unit of energy invested, the active sessile filter feeders gain much more food energy. At 20°C, the sessile sponges can filter an amount of water equivalent to their volume in < 20 seconds, and acquire bacteria and algae at 77 and 80% efficiency, respectively (see Frost, 1987). In other words, the volume filtered water for a day is as much as 4,320 times more than their own respective body volume. But the motile rotifer turns over only 10 times more than its body volume (Starkweather, 1987). Recalculated data reported by Fiala-Medioni (1987) reveal that the sessile filter feeding ascidian *Styela plicata* filters 210 l/g body weight/d to gain 25.2 mg C food/g/d, in comparison to the motile filter-feeding doliolid *Dolioletta gegenbauri* filtering just 175 ml/g/d to gain as little as 1.32 mg C food/g body

TABLE 4.4

Active and passive filter feeders: food size and volume of water filtered by filter feeders. The vacant columns indicate the need for data

Species/Taxa	Food size	Filtered volume
Active but sessile filter feeders: Porifera (Frost, 1987)		
Verongula sp	Marine	3.1 l/g dry weight/h
Mycale sp		12.4 l/g dry weight/h
Ephydatia sp	Freshwater	2.2 l/g dry weight/h
Spongilla lacustris		6.8 l/g dry weight/h
Bivalvia (Griffiths and Griffiths, 1987)		
Most bivalves	7–8 μm	~ 1.35 l/bivalve/h
Bryozoa (Gordon et al., 1987)		0.24–9.3 l/zooid/h
Cephalochordata (Fiala-Medioni, 1987)		
Branchiostoma lancelatum		4 ml/animal/h
Rotifera	4–17 μm	1–10 μl/rotifer/h
Cladocera	1–17 μm	
Active motile filterfeeders: Urochordata (Fiala-Medioni, 1987)		
Pelagic tunicates		2.5–8.8 l/g/h
Larvacea		0.0003–1.5 l/animal/h
Thaliacea		0.0012–15 l/animal/h
Pisces (Pandian, 1987b)		
Brevoortia, Engraulis	80–1,200 μm	360–900 l/fish/h
Amphibia (Seale, 1987)		
Xenopus tadpole	< 0.1, > 200 μm	0.31 l/g/h

Caddisfly larvae (condensed from Wallace et al., 1977)			
Species	Water velocity (cm/s)	Capturing mesh size (μm)	Diatom in gut (no)
Dilophilodes distinctus	19	41 × 6	3113
Arctopsyche irrorata	136	413 × 534	157
Hydropsyche incommoda	20–116	156 × 260	2454

C/d. A similar calculation of data reported for highly motile filter feeding fish *Brevoortia tyrannus* indicates that it is an energetically costlier strategy. (d) Hence, *with ~ 42,000 species, the active but sessile filter-feeding mode seems to facilitate and enrich species diversity. The other modes of active but motile filter feeding (~ 3,000 species) and passive filter feeding modes seem to limit the diversity.* (2a) Filter feeding fishes retain the food particle, whose size ranges from > 80 μm to 1,200 μm. In fact, they are more of 'biters' than 'filterers'.

Contrastingly, sessile filter feeders are capable of retaining food particle size of 1–17 μm; as ecosystem engineers, they play an important role in cleaning the water. (b) Irrespective of their huge number (~ 42,000 species), they do not incur any loss from competition. For example, sponges serve as refuge-substrata for many echinoderms (see Pandian, 2018); some entroprocts like *Loxosomella plakorticola* obligately require sponge to serve as substratum (Sugiyamo et al., 2010). (c) Contrastingly, competition among active but motile filter feeders eliminates a less competitive species. For example, the motile filter feeding rotifer *Brachionus calyciflorus* is a voracious feeder. A rotifer with 0.2 μg dry weight can ingest food equivalent to its body weight every 2 hours or at least 10 times a day (Starkweather, 1987); it steadily grows in number during the rearing period of 18 days. However, when the rotifer is reared with another motile filter feeder *Daphnia pulex*, the rotifer number is reduced on the 10th day and is outcompeted on the 16th day (see Wallace and Snell, 2010). Besides the size, the simpler structural organization of rotifers with limited number of cells and cell types may be a reason for their elimination against the structurally more complex daphnid.

About 29 and 70% of the earth's surface is covered by land and oceans, respectively (see Pandian, 2011). Nevertheless, the Primary Productivity (PP) on land (115 Gt C) is nearly two-times more than that (55 Gt C) of the oceans (*http://science.nasa.gov/earth-science/oceanography/living-ocean/remote-sensing/*). Hence, per unit area, the former is 6-times more productive than the latter. Though marine Primary Productivity (PP) is carried by 1 Gt biomass of phytoplankton (Carr et al., 2006), not more than 10% of their PP is consumed by herbivores in the pelagic realm and the remaining 90% of the production sink and sediment at the bottom of the oceans. While sinking, a major fraction of it is filtered and ingested by filter feeders. For example, the synconoid sponge *Leuconia aspersa* with 2.25 million flagellated chambers filters 22.5 l of water to acquire and satiate its daily food requirement (see Pandian, 1975). At the density of four individuals (indi)/l, *Artemia* is capable of filtering the entire water of Great Salt Lake once a day (see Reeve, 1963). Hence, the role played by the ecosystem engineering filter feeders by filtering enormous volume of water is admirable. But for filter feeders, the water in any aquatic system may remain dirtier and stinking. The unconsumed decomposing PP settles to enrich the organic composition of the sediment. For example, the sediment, composed of 0.1 mm sized particle may contain 0.02% organic nitrogen and 0.2% carbon, supports a rich flora of microbes. Sediments containing 0.06% (e.g. the sediment-feeding polychaete *Axiothella rubrocincta*, see Cammen, 1987) to < 0.3% carbon (sediment-feeding mullet *Mugil cephalus*, see Pandian, 1987c) are acceptable to sediment-feeding animals. The following two examples may reveal the massive role played by the bioturbulant sediment feeding ecosystem engineers. Sediment turbulation ranges from 114 g/animal/d in *Holothuria scabra* to 194 g/animal/d in *H. atra* (see Pandian, 2018). Inhabiting 1.6 km long, 3 m width and 30 cm depth of the intertidal zone of the Pacific coast, USA, the small (25 mm long) ophelid polychaete

Euzonus mucronata annually turbulate 14,600 ton (t) sediment (see Pandian, 2019). In all, ~ 23,500 species are assessed as aquatic sediment feeders (Table 4.2).

The same occurs on land too. But there are succinct differences: 1. Understandably, in the absence of filter feeders on land, the sediment feeders alone have to accomplish the decomposition of the leaf litter. 2. Mostly, bacteria serve to decompose the unconsumed sinking plants in aquatic habitats. But the arthropods—one reason for the enormous species diversity in them—and fungi do it on land. The former fragment the litter and render a relatively larger surface area for enzymatic decomposition by fungi. For example, the ruminant feces are degraded by coprophagous dung beetle and are decomposed completely by fungi. 3. Whereas the decomposing bacteria along with decaying plants serve as food for the suspension feeders and short-circuit the food chains in aquatic system, the nutrients and minerals arising from decomposition of the litter is directly returned to plants on land. Being ecosystem engineers, the service rendered by earthworms need no emphasis. In woodland forests, the leaf litter fall decreases from 5.5–15.0 t/ha to 2.5–3.5 t/ha and ~ 0.5 t/ha in tropics, temperate and alpine-arctics, respectively. In New South Wales, the UK, pastures containing no earthworm accumulate surface mats of 4 cm/y. Consuming 27 mg litter/worm/d, the earthworms decompose 3 t litter/ha/y (see Dash, 1987). Thus, earthworms accelerate further decomposition of decaying litter and render the nutrients, especially nitrogen (with additional inputs from nitrogen excretory products and mucous protein, see Dash, 1987) reusable by plants. Total production of mineral nitrogen by worms range from 30 to 50 kg/ha/y (see Pandian 2019).

The aquatic sediment feeders (~ 23,500 species) comprise Holothuroidea (1,000 species), living irregular spatangoids, acoelomates (1,000 species), pseudocoelomates (~ 1,000 species), hemocoelomates (409 species) and some polychaetes (8,335 species), fishes like mullets (975 species) and crustaceans (Anomura + mud crabs species) (Table 4.2). Along with the lumbricids (500 species) and a large number of litter fragmenting arthropods constitute terrestrial sediment feeders. In all, the number of sediment feeders may not exceed (~ 23,500 + 3,000) 26,500 species. Notably, the number of terrestrial sediment feeders constitute about eight times less than that of aquatic sediment feeders. *Arguably, the sediment feeders invest relatively more energy to acquire and satiate their daily food requirements than filter feeders. The turbulation of denser sediment substratum may demand relatively much more energy than filtering the less dense water medium. Hence, the sediment feeding mode—albeit required for cleaning the sediment—seems to limit the species diversity, especially in terrestrial habitat with less moisture content.*

Coprophagy: The fecal carbon content ranges from 2.0% in *Littorina planaxis* to 10–75% in *Hydrobia ulva* and nitrogen content from 0.02% in *H. ulva* to 4.48% in *Palaemonetes pugio* (Pandian, 1975). Likewise, the composition of feces of fishes ranges from 7.0 to 17.8% for protein, 2.6 to 33.3% for lipid,

3.5 to 55.9% for carbohydrate and 0.6 kJ/g dry weight to 3.5 kJ/g for energy content (see Pandian, 1987c). The intertidal populations of *Callianassa major* produce fecal pellets at a rate sufficient to provide 0.06 g organic carbon/ m^2/d for the coprophagic animals. With slow sinking rate (1,028 m/d) and low solubility (< 0.5%, see Pandian, 1987c), the nutritive feces of animals can be an important source of food for other animals not only in aquatic but also in terrestrial habitat; for example, there are 6000–8000 dung beetle species (Holter, 2016) feeding on ruminant dung. Among fishes alone, there are 45 coprophagic fish species ingesting feces of 65 fish species (Pandian, 1987c). At least nine species of shrimp and snails ingest feces of *Palaemonetes setiferus*. Their feeding rate ranges from 12 mg/g/d for *P. setiferus* feeding its own feces to 415 mg/g/d in *Palaemonetes pugio* ingesting feces of *P. setiferus*. The fraction of estimated fecal energy that meets metabolic requirement ranges from 9% in *Pagurus annulipes* to 86% in a 500 mg weighing *Glycera dibranchiata* ingesting feces of *Mugil cephalus*. In nine of 25 species investigated, the fecal energy provided 100–300% energy to meet the metabolic cost and to save the excess energy for growth. It is not known whether the feces-feeding species produce eggs from the excess energy gained. However, coprophagy seems not to have facilitated species diversity.

As indicated earlier, acquisition of adequate food is a prime determinant of progeny production and variations among the progenies are the raw materials for evolution and speciation. However, not all the food consumed by an animal is digested and absorbed, due to differences in the digestive ability; the ability varies from one food component to other and from species to species. For example, depending on algal species, the herbivorous fishes digest 43–81% of protein, 45–62% of carbohydrates, 21–41% of lipids and 22–87% of energy. In animals, energy budget is measured by C = F + U + R + P, where C is the food energy consumed, F, the Feces, U, nitrogenous excretory substances, R, the metabolism and P, the growth or production (see Pandian, 1987c). However, collection and quantification of F for aquatic animals pose problems like (i) filtration of large volume of water and (ii) dissolution (~ 0.5%) and disintegration (< 5%) of feces. In terrestrial habitats, Consumption (C) of incessantly feeding lepidopteran larvae involving the transpiratory water loss by the leaf poses a problem to the precise estimate of C. Not surprisingly, a number of indirect methods have been developed to estimate C from the readily measurable F in lepidopteran insects; hence, Absorption Efficiency (AbE), which is (C – F) ÷ C × 100, can also be easily estimated. For example, ^{14}C labeled *Chlorella* and chromic oxide are some markers to assess C, F and AsE (Assimilation Efficiency) or AbE (absorption efficiency). The ^{14}C labeling is a costlier and skilled method. Mathavan and Pandian (1974) discovered that for every unit of F + U egested, C was equal to 1.5 units for moths and 1.9 units for butterflies. The collection of fecal pellets and weighing them are easier and simpler. Thus, the method of estimation of C in lepidopteran larvae from F + U involving an error of

< 4% at a wide range of temperature (17–37°C) and food quality has proven as the easiest method to estimate C in lepidopterans (see Muthukrishnan and Pandian, 1987). Employing the F + U method, Muthukrishnan and Pandian (1987) went to the extent of predicting egg production from F + U in a number of lepidopterans.

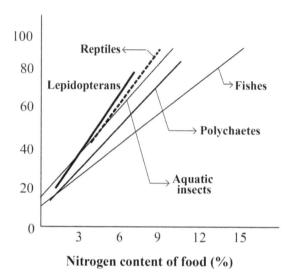

Nitrogen content of food (%)

FIGURE 4.1

Predicted trends for absorption (in lepidopterans, aquatic insects, fish), assimilation (insects and reptiles) efficiency of selected taxa as function nitrogen content of food (compiled from Pandian and Marian, 1985a, b, 1986a, b, c, d).

Chromic oxide (Cr_2O_3) does not dissolve in water and its presence in food can serve as a marker to quantify F. However, its use is limited to synthetic feed pellets only. Explaining the limitation of Cr_2O_3, Pandian and Marian (1986a) showed that determining N in food can be a simpler and widely practicable (in natural and synthetic foods) method to estimate F and AbE in fish. From over 100 values reported for 50 fish species, they reported that nitrogen content of food holds a positive significant correlation ($r = 0.9$) to AbE. Subsequently, they authors reported similar correlations of AbE for aquatic insects and polychaetes as well as AsE ($C - F + U \div C \times 100$) for lepidopterans and reptiles. Notable is that the maximum AbE ranges from < 80% for the lepidopterans consuming leaves with a maximum of 7.5% nitrogen to 92% in fishes eating food with nitrogen up to 13% (Fig. 4.1).

Attempts have also been made to predict C from the readily observable surfacing frequency in obligately air-breathing murrels and anabantids. In them, vertical swimming from deeper ponds may demand a larger quantum of food than those in a shallow pond. For example, a 4–5 cm long *Channa striatus* fingerling kept in 40 cm tall jar travels a distance of 1,503 m/d surfacing to breath air 1,879 time/d at the cost of 36 mg food/g fish/d than that kept at 2.5 cm shorter jar, in which the fish travels a distance of 65 m/d and surfaces 1,294 time/d at the cost of 20 mg/g/d (Pandian and Vivekanandan, 1976). From these observations, C can be predicted from the surfacing frequency with precision of ~ 95% in *C. striatus*, *C. punctatus* and *Anabas scandens* but with only 84% precision in *Macropodus cupanus* (see Pandian, 1987c). Like *C. striatus*, the Indian bullfrog *Rana tigrina* tadpoles also surface to breath air. Reared in a shallow (2.5 cm depth) aquarium, the tadpole surfaced 98 time/d travelling a distance of 5.0 m/d, but the tadpole reared in 5 cm deep aquarium surfaced 1,325 time/d travelling 14.6 m/d (Pandian and Marian, 1985e). In a subsequent elaborate study, Pandian and Marian (1985d) estimated the effect of feeds, ration, temperature and rearing density on food utilization by the tadpole. From these studies, they used Surfacing frequency (Sf) as a predictor of feeding rate (Cr) and Metabolic rate (Mr). According to them and Ponniah and Pandian (1981), Cr is a potent factor affecting growth (Pr) and Metabolism (Mr). Viewing Cr as the cause and Mr (= Sf) as its effect, Sf was shown as a predictor variable Cr. The computed regression equations to describe the relations between Sf and Cr, Sf and Mr, Cr and Mr as well as Cr and Pr were all linear and highly significant (p < 0.001). Hence, Pr of the tadpole can be predicted using the readily measurable Sf. This predictable model is described below:

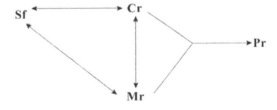

4.4 Macrophagy

With this lengthy preamble, this account returns to assess the number of macrophagous species. In the comparison to the microphagous filter feeding minimum duration of ~ 12 h/d in intertidal zone, macrophages require far less duration to predate and consume their prey. Due to page limitation, the description is limited to predation of mosquito (*Culex fatigans*) larvae, being the biological measure of controlling the pest. The number of larvae predated by larvivorous fishes ranges from 17/d by *Gambusia affinis* to 20/d by *Poecilia reticulata* (see Reddy, 1973). Surprisingly, it is higher for insects like dystiscid coleopterans (40/d) and hemipteran *Notonecta undulata* (56/d). To this

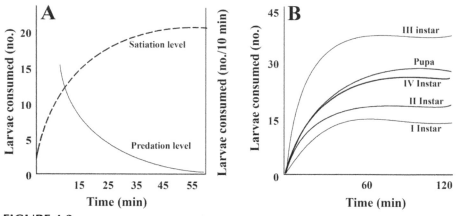

FIGURE 4.2

A. Levels of satiation and predation (*Culex fatigans*) of larvae by *Mesogomphus lineatus* as a function of time (compiled and redrawn from Mathavan, 1976). B. Effect of larval stage and density on larvae consumption (compiled and modified from Marian et al., 1983).

list, a frog's tadpole and dragonfly nymph are to be added. The predation rate of *Mesogomphus lineatus* nymph decreases from 15 larvae during the first 10 minutes to 2 larvae during the fourth 10 minutes, as satiation level is progressively increased (Fig. 4.2A). The rate of predatory attacted also decreases from 2.05/min to 0.4/min and successful capture from 1.33/min to 0.05/min during the first and fifth 10 minutes, respectively. The numbers of prey consumed by (*Rana tigrina*) tadpole increases from 15 to 40 for the third instar larvae and 15–30 for pupae (Fig. 4.2B). This contrasting relationship for appeptite vs time and stomach fullness vs time hold good for *Salmo gaidneri* (Grove et al., 1978). However, the duration required for satiation increases not only with increasing body size but also with increasing temperature in *S. trutta* (Eilliott et al., 1975). The dragonfly nymph requires 15 and 5 minutes to handle a pupa at the prey density of 100 and 800 mg, respectively. Hence, the predators seem to prefer larger prey, and predate and consume more number of them at higher prey density. Irrespective of prey density and species, *M. lineatus* consumes ~ 45 mg larvae of either *C. fatigans* or *Anopheles* sp (Mathavan, 1976).

While it has been possible to assess the number of fluid feeders and suspension feeders, repeated surveys revealed that for want of relevant information on food and the feeding mode in a vast number of macrophagy is not possible even to approximately assess the number of microphagic species. Therefore, the substraction of the cumulative number of fluid feeders (228,198) and suspension feeders (76,500) from the total number animal species (1,543,196, see Table 2.1) is perhaps the only means, by which the number of macrophagic feeders (1,543,196 – 304,698 = 1,238,500) can be assessed. Accordingly, the number of macrophagous feeders is 1,238,500 species, which is 80.3% of all animal species. Except for a few, not much information is yet available on the costs of feeding duration and food acquisition of the selected feeding modes. However, the number of species for each of the feeding modes may indirectly provide an idea of the required information and its implication to species diversity.

From Table 4.5, the following may be inferred: (1) The proportion of feeding modes decreases in the following order: macrophagous feeders (80.3%) > fluid suckers (14.2%) > filter feeders (3.2%) > sediment feeders (1.7%) > parasitic osmotrophs (0.6%) > free-living osmotrophs (0.006%). Therefore, macrophagy and osmotrophy are at the two opposite ends, suggesting that feeding duration and food acquisition cost of macrophagy are far less than that for osmotrophs. Acquisition of low molecular nutrients against osmotic gradient costs the osmotrophs a longer duration and perhaps the highest cost of food acquisition. *Not surprisingly, osmotrophy deters species diversity.* In comparison to the free-living gutless aquatic oligochaetes, the parasitic osmotrophs inhabit in richer and more concentrated liquid food. Not surprisingly, *the species number of parasitic osmotrophs is ~ nine times more than that of the free living gutless oligochaetes.* (2) The number of fluid suckers

TABLE 4.5

Food and feeding modes. Number of species arrived from Table 3.1

Feeding mode	Species (no)	As % of total no. of animal species
I. Fluid-feeders	228,198	14.7
1. Osmotrophs		
A. Free-living	100 (0.006%)	
B. Parasitic	9,400 (0.60%)	
Subtotal	9,500 (0.61%)	
2. Fluid-suckers		
C. Free-living	11,070 (0.72%)	
D. Pests	164,371 (10.65%)	
E. Parasites	43,257 (2.80%)	
Subtotal	218,698 (14.2%)	
II. Suspension-feeders	76,500	5.0
1. Filter-feeders	50,000 (3.2%)	
2. Sediment-feeders	26,500 (1.7%)	
a. Aquatic	23,500 (1.5%)	
b. Terrestrial	3,000 (0.2%)	
III. Macrophagers (1,543,196–304,698)	1,238,500	80.3

is more than that of the filter feeders. Notably, the fluid suckers, irrespective of their pest or parasitic nature, enjoy nutritionally richer food than the filter feeders. Further, their feeding duration may be relatively less than that of filter feeders. It may be recalled that leeches have a long interval between successive feedings. *In terms of food acquisition cost, fluid sucking may not be as costly as that of filter feeders.* (3) Within suspension feeders, *the number of filter feeders is almost two times more than that for sediment feeders and suggests that the cost of food acquisition by sediment feeders is higher, as they turn over denser sediment substratum, in comparison to the filter feeders turning over the less dense water medium* (despite their total absence in terrestrial habitat). Arguably, *the high cost sediment feeding mode does not foster or promote species diversity.* Apparently, *the costs of time and energy are the lowest for macrophages. Not surprisingly, macrophagy fosters and facilitates species diversity.*

Part B
Life History Traits

Of many life history traits, this account has identified (i) sexuality, (ii) gametogenesis and (iii) embryogenesis as some that profoundly influence species diversity.

B1
Sexuality

"Sex is a luxury and costs time and energy but ensures recombination to generate genetic diversity (Carvalho, 2003). As benefits accruing from genetic recombination outweighs the cost of time and energy, sex is successful and evolved as early as 1.6–2.0 billion years ago (Butlin, 2002) and has been successfully manifested in a wide range of microbes, plants and animals" (Pandian, 2012). In sexual reproduction, recombination during meiosis and gamete fusion at fertilization produce genetic diversity—the raw material for evolution; speciation is a by-product of evolution. In some animals, sexual reproduction is, however, supplemented by 'asexual' reproduction. There are three types of the so called 'asexual' reproduction. Progenies appearing from clonal multiplication or parthenogenic or unisexual reproduction tend to accumulate deleterious mutations and cause inbreeding depression; hence, many authors consider them as 'asexual' reproduction. Yet, the differences between them must be distinguished. The clonals arise from pluripotent/ multipotent or dedifferentiated stem cells but progenies appear from unreduced eggs in parthenogens and reduced egg diploidized by gynogenesis in unisexuals. The 'asexual' progenies exploit favorable conditions like abundant food supply and/or optimum temperature by rapid multiplication of the fittest clone(s). With difficulty of finding a mate in patchily distributed populations, some have switched preferably to hermaphroditism or parthenogenesis. With meiosis during gametogenesis, the hermaphrodites produce new gene combination(s) but miss them at fertilization, as gametes arising from the same individual are fused. Yet, they have developed an array of strategies to avoid selfing. But parthenogenesis is deprived of producing any new gene combination, as neither meiosis nor fertilization occurs in them. Not surprisingly, parthenogenesis is chosen by < 1% of all animals (Bell, 1982), whereas 5% of them have opted for hermaphroditism (Jarne and Auld, 2006). Nevertheless, meiosis occurs in meiotic parthenogens and males do appear sporadically in mitotic parthenogens, except in bdelloid rotifers; the sporadic occurrence of males results in meiosis and fertilization, and thereby eliminates the accumulated deleterious mutations. In clonals, the probability of elimination of these mutations is minimal.

5

Gonochorism and Males

Introduction

In gonochores, female and male sexes are expressed and functional in separate individuals. In them, motility is demanded, as male and female have to search for mutually acceptable mate(s). Surprisingly, cent percent gonochorism is manifested from the earliest Aorganomorpha, the Placozoa onward to hemocoelomate Pentastomida (Fig. 5.1), despite their simpler structural organization and low motility. However, with the breaking point commencing from the lophophorate Entoprocta, cent percent hermaphroditism occurs in Bryozoa (5,700 species), Chaetognatha (150 species) and Urochordata (3,000 species); it occurs also in Gastrostomulida (100 species) and pterobranchiate Hemichordata (25 species), which are more or less drop-outs (Fig. 5.1). *Incidentally, gnathostomulids are reciprocals and are functionally gonochores. In minor phyletics, evolution seems to have proceeded from a 'wrong combination' of gonochorism and structurally simpler low motility to that of structurally complex sessile hermaphroditism. This may partly explain species poverty (except in Rotifera and Nematoda) among the lower minor phyletics (Placozoa to Pentastomida) and species richness among higher (Bryozoa and Urochordata) minor phyletics. Contrastingly, major phyletics have commenced with the 'right combination' of 62% hermaphroditism and sessility in Porifera and proceeded toward gonochorism with increasing structural complexity and motility in higher major phyla like Arthropoda and Vertebrata. Not surprisingly, motility has been the driving force of evolution.*

Available evidence suggests the origin of hermaphroditism from gonochorism. For example, Jarne and Auld (2006) estimated from Heller's (1993) data that hermaphroditism has originated independently at least 40 times from the ancestral gonochoric gastropods. Yet, it must be noted that among the earliest diploplastic Porifera, most of them are hermaphrodites but function as behavioral gonochores. This is also true of the triplastic Platyhelminthes with the exception of cestodes. It is then difficult to comprehend the origin of ~ 5,500 hermaphroditic turbellarian species from about a dozen gonochoric fecampiids and acholadids, and

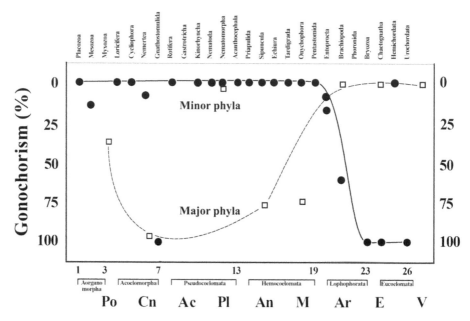

FIGURE 5.1

Contrasting trends (drawn from data listed in Tables 5.1, 6.3) showing proportion of gonochorism and hermaphroditism in minor and major phyla. Po = Porifera, Cn = Cnidaria, Ac = Acnidaria, Pl = Platyhelminthes, An = Annelida, M = Mollusca, Ar = Arthropoda, E = Echinodermata, V = Vertebrata.

12,012 speciose hermaphroditic digeneans to originate from 30 and odd gonochoric schistostomatids. This is true for sponges (Sara, 1974). Hence, it is likely that *the metazoans have originated from functionally gonochoric hermaphrodites, as in sponges and flukes. To reduce the cost of manifestation and maintenance of dual sexuality, the proportion of the gonochorics increases as in annelids and molluscs (see Fig. 1.1)*; with increased motility, almost cent percent gonochorism emerged subsequently in echinoderms and vertebrates. Of course, hermaphroditism may have secondarily appeared from the gonochoric gastropods and others.

5.1 Quantification

A vast majority of animals are structural gonochores (Table 5.1). Surprisingly, none of the major phyla are comprised of only gonochores. Conversely, many minor phyla consist of only hermaphrodites (e.g. Gnathostomulida, Bryozoa, Chaetognatha and Urochordata). Remarkably, all pseudocoelomates and hemocoelomates are gonochores with a very few exceptions. The values for gonochoric species in Porifera, Cnidaria, Annelida, Arthropoda,

TABLE 5.1

Number of gonochoric species in major and minor phyla. 1 = Van Soest et al. (2012), 2 = Ereskovsky (2018), 3 = Mapstone (2014), 4 = Shikinova and Chang (2018), 5 = Carlon (1999), 6 = Hyman (1940), 7 = Pandian (2020), 8 = Pandian (2019), 9 = Zhang (2011), 10 = Pandian (2017), 11 = Pandian (2018), 12 = Pandian (2016), 13 = Pandian (2011a), 14 = Pandian (2021)

Major phyla			Minor phyla[14]		
Phylum	Species (no.)	Gonochore (no.)	Phylum	Species (no.)	Gonochore (no.)
Porifera[1,2]	8,553	3,260	Placozoa	3	3
Cnidaria[3,4,5]	10,856	10,834	Mesozoa	150	108
Acnidaria[6]	166	0	Myxozoa	2200	2200
Platyhelminthes[7]	27,700	45	Loricifera	34	34
Annelida[8]	16,911	12,855	Cycliophora	2	2
Arthropoda[9]	1,242,040	1,240,988	Nemertea	1,300	1,299
Mollusca[10]	118,451	91,378	Gnathostomulida	100	0
Echinodermata[11]	7,000	6,992	Rotifera	2,031	2,031
Vertebrata[9]	64,832	64,695	Gastrotricha	813	[††]813
Total	1,496,509	1,431,047	Kinorhyncha	200	200
			Nematoda	27,000	[†]26,990
			Nematomorpha	360	360
			Acanthocephala	1,100	1,100
Major Phyla	1,496,509	1,431,047	Priapulida	19	19
Gonochores (%)	95.6		Sipuncula	160	160
Minor phyla	46,687	37,045	Echiura	230	230
Gonochores (%)	79.3		Tardigrada	1,047	*605
Grand total	1,543,196	1,468,092	Onychophora	200	200
(%)	95.1		Pentastomida	144	144
			Entoprocta	200	164
			Phoronida	23	14
Crustacea[12]	54,384	53,241	Brachiopoda	391	261
Teleost fish[13]	32,510	32,373	Bryozoa	5,700	0
[†] only ~ 10 species hinted			Chaetognatha	150	0
[††] commence as parthenogens			Hemichordata	130	108
* 164 ♂ + 50% ♀ = 441 species			Urochordata	3,000	0
			Total	46,687	37,045

Mollusca and Echinoderamta were taken from Table 6.3. For example, the 3,849 speciose Clitellata are all hermaphrodites. The number for polychaetes was reached after substracting another 207 hermaphroditic polychaete species (see Pandian, 2019). Among minor phyla, a compromise was made only for the 1,047 speciose Tardigrada, in which 164 species are identified as gonochores. For the remaining 882 species, 50% was taken as gonochores and the other 50% as hermaphrodites, as no information is available for them (see Tables 18.1, 18.2 of Pandian, 2021). *Of 1,496,509 major phyletics, 95.6% species are gonochores. However, the corresponding value for the minor phyletics is far less at 79.3%. On the whole, 95.1% of the animals are structural gonochores.*

5.2 The Males

Maintenance of males is indeed a luxury, as their main function is to provide sperms. Nevertheless, the range of morphotypes encountered within males is astounding, for example, teleosts: primary and secondary males, initial and terminal males, territorial, sneaker and satellite males in scarid and labrid fish (Pandian, 2011a), copulatory and non-copulatory males in *Orconectes immunis* (Pandian, 2016), insects: digger and hover males, reproductive males and non-reproductive drones (Muthukrishnan, 1994), molluscs: sneaker and consort males in *Loligo bleekeri* (Pandian, 2017) as well as hermaphrodites and dwarf males in androdioecious crustaceans and bivalves. In teleosts, the described morphotypes represent the different phases in the ontogenetic pathway of sexualization. In gonochoric *O. immunis* too, the occurrence of copulating and non-copulating males is a season-dependent event. Conversely, the dwarf males occur mostly in the hermaphroditic cirripede crustaceans, molluscs and in two speciose Cycliophora.

5.2.1 Size Reduction

With inclusion of the incredible wastage of sperms, gonochores allocate up to a quarter of their body weight on semen production (e.g. 27% in arctic cod: *Boreogadus saida*, 35% in *Galaxias maculatus*, see Pandian, 2011a; 28% in *Ophiocoma aethiops*, see Pandian, 2018). To reduce the cost of semen production, gonochorics have adopted the following strategies: (1) Reduction in male ratio at population level and (2) reduction in somatic size of males and/or production of semelparous males (see Section 15.4). Of them, the former is more widely distributed than the latter.

Accordingly, male size reduction occurs in prosobranchs, cephalopods (see Pandian, 2017), rotifers, sipunculids and onychophores (see Pandian, 2021). For example, the prosobranch *Lacuna pallidula* male is one third the length and one tenth of the weight of a female (Hyman, 1967). In cephalopods, sneaker males are present in three families: Trematopodidae, Ocythoidae and Argonautidae. The pygmy sneaker male of *Argonauta* sp not only lacks a shell (Finn, 2009) but also measures one 600th the size of the female (Finn and Norman, 2010). In the ascothoracid barnacle crustacean, *Trypetesa lampas*, the dwarf male weighs one 500th of the females and up to 15 dwarf males may be carried by a female (see Rinkevich, 2009). In Onychophora, a male measures half the length of a female and weighs one third of a female (Menge-Nagera, 1994). A pygmy male of *Bonellia viridis* measures only 1–3 mm in length, in comparison to 1 m length of a female (see Pandian, 2021). In monogonont rotifers also, the smaller males lack the gut and other organs, and the degradation leads to such an extent that the *Asplanchna sieboldi* male is a sac filled with ~ 50 sperms (Fig. 5.2C–F). In members of the Orders Flosculariacea and Collothecacae, the males are one tenth or less the size of females. They appear from smaller eggs, do not grow after hatching and often resemble the juvenile females. In the parasitic entoconchid prosobranchs, parasitism has not only 'sacculinized' the male into a 'testis' but also holds it within the female's body cavity (Fig. 5.2A–B). This reduction in somatic body size reaches a climax during the development of 'dwarf males' in cycliophores; in *Symbion pandora*, the differentiation involves miniaturization by massive reduction in the size of epidermal and mesothermal cells from 3,900 to 900 μm³. As a result, the male measures 4 μm in length and has only 50 nucleated somatic cells (Neves and Reichert, 2015). The massive enlargement of testes in dwarf males housed in the pallial or branchial chamber of hermaphroditic bivalves result in progressive loss of shells, excretory system, labial palps, inhalant and exhalent siphons (see Pandian, 2017).

Labile sex differentiation has been demonstrated in the echiuran *Bonellia* and cirripedes. In cirripedes, the cyprid larva settling on the mantle rim of conspecific hermaphrodite metamorphoses into a dwarf male but that on a substratum becomes a hermaphrodite. On exposure to surplus of hermaphrodites, not more than 50% cyprid larvae of the *Scalpellum scalpellum*

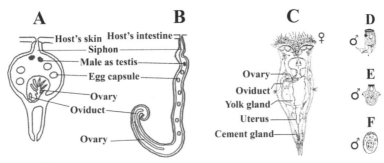

FIGURE 5.2

Anatomy of entoconchid prosobranch gastropods A. *Entocolax* and B. *Thyomicola* (free hand drawings from Lutzen, 1968). C. monogonontan female and males showing progressive reduction in D. *Mytilina mucronata*, E. *Pedalia mira* and F. *Asplanchna sieboldi* (free hand drawings from different sources).

develop into dwarf males. Hence, a genetic mechanism is suggested. With precocious sexual maturity, the cirripede dwarf males enjoy a higher fertilization success per sperm. The number of dwarf males per hermaphrodite ranges from one in *Paralepas klepalae* to 13 in *Octolasmis warwiokii*. A positive relationship between the number of dwarf males and body size of hermaphroditic *Koleolepas owis* and *Vernum brachiumcancri* has been reported. Apparently, the larger hermaphrodites receive more sperms. To facilitate cross-fertilization, the hermaphrodites settle within a distance of 50 mm from one another.

5.2.2 Ratio Reduction

To compensate the high cost of semen production, gonochores may reduce the male ratio. Male ratio is reduced by (i) parthenogenesis, (ii) protogyny (see Table 6.2), (iii) androdioecy, (iv) quality and/or quantity of food supply, and (v) parasitic castration (see Table 23.4). Male ratio is 0.07 in the deep sea copepod *Stenhelia*. Reduction in the life span of the male is another strategy. Males are found only up to the size of 17 mm (carapace length) but females up to 44 mm in *Potimirium brasiliana* (for more examples see Pandian, 2016). In Onychophora *Euperipatoides rowelli*, males weighing > 200 mg do not survive, while females survive up to 600 mg size (Menge-Najera, 1994). Among parthenogens, males do appear rarely; when they appear sporadically, their ratio does not exceed 0.01 (e.g. *Pleuroxus humulatus*, see Pandian, 1994). Androdioecy is constituted by the presence of hermaphrodites and males. It is common among notostrocan (12 species) and spinicaudatan (40 species) crustaceans. In them, the male ratio is 0.13 for *Triops* sp (see Pandian, 2016) and 0.21 for 14 species belonging to the genus *Eulimnadia* (Weeks et al., 2008). Food quality and quantity exert a profound effect on the sex differentiation process. Calanoid copepods pass through C_1–C_5 copepodid stages prior to becoming adults. When food supply is low or scarce, the presumptive males transdifferentiate into females and thereby reduce male ratio (Gusamao and McKinnon, 2012). In the digynic protandric hippolytid *Hippolyte inermis*, the diatom *Coconeis neothumens* is supplemented by a diet to zoeal and post-larval stages eliminating the male phase and facilitating direct differentiation to the female phase (Zupo, 2000, 2001). Whereas a cereal diet does not change the genetic sex ratio of 0.5 ♀ : 0.5 ♂ in the polychaete *Dinophilus gyrociliatus*, the tetranin diet reduces male ratio to 0.33 (Prevedelli and Vandini, 1999).

In all, the incidence of dwarf males may not exceed 1,000 species. With increasing food supply, male ratio of calanoid copepod was increased for selection of the fittest male to mate (Guasamao and McKinnon, 2012). In the lepidopteran *Pontia protodice*, the male ratio increased from 54% at the density of 133 no/ha to 92% at the density of 1,019 no/ha (Shapiro, 1970) for selection of the fittest male to mate. Therefore, alteration of sex differentiation to produce changes in the sex ratio seems to be the more preferred strategy than the strategy of reducing somatic growth including those in dwarf males.

6

Hermaphroditism and Selfing

Introduction

Sexual reproduction introduces new gene combinations during meiosis and at fertilization and thereby increases heterozygosity at the population level. Yet, it levies a heavy premium on time and energy budgets of animals, especially in those with a short life span, and causes an incredible waste of gametes. Hence, the gonochoric sexual reproduction is indeed a luxury. Not surprisingly, animals display the most divergent expression of sex. In them, sexuality ranges from gonochorism to unisexualism (e.g. Lepidoptera: *Luffia lapidella*, see Muthukrishnan, 1994; Teleostei: *Poecilia formosa*, see Pandian, 2011, Section 7.2) and hermaphroditism. The expression of female and male reproductive functions in a single individual results in hermaphroditism. Within hermaphroditism, sexuality ranges from simultaneous to sequential and serial hermaphrodites. The sequentials undergo natural sex change once in their lifetime in a single direction but the serials do it more than once in either direction. The sequentials are divided into protandrics changing sex from male to female and the reverse is true of protogynics. In fishes, the protogynics are further divided into monandrics and diandrics. In the former, the secondary males appear indirectly from sex-changing females but in diandrics, primary males arise directly, in addition to the indirectly appearing secondary males. A similar dichotomy is recognized in protandric species and the corresponding terms are monogyny and digyny. In sequential teleosts, the monandrics are again divided into monochromatics with all mature initial males (IP) displaying the same dull body color as the females and dichromatics, in which the body color is bright in Terminal Males (TP). The serials may also be divided into groups: 1. Bidirectional hermaphrodites, in which the undifferentiated juvenile always matures into a female followed by the usual serial sex change, as in *Gobiodon histrio* and 2. In cyclic hermaphroditism, the juvenile differentiates into either a female or a male and then undertakes the usual serial sex change, as in *Paragobiodon echinocephalus* (Pandian, 2011). The bivalves, *Teredo navalis* and *Ostrea edulis* commence as males and then cyclically change sex (Pandian, 2017). In a

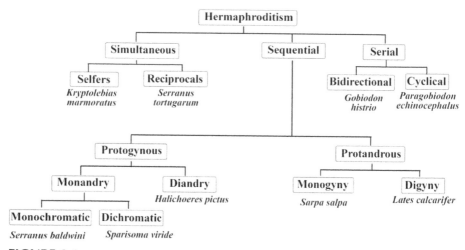

FIGURE 6.1

Patterns of functional hermaphroditism in teleosts (from Pandian, 2011b, modified).

5-year study, Galtsoff (1964) found that in *Crassostrea virginica*, an individual changed sex four times. Considering teleosts as an example, the patterns of functional hermaphroditism is illustrated in Fig. 6.1.

6.1 Gonadal Diversity

The hermaphroditic platyhelminths, annelids, cirripede crustaceans, lophophorates, chaetognaths and ascidians possess distinctly separate female and male reproductive systems. Conversely, the two reproductive systems commence with one or more ovotestes with separate (e.g. crustaceans, fishes) or combined (e.g. pulmonates) gonadal ducts. The ovotestis occurs in pulmonates (see Pandian, 2017), decapod crustaceans (see Pandian, 2016), some fishes (see Pandian, 2017) and a few holothuroids (e.g. *Amphiura monorina, Ophiolebella biscutifera,* see Pandian, 2018) and nemerteans (see Pandian, 2021). Davison (2006) considers ovotestis as an underdeveloped organ in evolution. Interestingly, the monoaulic ophisthobranchs like *Okadaia* have separate ovarian and testicular follicles, whereas the diaulics like *Lobiger* have an undelimited ovotestis. The structural diversity in ovotestis indicates that shifts have occurred at least 40 times from the ancestral gonochoric state to hermaphroditism (see Jarne and Auld, 2006). Figure 6.2 illustrates the ovotestis and its structural diversity in some hermaphrodites. The ovotestis may be grouped into two types: 1. In delimited type, the ovary and testis are located in separate chambers, separated by a connective tissue, as in the sparid fish *Pagurus pagurus* (Fig. 6.2A) and shrimp

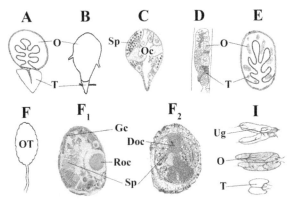

FIGURE 6.2

Structural diversity of ovotestis in some hermaphrodites. Delimited type: A. *Pagurus pagurus*, B. shrimp (from Pandian, 2014, 2016). Undelimited type: C. *Prostoma rubrum* (after Coe, 143), D. alternate locations in a hermaphroditic nematode (after Potts, 1910), E. *Epinephelus guttatus* (from Pandian, 2014), F. pulmonate, F_1. section through the ovotestis of *Biomphalaria glabrata* and F_2. *Pseudunela cornuta* (modified and redrawn from Rivero-Wendt et al., 2014, Neusser et al., 2009). I. Protogynic change from undifferentiated to ovary and then to testis in *Safflamen chrysopterus* (from Pandian, 2014). O = ovary, Doc = differentiating oocyte, Roc = ripe oocyte, Oc = oocyte, T = testis, OT = ovotestis, Gc = germ cells, Ug = undifferentiated gonad, Sp = sperms.

(Fig. 6.2B). 2. But they are located in a single chamber in the undelimited type (e.g. nemertean *Prostoma rubrum*, nematode *Diplogaster maupasi* and grouper fish *Epinephelus guttatus*, Fig. 6.2C–E). In them, oogenesis and spermatogenesis may temporally and transiently be separated. For example, the former occurs in the ovotestis of *Biomphalaria glabrata*, when the latter remains inactive; the reverse occurs in *Pseudunela cornuta* (Fig. 6.2F$_1$, F$_2$). It is likely that the pulmonates adopt the temporal separation strategy of oogenic and spermatogenic activity to escape from selfing. More importantly, it is from the same anlage of the gonad, the ovary functions initially and then testis in the protogynics like *Safflamen chrysopterus*, (Fig. 6.2I) and the reverse is true in the protandric bryozoans like *Alcyanidium duplex* (see Pandian, 2021). In protogynics, protandrics and serials as well as Simultaneous Hermaphrodites (SHs), the Primordial Germ Cells (PGCs) retain bisexual potency beyond sexual maturity.

6.2 Inbreeding and Consequences

Simultaneous Hermaphroditism (SH) provides a theoretical option to self, although all SH need not necessarily opt for it (Heller, 1993). Among SH, especially selfing SH, inbreeding depression is measured using the following parameters: (i) parental fecundity, (ii) survival and (iii) fecundity of F_1 generation (e.g. Jarne et al., 2000). As it is difficult to estimate these parameters in natural populations, information is assembled from experiments (e.g. Escobar et al., 2011). When cross breeders face a dilemma in the absence of mate(s), they can self and pay the cost of selfing or wait for the arrival of a potential mate. Hence, the waiting time is also added as another parameter, which results in delayed sexual maturity, i.e. longer generation time.

Survival: For relevant information on reduced survival of F_1 progenies arising from SH, only a few examples could be assembled. At 10°C, survival of miracidium larva decreased from 40 hours in gonochoric *Schistosoma mansoni* to ~ 22 hours in *Fasciola hepatica* SH. The corresponding values are 12 hours and 6 hours for them at 30°C (Fig. 6.3A). These values suggest that the fitness of SH is ~ half of that of gonochores. In them, not only survival but also swimming speed is reduced from 65 mm/second (s) to 50 mm/s (Pandian, 2020, Chapter 3). In *Ophryotrocha*, adult survival decreases at a faster rate in *O. diadema* SH than that in gonochoric *O. labronica* (Fig. 6.3B). Notably, a small fraction of *O. diadema* survives longer (45–50 days) than *O. labronica* (40 days), indicating that the former waits for a potential mate. Within *Helobdella robusta*, cross fertilizing leeches survive longer than selfers; the trend for the survival is more gradual for cross fertilizers than that for the selfers (Fig. 6.3C).

Parental fecundity: In snails, the oviposition frequency increases, as age advances. With facultative or obligate selfing treatment, the frequency is almost equal in *Pseudosuccinea columella*. But the difference between the treatment widens in others; (i) *Planorbula armigera* is driven to obligate cross breeding but *Heliosoma duryi* to facultative selfing up to the age of 50 days and then 40% of them switching to obligate cross breeding between the age of 85 and 260 days (Fig. 6.3D). Notably, the age at first sexual maturity and generation time are approximately equal in cross breeders. But these are delayed or prolonged in facultative selfers due to the addition of waiting time. Among 29 bassommatophoran pulmonate species, the waiting time is prolonged to a maximum of 32 days in *Heliosoma trivolvis* (Escobar et al., 2011). In the terrestrial bulimullid pulmonate *Bulimulus tenuissimus*, the enforced isolation extends generation time up to 475 days and prolongs the waiting for the arrival of a potential mate. Lifetime fecundity is also reduced from 48 eggs in group reared snails to only 9 eggs in snails reared in isolation (Table 6.1). Relevant data available for reduction in fecundity in SH is scarce. Perhaps, the best example for fecundity reduction in SH is reported from the polychaete genus *Ophryotrocha*, in which some species are gonochores and

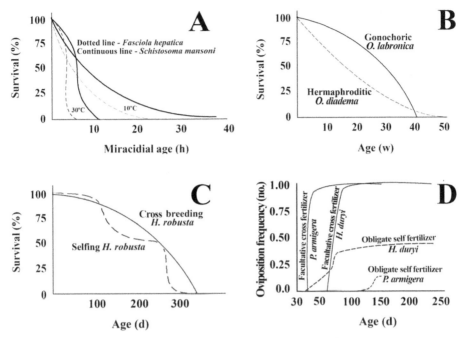

FIGURE 6.3

A. Effect of temperature on survival of miracidium of gonochoric *Schistosoma mansoni* and hermaphroditic *Fasciola hepatica* (compiled and redrawn from Anderson et al., 1982, Smith, and Green, 1984). B. Survival as a function of age in gonochoric *Ophryotrocha labronica* and hermaphroditic *O. diadema* (compiled from Akesson, 1976, Simonini and Prevedelli, 2003). C. Survival as a function of age in selfing and inter-breeding *Helobdella robusta* (compiled and redrawn from Iyer et al., 2019). D. Oviposition frequency as a function of age in *Planorbula armigera* and *Heliosoma duryi*. Continuous lines indicate facultative cross fertilizing treatment and dotted lines obligate self fertilizing treatment (compiled and redrawn from Escobar et al., 2011).

others are SHs. Table 6.1 shows that fecundity is reduced from 131 egg/♀ or 9.4 egg/♀/d in gonochoric species to 28 egg/♀ or 4.5 egg/♀/d, as if to confirm that the fitness of SH is half of that of gonochores. A second example comes from the gonochoric branchiopod *Branchipus schaefferi* and cirripede SH *Bendria purpurea*. The batch fecundity is reduced from ~ 200 eggs in the gonochoric branchiopod to ~ 75 in the cirripede SH. In *Helobdella octatestisaca*, batch fecundity is reduced from 91 eggs in a cross breeder to 49 eggs in a selfer.

Fecundity of F_1 *progeny*: Available information on F_1 fecundity of SH is scarce. With inability to copulate and donate sperm, aphallic (see below) snails allocate more resources for the female function. As euphallics have to develop dual reproductive systems, they allocate relatively less for female development than that of aphallics (Table 6.1). As a result, the aphallics are more fecund not only in F_1 but also F_2. For example, the F_1 'gonochoric'

TABLE 6.1

Comparative details on fecundity and other parameters in gonochoric and
simultaneous hermaphrodites

Bulimulus tenuissimus (Silva et al., 2013)		
Parameter	**Reared in**	
	Group	**Isolation**
Life span (d)	880	725
Generation time (d)	250	475
Reproductive life span (d)	230	30
Oviposition frequency (no.)	3	1
Lifetime fecundity (no./♀)	48	9

Gonochorics		**Simultaneous hermaphrodites**	
Species	**Fecundity (egg no.)**	**Species**	**Fecundity (egg no.)**
Ophryotrocha (Premoli and Sella, 1995)			
O. macrovifera	80	O. gracilis	11
O. notoglandulata	115	O. diadema	25
O. labronica	130	O. socialis	25
O. robusta	200	O. hartmanni	30
		O. maculata	50
131 egg/♀ or 9.4 egg/♀/d		28 egg/♀ or 4.5 egg/♀/d	
Branchipus schaefferi (Beladjal et al., 2003a)	200 egg/ batch	Bendria purpurea (Utinomi, 1961)	75 egg/batch

Bulinus contortus (Delay, 1992)		
Parameter	**Euphallic (SH) (no./♀/d)**	**Aphallic (no./♀/d)**
F₁ Offspring		
Egg (no./capsule)	0.83	0.87
Capsule (no.)	4498	5609
Fecundity (no./snail/d)	5438	6296
F₂ Offspring		
Hatchability (%)	95.4	97.5
Hatching size (mm)	1.21	1.07
Survival on 30th d (%)	68.7	61.4

aphallic *Bulinus contortus* are 14% more fecund than that in SH euphallics. Due to differences in F_1 egg size, F_2 hatchling size is also 13% larger in the aphallics than that in the euphallics.

6.3 Sequential Hermaphroditism (SQH)

Not surprisingly, a large number of hermaphroditic species have devised an array of strategies to adopt functional/behavioral gonochorism. To escape from inbreeding depression, they have chosen two escape routes: (i) the costlier and less preferred Sequential Hermphroditism (SQH) involving structural changes in reproductive and associated morphological systems including color and size changes and (ii) the less costly and more preferred behavioral change to function as gonochores. A description of the subgroups in SQH has already been described (see Fig. 6.1). With a flexible shift from one to another subgroup, the assessment of the number of SQH and their subgroups namely protandrics (PAH), progogynics (PGH) and serials (Ser) has proved to be a task. The following are some intriguing examples for their flexibility. (1) In chondrinid gastropod *Solatopupa similis*, male and female gametes differentiate simultaneously, but maturation commences early for spermatocytes in summer, but that for oocytes, it is delayed until winter. (2) The salt-marsh dwelling ellobiid *Ovatella* commence as males, continue as females and terminate as Simultaneous Hermaphrodites (SH). (3) In polyplacophore *Cyanoplax dentiens*, the chiton is a gonochore, hermaphrodite and female during autumn, winter and spring, respectively. (4) But the stylommatophoran *Arion* pass through the male, female and hermaphrodite phases, as the snail grows and ages. (5) A major complication is the occurrence of aphallism and euphallism in some snails. The euphallics develop an ovotestis with a fully functional female and male reproductive tracts. In contrast, aphallics do not develop the distal portions, i.e. the copulatory organ, prostate gland and vas deferens of the male tract but functional sperms are produced in the testicular tissues. Hence, they cannot donate sperm but function as females only, whereas euphallics can function as SH. Phally is irrevocably developed prior to sexual maturity. Aphally occurs in *Pyramidula*, *Pupilla*, *Acanthinula*, *Vallonia*, *Prophysaon*, *Arion*, *Zontinoides*, *Deroceras* and *Thysanophora*. However, some species among the vertiginids like *Vertigo*, *Trunctallina*, *Columella* are euphallics, while others are aphallics. Furthering the complication, the proportion of aphallics is increased from 0% in *Vertigo pygmaea* to 95% in *V. pusilla*. In *Columella*, it ranges from 7% in *C. edentula* to 69% in *C. columella*. In some of them, all are aphallics during spring but subsequently euphallics dominate, indicating the regulation of phally by temperature and photoperiod (see Heller, 1993). Among sequential Teleostei also, (i) within a protogynic species, some females do not change sex, while others do (e.g. Scaridae: *Sparisoma radians*;

Sparidae: *Pagurus pagurus*; Serranidae: *Mycteroperca bonaci*). (ii) within a species, some are haremics, while others are pair-spawners (e.g. *Sparisoma cretense*) and (iii) within a haremic species, mating system changes from place to place; *Helichoeres maculipinna* is haremic in the Panama but not in Florida (see Pandian, 2011). In pulmonates, anthozoans, teleosts and possibly others, the extraordinary flexibility at individual, population and species level do not easily permit identification and quantification of selfers and outcrossers as well as different types of sequentials. Having chosen the 'wrong' route to escape from selfing, the SQH crustaceans have explored almost all available routes for sex change (Fig. 6.4).

In Platyhelminthes, no sex change occurs; their protandrics simply add the female component, as they grow in size or as age advances. This is also true for the higher invertebrate major (e.g. Annelida, decapod crustaceans) and minor (Chaetognatha) phyla. Similarly, the protogynics may add the male component, as in Gastrotricha and crustaceans (e.g. *Chelonibia patula*). In a few others, sex is changed from one to another; for example, from male to female (e.g. echinoderms: *Ophiolepis kieri*, *Holothuria atra*: polychaetes: *Hydroides elegans*, Crustacea: see Pandian, 2016, Table 3.8 for a list). In some others, the androdioecy with hermaphrodites and males (e.g. spinicaudate crustaceans, *Anodonta* in bivalves) or hermaphrodites with females (e.g. the freshwater bivalve *Elliptio complanata*) is established. Arguably, sex change from one sex to another is costlier than the simple addition of another sex

Sexuality	Ontogeny
1. Gonochory	Juvenile → male of female
2. Monogynic protandry *Anilocra fontinalis*	Juvenile → male → female
3. Digynic protandry *Processa edulis*	Juvenile → male → secondary female Juvenile → primary female
4. Protandry (with EM and primary females)	Juvenile → male → secondary female ↘ EM female
5. Protandric simultaneous hermaphroditism *Cheloniba patula, Lysmata boggessi*	Juvenile → male → simultaneous ♀⚲ Juvenile → female → male
6. Monandric protogyny *Gnorimosphearoma*	Juvenile → female → male
7. Diandric protogyny *Hageria rapax*	Juvenile → female → secondary male Juvenile → primary male

FIGURE 6.4

Ontogenetic pathways of sex changing hermaphroditic crustaceans (from Pandian, 2016).

component. The costliest may be the serials, in which sex is changed serially. *Not surprisingly, none of the minor phyletics has chosen the serial route. Even among the major phyletics not more than 1229 species (or 0.08% of all animal species) have opted for the SQH route* (Table 6.2). *In fact, the cumulative (major + minor phyletics) number and percentage of SQH is only 1,740 species and 0.11%,*

TABLE 6.2

Number of sequential hermaphroditic (SQH) species in major and minor phyla. 1 = Van Soest et al. (2012), 2 = Ereskovsky (2018), 3 = Hyman (1940), 4 = Pandian (2020), 5 = Zhang (2011), 6 = Pandian (2017), 7 = Pandian (2019), 8 = Pandian (2018), 9 = Pandian (2016), 10 = Pandian (2011a), 11 = Pandian (2021), PAH = Protandric ♀, PGH = Protogynic ♀, SER = Serial ♀

Major phyla				Minor phyla[11]			
Phylum	**Species (no.)**			**Phylum**	**Species (no.)**		
	PAH	**PGH**	**Ser**		**PAH**	**PGH**	**Ser**
Porifera[1,2]	+	+	6	Placozoa	0	0	0
Cnidaria[3]	+	+	0	Mesozoa	0	0	0
Acnidaria[3]	-	-	0	Myxozoa	0	0	0
Platyhelminthes[4]	275	0	0	Loricifera	0	0	0
Arthropoda[5]	76	14	0	Cycliophora	0	0	0
Mollusca[6]	646	1	9	Nemertea	116	0	0
Annelida[7]	63	2	6	Gnathostomulida	0	0	0
Echinodermata[8]	7	3	0	Rotifera	0	0	0
Vertebrata[6]	29	84	8	Gastrotricha	0	100	0
Subtotal	1,096	104	29	Kinorhyncha	0	0	0
Total	1,229			Nematoda	0	0	0
				Nematomorpha	0	0	0
				Acanthocephala	0	0	0
Major Phyla (no.)	1,096	104	29	Priapulida	0	0	0
(%)	89.0	8.5	2.3	Sipuncula	0	0	0
Subtotal (no.)	1,229			Echiura	0	0	0
Minor phyla (no.)	336	175	0	Tardigrada	0	0	0
(%)	65.6	34.3	0	Onychophora	0	0	0
Subtotal (no.)			511	Pentastomida	0	0	0
Total (no.)	1432	279	29	Entoprocta	0	0	0
(%)	75.0	23.3	1.7	Phoronida	0	0	0
Grand total (no.)	1,740			Brachiopoda	0	0	0
				Bryozoa	0	0	+ 0
				Chaetognatha	150	0	0
				Hemichordata	0	0	0
Crustacea[9]	76	14	0	Urochordata	70	75+	0
Teleost fish[10]	29	84	8	Subtotal	336	175	0

respectively. More interestingly, 89% of major phyletics have opted for protandry among SQH, indicating that the addition of the female component to the existing male component or changing male to female may be less costly and a larger body size can provide relatively more space for maturing oocytes. Understandably, females do require more resources to produce the costlier eggs. It is notable that *whereas more number of vertebrate teleosts species have preferred protogyny (84%) but all the invertebrates have opted for protandry.*

6.4 Behavioral Gonochorism

Hermaphrodites (H) include Simultaneous Hermaphrodites (SH) and sequential hermaphrodites (SQH) and Functional/Behavioral Hermaphroditism (F/BH). For, the more preferred, less costly escape route, an estimate on the number of behaviorally functioning gonochoric hermaphrodites (F/BH) has been made from (Table 6.3). Porifera are known

TABLE 6.3

Number of structural (SG) and functional/behavioral (F/BH) gonochores and selfers in major phyla. H – hermaphrodites. 1 = Van Soest et al. (2012), 2 = Ereskovsy (2018), 3 = Mapstone (2014), 4 = Shikina and Chang (2018), Carlon (1999), 5 = Hyman (1940), 6 = Pandian (2020), 7 = Pandian (2019), 8 = Pandian (2016), 9 = Pandian, (2017), 10 = Echinoderamta (Pandian, 2018), 11 = Zhang (2011, 2013), * = Heller (1993), H*: Hermaphrodite = SG + F/BH

Phylum	Species (no.)	SG (no.)	F/BH ♀ (no.)	Selfer (no.)	H* (no.)
Porifera[1,2]	8,553	3,260	5,293	0	5,293
Cnidaria[3,4]	10,856	10,834	21	1	22
Acnidaria[5,6]	166	0	166	0	166
Platyhelminthes[6]	27,700	45	23,008	4,647	27,655
Annelida[7]	16,911	12,855	3,956	100	4,056
Arthropoda[8,11]	1,242,040	1,240,988	881	171	1,052
Mollusca[9]	118,451	91,378	19,393	7,680	27,073
Echinodermata[10]	7,000	6,992	7	1	8
Vertebrata[11]	64,832	64,695	134	3	137
Major phyla (no.)	1,496,509	1,431,047	52,859	12,603	65,462
%		95.6	3.5	0.9	4.4
Minor phyla (no.)	46,687	36,232	9,414	1044	10,462
%		76.6	20.2	2.2	22.4
Grand total (no.)	1,543,196	1,467,279	62,273	13,647	75,924
%		95.08	4.02	0.88	4.91

as Simultaneous Hermaphrodites (SH) (Fell, 1993). However, Ereskovsky (2018) reported the occurrence of gonochorism in at least ~ 10 orders across all over the major classes. WoRMS put the species number in these 10 orders as 3,441 but also the number of Porifera as 9,026. Considering the estimated number of 8,553 species for Porifera by Van Soest et al. (2012), the gonochoric species number for Porifera was taken as 3,260 (Table 6.3). In both oviparous and viviparous sponges, "the freely released conspecific spermatozoa enter the egg-containing individual via the incurrent water system and pass through one or more barriers to reach the eggs ..... scattered in the mesohyl" (Ereskovsky, 2018). Hence, sponges are not selfers, irrespective of gonochoric and hermaphroditic vivipares. The cnidarians are reported as gonochores (Shostak, 1993). However, Shikina and Chang (2018) reported the incidence of hermaphroditism in 70% of the stony scleractinid corals. Using morphological (e.g. pigments), biochemical (e.g. enzymes) and/or DNA markers, it has been possible to identify and quantify the number of selfers in these broadcast spawning SH; some of them broadcast both male and female gametes, while brooders with internal fertilization broadcast male gametes only. Fortunately, Carlon (1999) summarized the experimental evidences for selfers within each of 23 scleractinid species; from Carlon, the following may be inferred: (1) In coral species with internal fertilization, selfing ranged from 0.0% in *Montipora spumosa* to 13.0% in *Acropora valida* among acroporids and from 1.4% in *Platygyra* sp to 65.8% in *Goniastrea favulus* among favids. (2) None of the experimental individual coral included cent percent selfers. (3) In 3 species of 12 acroporids and 2 species of 11 favids, none of the tested individuals was a selfer. (4) Within the other tested species, the proportion of the selfer ranged from 0.5% in *Acropora tenuis* to 65.8% in *G. favulus* and it averaged 2.8% for Acroporidae and 9.3% for Favidae. (5) On the whole, it exceeded 50% only in *G. favulus* with 65.8%, which may be considered as a selfing species, and all the remaining 22 species are cross fertilizers, as the proportion of selfers in them was < 13%. Some 166 species of Acnidaria are cross fertilizers (Hyman, 1940) and the 10,856 speciose cnidarians are also cross fertilizers (see Hyman, 1940).

The 27,700 speciose Platyhelminthes are SH. However, most members of Acoela (150 species) and Catenulida (120 species) are protandrics (Hyman, 1951a). Pandian (2020, Table 6.1) lists 33 and odd gonochoric including 23 schistostomatid and gynogenic species strewn across the three major classes of Platyhelminthes. In turbellarians, the female and male gametes are exited through the mouth in prolecithophorans. All species belonging to the parasitic turbellarian families Fecampiidae (10 species) and Acholadidae (2 species) are also gonochores. Hence, 33 + 12 = 45 species are all gonochores among the flukes. In all others, there are separate exits for the gametes (Pandian, 2020, Fig. 2.2D). They possess a penis and reciprocally inseminate each other (e.g. *Dugesia* spp, Vreys and Michiels, 1988). For Monogenea, available data on intensity of infection ranging from 2.9 fluke/host in *Discocotyle sagittata* to 210 fluke/host in *Diplectanum aequans* on the sea bass

Dicentarchus labrax (Pandian, 2020, Table 3.2) suggest the scope for cross fertilization. In the monogenean genus *Diplozoon*, two diporpa larvae, on successful infection, unite to form an X-shape position; in the position, the sperm duct of each fluke opens into the copulation canal of the other and thereby ensures monogamic cross fertilization. Among other flukes, a few are reported as protandrics (e.g. *Benedenia seriolae*, Lackenby et al., 2007). In *Entobdella solea*, Whittington and Horton (1996) experimentally showed that selfing does not occur in this fluke. Hyman (1951a) asserted that as a rule, monogeneans are all cross-fertilizers. In the 12,012 species of digeneans, the penis and Laurer's canal ensure cross fertilization (Hyman, 1951a). Nevertheless, the 4,647 speciose cestodes are selfers. Even with experimental offering of mating pairs, some pairs of *Schistocephalus solidus* opted for selfing (Scharer and Wedekind, 1999). Considering all these facts, some 4,647 cestode species and 23,053 other fluke species are considered as selfers and functional gonochoric SH, respectively (Table 6.3).

In crustaceans, the listed number of SH was arrived from Pandian (2016). Simultaneous hermaphroditism occurs among Notostraca (12 species), Spinicaudata (40 species), and Cirripedia (~ 1,000 species). In them, selfing is minimized or eliminated by the existence of androdioecic mating system or dwarf males. A single male of spinicaudate *Eulimnadia texana* may inseminate 12 hermaphrodite/day. The male engages its claw-like claspers to hold the hermaphrodite during copulation. Possessing no claspers, the hermaphrodites may not mate with each other. To facilitate cross fertilization, the hermaphroditic barnacles settle within a distance of 50 mm from one another (Yusa et al., 2001). Of 1,052 crustacean SHs, this analysis noted the occurrence of selfing in < 170 species (Table 6.3). In Insecta, a very rare incidence of selfing is hinted in *Icerya purchasi* SH (Royer, 1975).

According to Heller (1993), 27,073 molluscan species are SH (Polyplacophora = 100%, 263 species, Prosobranchia = 3% of 1,560 species, Ophisthobranchia = 100%, 2,000 species, Pulmonata = 100%, 24,000 species and Bivalvia = 9% of 810 species, Table 6.3). In his detailed and interesting review, Heller (1993) reported that selfing is a rarity in prosobranch SH. Except for three selfing species, cross fertilization seems to be a general rule in ophisthobranch SH. In bivalves, 13 of 117 families constitute ~ 810 hermaphroditic species. In them, 75 genera of 85 are SH. Their simple reproductive system consists of a single gonoduct appearing from the ovotestis; the duct is then paired and opens directly into the exterior. The simple system does not include any specialized area or organ for collection, storage of gametes and their fertilization. Ostrovsky et al. (2016) reported the existence of 42 brooding matrotrophic bivalve species; of them, oviparous and larviparous bivalves constitute an equal number of 21 species each. Arguably, the matrotrophic brooding bivalves collect the fertilized eggs to brood them. Heller also noted that brooding need not necessarily involve hermaphroditism. Therefore, the probability of selfing in bivalves is almost zero.

However, it has been indeed a challenge to assess the number of selfing pulmonate species, which have devised structural and behavioral routes to escape from selfing. Some pulmonate snails display dimorphic phallism, namely euphally and aphally. As described earlier, aphallics, with no copulatory organ and others, cannot donate sperm and function as females only, whereas euphallics can function as SH. In some of these phallics, the dimorphism has appeared 14 independent times (Johnson et al., 1995). Among Stylommatophora, phallism is reported from 12 genera in 8 families (Heller, 1993). With phallism, some pulmonate species have almost attained the structural gonochoric status. The other pulmonates have developed a behavioral strategy to act as either a gonochore or selfing SH. In fact, SH can be obligate selfers (e.g. *Lasaea subviridis*), or obligate cross breeders (e.g. *Ancylus fluviatilis*, which, on enforced isolation, becomes sterile). Still others are facultative selfers; among them, there are bilateral reciprocals, i.e. the mating partners inseminate each other (e.g. *Bathyomphalus contotrus*) or unilaterals, i.e. one of the partners that acts as a male, while the other as a female (e.g. *Segmentina oelandica*). Among the ophisthobranchs, the sea hares copulate in a chain, in which a mate acts as a male to the preceding one and the female to the succeeding one (e.g. *Aplysia californica*, Angeoloni et al., 2003). However, of 108 pairs observed, three pairs mated both reciprocally and unilaterally. Among the freshwater pulmonates too, *Biomphalaria glabrata* and *Potamopyrgus antipodarum* are capable of self- and cross-fertilization. Hence, selfing, or unilateral or bilateral insemination is a highly flexible behavioral trait. The freshwater pulmonates belonging to the families Acroloxidae, Ancylidae, Lymnaeaidae, Planorbidae and Physidae, on enforced isolation, self-fertilize themselves. Hyman (1967) described that the selfing *Lymnaea columella* produced viable egg capsules for 47 generations over a period of nine years. But Heller (1993) concluded that "in pulmonates, cross fertilization is the preferred mode of breeding, when a mate is available. Many pulmonates are, indeed, so abundant that is hard to believe that finding a mate really poses a serious problem". However, some enthusiastic authors enforced isolation in many pulmonates and reported experimental evidences for selfing in them. Heller assessed that some 7,680 pulmonate species are selfers (Table 6.3).

In the 16,911 speciose Annelida, Simultaneous Hermaphroditism (SH) is restricted to 4,056 species, i.e. Polychaeta: 207 species; + Oligochaeta = 3,165 species; + Hirudinea = 684 species (Table 6.3). In terrestrial oligochaete SH, spatial separation of the female and male gametogenic segments and location of metanephredia, which serve as exits for the gametes, eliminate selfing. Of 56 earthworm species, 42 are parthenogens and in most of them, the male reproductive system is degenerated either partially or wholly (see Pandian, 2019). The others reciprocally inseminate each other (e.g. *Eisenia fotida*, Monroy et al., 2005). Being more of clonals, sexual reproduction is very rare in aquatic oligochaetes. In very few sexually reproducing aquatic oligochaetes, gonoducts are lacking and sperms are discharged through metanephridea and eggs are shed through a porous body wall. Hence, they are all cross breeders. In two hirudinean species (*Helobdella robusta* and *H. octatestisaca*), selfing can be experimentally enforced (Iyer et al., 2019). The SH polychaetes discovered different escape routes from hermaphroditism: (i) In some of them, the ovarian and testicular segments are located in different segments (e.g. *Aracia sinaloae, Neanthes limnicola*). (ii) In others like *Branchiomma bairdi,* the anlage for the female and male gametogenic segment is the same but they are topographically separated. (iii) In *Laonome albicingillum*, fertilization success is reduced to ~ 60% in self fertilizers, in comparison to that of cross fertilizers. (iv) *Diopatra marocensis* is unique in that it switches from SH to protoandric and gonochoric with increasing population density from 2 no/m^2 to 4 no/m^2 and > 10 no/m^2, respectively (see Pandian, 2019, Table 2.1). In all, not more than 100 annelid species inclusive of facultatives may be regarded as selfers.

In echinoderms, the few hermaphroditic species are all cross breeders, however with the exception of the viviparous *Amphipholis squamata* (Pandian, 2018). In teleosts, 137 species are confirmed as hermaphrodites (Pandian, 2013); of them, three SH [e.g. *Kryptolebias* (=*Rivulus*) *marmoratus*] alone are selfers; in them too, clonal diversity has been demonstrated. All other SH are reciprocals or change sex to the female (e.g. *Sarranus fasciatus*, see Pandian, 2011, Chapter 6). On the whole, only three teleostean species are selfers.

Among minor phyletics, the pseudocoelomates are all gonochores, except for the 813 speciose Gastrotricha, which are parthenogens and terminate as SH. As their SH produce < 33% of lifetime fecundity, they are considered as parthenogens (Table 6.4). In the gastrotrichan order Cheetodontoidea, the male system is degenerated (except in *Xenotrichula*); hence, they are all functional females (Hyman, 1951b, p 164). Among Mesozoa, the dicyemids (42 species), Gastrostomulida (100 species), some tardigrade species, Bryozoa (5,700 species), Chaetognatha (150 species), a few hemichordate (25) species and Urochordata (3,000 species) are all SH. The incidence of SH is reported in six nemerteans (*Geonemertes palaensis, Neonemertes agricola, Tetrastemma caecum, T. hermaphroditicum, T. kefersteinii, T. marioni,* Riser, 1974) but selfing only in *Prosorhochnus americanus* SH (Caplins and Turbeville, 2015). From values reported by Pandian (2020, Tables 18.1, 18.2), many tardigrade species are identified as selfing SH. However, Nelson et al. (2010) found that selfing in them is limited to seven species only, i.e. *Bertolanius weglarskae, Macrobiotus joannae, Parhexapodibius pilatoi* and four species belonging to the genus *Isohypsibius*. Subsequently, Bertolani (2001) confirmed selfing only in the isolated *I. minicus* and *M. joannae*. Hence, < 7 tardigrade species are considered as selfers. In Entoprocta, the gametes are shed through a common gonoduct; it is not known whether the female and male gametes are simultaneously released or at temporally separated events. Hence of 36 SH entoproct species, 18 are considered as selfers. Spawning is continuous in the SH phoronids of *Phoronis hippocrepia, P. mulleri* and possibly *P. ijimai, P. viridis* and *Phoronis australis*. In them, the gametes are shed

TABLE 6.4

Number gonochores (G), hermaphridates (H), functional/behavioral gonochoric (F/BG/SH) and selfers in minor phyla (compiled from Pandian, 2021)

Phylum	Species (no.)	G (no.)	H (no.)	F/BG (no.)	Selfer (no.)
Placozoa	3	3	3	0	3
Mesozoa	150	108	0	0	0
Dycyemida	-		42	-	42
Myxozoa	2,200	2,200	0	0	0
Loricifera	34	34	0	0	0
Cycliophora	2	2	0	0	0
Nemertea	1,300	1,299	1	0	1
Gnathostomulida	100	0	100	100 Reciprocals	
Rotifera	2,031	2,031	0	0	0
Gastrotricha	813	813	Parthenogens → ♀ 813		
Kinorhyncha	200	200	0	0	0
Nematoda	27,000	26,990	10	10	0
Nematomorpha	360	360	0	0	0
Acanthocephala	1,100	1,100	0	0	0
Priapulida	19	19	0	0	0
Sipuncula	160	160	0	0	0
Echiura	230	230	0	0	0
Tardigrada	1,047	605	442	435	7
Onychophora	200	200	0	0	0
Pentastomida	144	144	0	0	0
Entoprocta	200	164	36	18	18
Phoronida	23	14	9	4	5
Brachiopoda	391	261	130	0	130
Bryozoa	5,700	0	5700	5700	0
Chaetogantha	150		150	150	PAH
Hemichordata	130	108	22	0	22
Urochordata	3000	0	3000	2997	3
Total (no)	46687	36,232	10,462	9,414	1,044
%		77.6	22.4	20.2	2.2

into the coelom, where fertilization occurs. In Bryozoa, the long life span of sperms (Manriquez et al., 2001) and high efficiency of sperm capture by lophophores (Pemberton et al., 2003, 2007) is reported. Hughes et al. (2009) found that the intra-colonial self fertilization resulted in aborting embryos, reduced production and fitness of larvae. Hence, all bryozoans are assessed as cross fertilizers. Among brachiopods, of 18 species, 6 are SH (Pandian, 2021, Table 24.1). Hence, the number for selfing SH is taken as 130 species. The SH ascidians like *Ascidia callosa, Corella inflata* and *Phallusia mammillata* are reported as selfers (Cloney, 1990). However, Gasparini and Ballarini (2015) noted that there are no selfers among urochordates.

In a review, Jarne and Chalresworth (1993) summarized the then available experimental evidence for the estimates on the number of selfers, crossers and mixers (facultative SH) (Table 6.5). The values reported no selfing in Urochordata and one selfing species in Cnidaria confirming those arrived from the present analyses. The reviewers reported 14 of 44 species in two major pulmonate taxa or 32% are selfers. This 32% for 24,000 speciose pulmonates is 7,680 selfing species, which agrees completely with that of Heller. According to the reviewers, selfing does not occur in polychaetes. However, the present estimate has put it as 100 species for the phylum Annelida including some oligochaete and hirudineans, which may be selfers. For platyhelminths, the present analysis has given adequate evidence to show that except for the 4,647 speciose selfing cestodes (i.e. 17% flukes), all other flukes are cross fertilizers; however, the value reported by reviewers is 33%. On the whole,

TABLE 6.5

Number of selfing species in five hermaphroditic taxa (modified from Jarne and Charlesworth, 1993)

Phylum/Taxa	Breeding system (no.)			Total (no.)
	Outcrosser	Mixed	Selfers	
Cnidaria	0	1	1	2
Platyhelminthes	4	2	3	9
Pulmonates				
Stylommatophora	16	2	5	23
Basommatophora	11	1	9	21
Annelida				
Polychaeta	5	0	0	5
Urochordata	2	2	0	4
Total	38	8	18	64
As % of total 64 species	59.4	12.5	28.1	
	6.5	←⌐⌐→	6.0	
	(66%)		(34%)	

~ 33–34% of all the investigated species are selfers. But this value is diluted to < 1%, when cumulative number of species from 9 major and 26 minor phyla is considered together; incidentally, the most speciose Arthropoda are cross fertilizers except for 170 crustaceans.

Table 6.6 provides for the first time a complete picture on the number and proportion of gonochores and hermaphrodites inclusive of their subgroups, from which the following may be inferred: 1. *Of 1,543,196 species, 95.1% animals are structural gonochores. 2. Only 4.9% animals are hermaphrodites. This value is closer to that of 5.4% arrived by Jarne and Auld (2006), who had estimated ~ 65,000 hermaphroditic species among 1.2 million of the total number of animal species considered by them. 3. Of them, 3.92% and 0.11% are behavioral and sequential hermaphrodites, respectively; the 3.92% was arrived after subtraction of 0.11% from 62,273 species for behavioral gonochorics (see Table 6.6). This value is higher than that (24,000 species or 0.2%) estimated by Jarne and Auld (2006). However, it must be noted that the present account has considered 1,543,196 species and also taken into consideration the sequential hermaphrodites. 4. Of the two escaping routes from selfing hermaphroditism, more number of species have chosen the less costly behavioral gonochoric strategy, as that of sequential hermaphroditism is costlier. 5. For the minor phyla, the value for the Structural Gonochorics (SG) is 77.6%, and the remaining 22.4% for simultaneous hermaphroditism (Table 6.3), which is three-four times higher than that of major phyletics. However, 20.2% minor phyletics function as behavioral gonochoric hermaphrodites* (Table 6.4). *Thereby, minor phyletics have more effectively reduced selfing. 6. Selfing occurs in < 0.88% of all animal species, indicating that it assures reproductive multiplication under unusual conditions but at the cost of inbreeding depression.*

TABLE 6.6

Approximate species number of gonochores, hermaphrodites, functional/behavioral gonochoric hermaphrodites (F/BG/SH) and Sequential hermaphrodites (SQH)

Particular	Species (no.)	%
Animal species (Table 4.1)	1,543,196	100.0
Structural gonochorism (SG)	1,467,279	95.08
Behavioral gonochoric SH*	60,533	3.92
SQH (Table 6.2) (2.2% of H)	1,740	0.11
Selfing hermaphrodites	13,647	0.88
Hermaphrodites (Table 6.3) (H)	75,924	4.91

* 62,273 – 1,740 SQH = 60,553

6.5 Meiosis vs Fertilization

The findings reported in Table 6.6 reveal that (i) *gonochorism accelerates species diversity but hermaphroditism decelerates it, (ii) within the limitation of the diversity in hermaphroditism (expressed in different forms), sequential hermaphroditism deters and reduces the scope for diversity, whereas behavioral gonochoric hermaphroditism enhances the scope for diversity.* In gonochores, as well as sequential and functional/behavioral gonochorics, the gametes arising from different individuals are fertilized. Interestingly, the findings reported in Table 6.5 have also provided a unique opportunity for the first time to know whether meiosis or fertilization plays a bigger role in fostering species diversity. As in gonochores, male and female gametes are also produced in all hermaphrodites. However, fertilization is restricted to the male and females appearing from the same individual among selfing hermaphrodites.

As stated elsewhere, new gene combinations are generated during meiosis and at fertilization of gametes arising from two individuals. *Missing cross fertilization, the selfing hermaphrodites suffer a steep reduction in species diversity to ~ < 1%, in comparison to 99% of the diversity in (structural and behavioral) gonochores. Interestingly, the cytological events show that during meiosis, only small fractions are crossed over and recombined between homologous chromosomes. But fertilization brings a whole set of new haploid chromosome from another individual. Hence, the scope for the generation of new gene combinations is nearly 99-times greater at fertilization than during meiosis. Amazingly, fertilization plays a greater role in fostering species diversity than that of meiosis.*

7

Parthenogenesis and Unisexualism

Introduction

Parthenogenesis is a peculiar mode of sexual reproduction, in which fertilization does not occur. Based on the chromosomal behavior in maturing oocytes, it is classified into three major types (Suomalainen, 1962): (i) ameiotic or apomictic, (ii) meiotic or automictic and (iii) generative or haplo-diploid inclusive of Paternal Genome Elimination (PGE, see Blackmon et al., 2017). In ameiotic parthenogenesis, females produce unreduced eggs, which develop without the need for a trigger from auto- or allo-sperm (as in gynogenesis). During maturation of oocytes, synapsis between homologous chromosomes does not occur; as a result no segregation or crossing over takes place. Hence, no new gene combination is generated. In laboratory cultures, not a single male appeared up to 180 generations in *Moina brachiata* and *Daphnia magna* as well as up to 666 and 780 generations in *D. pulex* and *M. macrocopa*, respectively (Banta, 1939). However, sexual reproduction may be restored in them following sporadic appearance of males (e.g. cladocerans, see Pandian, 2016). In contrast to the ameiotic type, meiotic parthenogenesis retains the normal meiosis. But the egg nucleus restores diploidy by one of the mechanisms described in Table 7.1. In generative or haplo-diploid parthenogenesis too, normal meiosis is retained in the oocytes. However, a fertilized egg gives rise to a diploid female but an unfertilized egg develops into an arrhenogenic male. In rotifers, the fertilized diploid mictic eggs develop into dormant eggs and unfertilized haploid eggs hatch as males. Generative parthenogenesis is common among rotifers (Serra et al., 2018), hymenopteran wasps, bees and ants, a few thysanopteran thrips, coleopterans (e.g. *Micromalthus*, scale insects and blood fluke *Schistosomatium douthitti*, see Pandian, 2020). Whereas most gall wasps and a few weevils have resorted to parthenogenesis alone, others such as aphids, cynipid hymenopterans, cecidomysid dipterans and a few nematodes (e.g. *Deladenus siricidicola*, Pandian, 2021, Fig. 12.12) undergo a series of advantageous parthenogenic homogonic cycles during a favorable condition and alternate sexual reproduction, when there is a need, for example, for dispersal. Consult Muthukrishnan (1994) for more information

TABLE 7.1

Modes of restoration of diploidy in meiotic parthenogenics (condensed and modified from Muthukrishnan, 1994)

Restoration mode	Remarks	Examples
Fusion of homologous egg nucleus and II polar body nucleus	Perpetuation of maternal homozygosity	Hemiptera: *Guerinella serratula*
Fusion of non-homologous secondary oocyte nuclear and I polar body nucleus	Perpetuation of maternal heterozygosity	Lepidoptera: *Solenobia lichenella*
Fusion of non-homologous nuclear I and II polar bodies	Perpetuation of maternal heterozygosity	Lepidoptera: *S. triquetrella*
Unusual premeiotic endomitosis producing homologous 2n eggs	Perpetuation of maternal heterozygosity	Orthoptera: *Warramba virgo*
Multiple conversions of n to 2n nucleus during or after blastoderm stage	Increase in homozygosity. Males are rare	Orthoptera: *Euhadenoecus insolitus*

on insects. In parasitic nematodes and free-living tardigrades, an increasing tendency for meiotic parthenogenesis occurs. For example, of 45 nematode species, 21 and 24 are mitotic (ameiotic) and 24 meiotic parthenogens, respectively (see Pandian, 2021, Table 12.3).

7.1 Taxonomic Distribution and Quantification

Among minor phyletics, parthenogenesis appears first in the structurally simple acoelomate Loricifera and peaks in the pseudocoelomate Rotifera; subsequently, it decreases progressively from Nematoda to Nematomorpha and from Tardigrada to Sipuncula among hemocoelomates. Notably, it is not reported from any eucoelomate minor phyletic (Table 7.2). Among major phyletics also, it peaks in hemocoelomate Arthropoda. Almost no sponge species is reported to reproduce parthenogenically (Fell, 1993). Shostak (1993) listed three species namely *Margelopsis haeckeli*, *Cunina probroscidea* and *Stygiomedusa fabulosa* as parthenogens. In Annelida, 66 species are parthenogens; of them, 56 are earthworms. It is not clear whether their low motility amidst denser substratum and consequent inability to readily find a mate has driven them to parthenognesis. For Annelida and Mollusca, more or less precise information is available (Table 7.2). Available information on the number of parthenogens in other major phyla is scarce, widely scattered and is not reliable. In the absence of the gonad or sperm, the turbellarians, *Tetracoelis marmorosa* and *Gyratrix hermaphroditus* were assumed as parthenogens. Despite providing cytological evidence for their ploidy status, no cytological evidence is provided for the occurrence of parthenogenesis in these flukes. *Bothrioplana semperi* is protandric, as its male gonad is either

TABLE 7.2

Number of parthenogenic species in major and minor phyla. 1 = Hyman (1940), 2 = Shostak (1993), 3 = Pandian (2020), 4 = Pandian (2017), 5 = Zhang (2011, 2013), 6 = Pandian (2016), 7 = Blackmon et al., 2017, 8 = Pandian (2019), 9 = Pandian (2018), 10 = Pandian (2021)

Major phyla		Minor phyla[10]	
Phylum	Species (no.)	Phylum	Species (no.)
Porifera[1]	-	Placozoa	0
Cnidaria[2]	3	Mesozoa	0
Acnidaria[1]	-	Myxozoa	0
Platyhelminthes[3]	1	Loricifera	34
Acoelomorpha[3]	1	Cycliophora	0
Mollusca[4]	7	Nemertea	0
Gastropoda	7	Gnathostomulida	0
Arthropoda[5]	6,075	Rotifera	2,031
Crustacea[6]	1,525	Gastrotricha	813
Insecta[7]	3,106	Kinorhyncha	0
Myriapoda	119	Nematoda	45
Chelicera	1,558	Nematomorpha	11
Annelida[8]	66	Acanthocephala	0
Echinodermata[9]	2	Priapulida	0
Vertebrata	0	Sipuncula	1
Total	6,156	Echiura	0
		Tardigrada	287
Major phyla (no.)	6,156	Onychophora	0
%	0.41	Pentastomida	0
Minor phyla (no.)	3,222	Entoprocta	0
%	6.9	Phoronida	0
Total	9,376	Brachiopoda	0
%	0.61	Bryozoa	0
		Chaetognatha	0
		Hemichordata	0
		Urochordata	0
		Total	3,222

poorly developed or atrophied at the time of ovarian maturation. In this functionally gonochoric female *B. semperi*, unusual dioogonic parthenogenesis is described (see Benazzi and Benazzi-Lentati, 1993). In the asteroid *Ophidiaster granifer* of 322 specimens examined, only 64 were found to have neither a gonad nor a sperm, and the asteroid was assumed as parthenogenic. Finding 2 of 25 specimens packed with blastulae in *Diadema antillarum*, which suffered a sudden and extreme decrease in population density (to 0.5%), *D. antillarum* was also assumed as parthenogenic. Whereas cytological evidence was cited for the existence of parthenogenesis in terrestrial earthworms and aquatic enhytraeids (Christensen, 1961), no evidence was provided for the assumed incidence of parthenogenesis in these echinoderms.

It was earlier shown that (i) hermaphrodites may incur inbreeding depression and may not be as fecund as gonochores, owing to the need for manifestation and maintenance of dual sexes in a single individual. For example, fecundity of many hermaphroditic polychaetes like *Ophryotrocha* is 50% that of gonochoric *Ophryotrocha* spp (see Table 6.1). However, the sequential copulatory (hence functionally gonochoric) hermaphrotidic sea hare *Aplysia californica* (Angeoloni et al., 2003) produces 478 million eggs in 18 weeks (MacGinitie, 1934) or 3.8 million egg/d. Hence, some other factor(s) may also regulate fecundity in hermaphrodites. Selfing hermaphrodites may suffer inbreeding depression but not the cross-fertilizers like *A. californica*.

In insects, the incidence for mitotic/meiotic parthenogenesis is reported for 3,106 species from 20 orders, of which it is more prevalent in Hemiptera (467 species) and Coleoptera (326 species). Haplo-diploidic parthenogenesis occurs in 1,891 species from six orders and is more prevalent in Hymenoptera (1,591 species) and Hemiptera (255 species). In hymenopterans, 1,591 (1.6%) of 100,000 species are haplo-diploid parthenogens. Of 90,000 hemipterans, 255 (0.28%) species undergo a modified haplo-diploidy namely Paternal Genome Elimination (PGE). Under PGE, males develop initially from fertilized eggs. However, the paternal genome is eliminated from fertilized eggs during embryonic development. Of 90,000 species, 722 (or 0.8%) hemipteran species are parthenogens. However, this value of 0.8% is still lower than that (1.0%) reached by Bell (1982). Despite its rarity among 1,020,007 speciose Insecta, parthenogenesis has evolved independently hundreds or even thousands time. Two reasons are attributed for the emergence parthenogenesis in insects. The endosymbiotic/parasitic *Wolbachia* eliminates the testis. Nevertheless, the probabilities of *Wolbachia* causing the transition to parthenogenesis are eliminated, as majority (71%) of parthenogenic taxa is nested within diploid clades. The second reason attributed to the repeated emergence of parthenogenesis is the loss of flight, which reduces the motility of parthenogens for the search of a mate. Increasing levels of flightlessness of females or males or both sexes has contributed to the increasing incidence of parthenogenesis in insects (Blackmon et al., 2017). At the phylum level, Loricifera, Rotifera and Gastrotricha are all parthenogens. In Nematoda 45 species are identified as parthenogens. Among the limnoterrestrial macrobiotid, cohypsibiid and hypsibiid tardigrades, some 287 species are approximately assessed as parthenogens (see Pandian, 2021).

The existence of haploids merits a few lines of description. It is known in the eusocial hymenopterans (1,591 of 100,000 species = 1.6%) and (paternal genome eliminated) hemipterans (255 of 90,000 species = 0.28%) and in a few schistostomids. In teleosts also, Yamaki et al. (1999) reported the existence of haplodiploid mosaic charr *Salvelinus leucomaensis*, in which haploid organs are sustained. In the Indian catfish *Heteropneustes fossilis*, the haploids occur at 1.7% frequency with sex ratio of 1 ♀ : 1 ♂. The haploid males produce haploid sperms (apparently with meiotic reduction) with 30n chromosomes. On fertilization by a haploid sperm, the development is initiated in the 2n fertilized egg to produce 2n female (Pandian and Koteeswaran, 1999). Nevertheless, the rarity of their incidence and short life span confirm that the haploids cannot and do not contribute to species diversity.

The present analyses have led to a couple of generalizations. 1. Pandian (2021) has brought convincing evidence to show that parthenogenesis and hermaphroditism among minor phyla do not coexist and mutually eliminate each other (Fig. 7.1). In tardigrades, family level analysis confirms the mutual elimination. Regarding the overlap between parthenogenesis and hermaphroditism in some major phyla, the following may have to be noted: (a) Shostak (1993) hinted that the incidence of parthenogenesis in three species but no cytological evidence is available to confirm it (see Pandian, 2019). (b) This is also true for seven gonochoric prosobranch molluscs but no report is available for parthenogenesis in hermaphroditic 2,000 speciose opisthobranchs and 24,000 speciose pulmonates (see Pandian, 2017). (c) Among crustaceans, hermaphroditism occurs in the 1,000 speciose cirripedes, spinicaudates and notostracans. In them also, the androdioecious mating system and long protrusible penis in gregariously sessile balanids ensure behavioral gonochorism (see Pandian, 2016). (d) Parthenogenesis is reported in four tubificid species, six enchytraeid species and 53 lumbricid species. Tubificids are known as protandrics (Poddubnaya, 1984). In most aquatic oligochaetes, gonadal organs are either missing or not functional, when present. In them, copulation has never been observed. In this context, more information is required to know whether all or some of these 68 hermaphroditic and parthenogenic annelids can be selfers or crossers. Of 33 North American lumbricids, 17 of them reproduce exclusively by parthenogenesis. In many of them, testis remains degenerated or sperms are not present, when it is present (Gates, 1971). In fact, Reynolds (1974) named them as 'female hermaphroditics'. In a few others, in which hermaphroditism is functional, they inseminate reciprocally and act as behavioral gonochores (e.g. *Eisenia foetida*, Monroy et al., 2005). These findings tend to confirm the fact that hermaphroditism and parthenogenesis mutually eliminate each other in major phyletics also. Hence, this new finding of mutual elimination between parthenogenics and hermaphroditism holds good for major phyletics also. 2. There is a remarkable increasing evolutionary tendency to switch from homozygous mitotic parthenogenesis in all species of Rotifera and Crustacea to the less homozygous meiotic parthenogenesis in free-living Tardigrada and Hymenoptera and parasitic Nematoda. 3. From Table 7.2, it may be noted that *the incidence of parthenogenesis is limited to 0.41% in major phyletic*

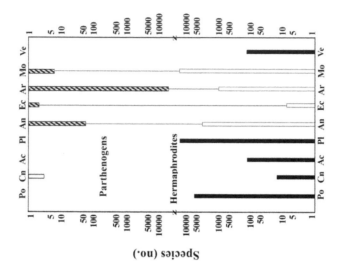

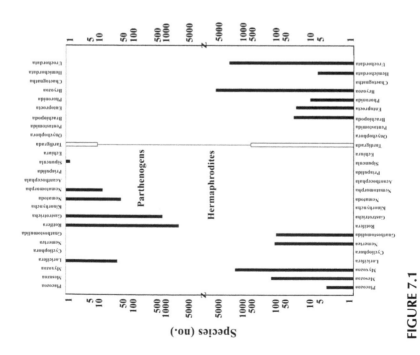

FIGURE 7.1

Parthenogenesis and hermaphroditism mutually eliminate each other in minor (left panel, Pandian, 2021) and major (right panel) phyla. For overlapping major phyla explanation is provided in the text. Po = Porifera, Cn = Cnidaria, Ac = Acnidaria, Pl = Platyhelminthes, An = Annelida, Ec = Echinodermata, Ar = Arthropoda, Mo = Mollusca.

species but it is as high as 6.9% for the minor phyletic species, i.e. nearly 17-times more than in major phyletics. Obviously, the preponderance of parthenogenesis has limited species diversity in minor phyletics. In all, parthenogenesis is limited to 0.61% of all animals. Hence, the value of 1% parthenogens reported by Bell (1982) may be an overestimate. Further, parthenogenesis and hermaphroditism mutually eliminate each other in both major and minor phyletics.

7.2 Unisexualism

Unisexualism is a modified version of invertebrate parthenogeneses in poikilothermic vertebrates. With no males, the rare unisexuals are all hybrids between them and one or more biparentally reproducing sexual species. Their incidence is limited to 30 species or rather biotypes: fishes (seven biotypes), amphibians (three biotypes) and lizards (20 biotypes) (Vrijenhoek et al., 1989). The rearing difficulty has eliminated the scope for experimental investigation in lizards. With amenability for rearing and experimentation, more or less adequate literature is available for fishes, albeit the number is limited. Hence, the description in this account is limited to fishes.

Defying gonochorism, the unisexual fishes seem to have entered into more complicated reproductive modes, which are summarized in five headings (Table 7.3): 1. In the simplest type *Poecilia formosa*, following activation of embryonic development, the donor genome of *P. mexicana* sperm is eliminated and diploidy is restored by gynogenetic retention of its own polar body. 2. To alleviate the negative clonal* effect, a small amount of genome leaked from the donor *Poecilia latipinna* sperm is incorporated into activated eggs of *P. formosa*. 3. In contrast to gynogenesis, hybridization is a hemiclonal mode of reproduction that changes the male genome at every generation in accordance with the mating male species. For example, the MP hybrid between *Poeciliopsis monacha lucida* ♀ × *P. occidentalis* ♂ produces an egg containing an irreplaceable M maternal genome but the male's P genome, representatively expressed by a dark spot on the dorsal fin, is replaced in each generation by a mating sexual partner species (Fig. 7.2). 4. In a little more complicated *Menidia clarkhubbsi*, progenies are triploidized by retaining the genomes of its own egg and its own polar body, and that of the donor *M. penninsulae* sperm.

The triploidization process is more complicated by the hybrid producing unreduced clonal sperm in *Squalius alburnoides*. 5. More complication is introduced by tetraploidization in *Carassius auratus sugu* and *Cobitis granoei taenia* complex. In *C. auratus gibelio*, the tetraploid males are sterile but

---

* This term is not appropriate but is used here, as it is familiar among biologists specializing on unisexuals.

TABLE 7.3

Polyploidization and reproductive modes in unisexual fishes (from Pandian, 2012)

Unisexual	Sexual pattern	Mode of reproduction	Sperm source
Menidia clarkhubbsi Athernidae Capano Bay, Taxas	2n ♀ only	Gynogenesis	*M. beryllina* *M. peninnsulae*
Fundulus sp Fundulidae Nova Scotia, Canada	2n ♀ only but 3n rarely	Gynogenesis?	*F. heteroclitus* *F. diaphanus*
Poeciliopsis spp Poeciliidae, Mexico	2n ♀ 3n ♀	Hybridogenesis Gynogenesis	*P. lucida, P. monacha* *P. occidentalis, P. infans*
Poecilia formosa Poeciliidae Texas, north east Mexico	2n ♀, ♂ extremely rare, 3 ♀, ♂ rare	Gynogenesis Gynogenesis	*P. mexicana* *P. latipinna*
Phoxinus eos neogaeus Cyprinidae East Inlet pond, New Hempshire	2n ♀ only 3 ♂; 0.5 % sterile ♂ 3 ♀ clonal; non clonal	Gynogenesis Variant in hybridogenetic direction	*P. eos, P. neogaeus*
Carrasius auratus Cyprinidae *gibelio*, north China *langsdorfi*, Japan *sugu* south China	2n ♀, 50% sterile ♂ 3n ♀, 50% sterile ♂ 4n ♀, 50% sterile ♂	Gynogenesis	*Cyprinus carpio*, other cyprinids
Squalius alburnoides Cyprinidae West Spain, Iberia	2n ♀, ♂ 3 ♀ only 4n ♀, ♂	Gynogenesis Hybridogenesis Gonochorism	*Squalius pyrenaicus* *S. carolitertii*
Cobitis granoei taenia Cobitidae Moscow River	3n ♀ only 4n ♀ only 4n ♀, ♂	Gynogenesis Gynogenesis Gonochorism	*C. granoei, C. taenia*

the female lays unreduced tetraploid eggs. The *C. granoei taenia* complex produces (i) unreduced 4n eggs, (ii) triploid eggs and (iii) mosaic eggs.

Having arisen as mostly interspecific hybrids on multiple occasions, each unisexual fish species has explored its own uncommon mode of reproduction. Gynogenesis has been a chosen mode of reproduction by almost all of them to escape the consequent clonal negative effects, but each one has explored alternative avenues. Perhaps, *Poecilia formosa* and *Menidia clarkhubbsi* have adopted gynogenesis as the only mode of reproduction and exhibit it in its simplest form. To introduce clonal diversity *Cobitis* and *Carassius* have included 'genome addition' and polyploidization in their gynogenic mode of reproduction. Besides gynogenesis, *Poeciliopsis* have opted for hybridogenesis, which provides them an opportunity to try and test a new genome at every generation. While most unisexual clones remain frozen and isolated, *Phoxinus eosneogaeus* has chosen to link them together and introduce

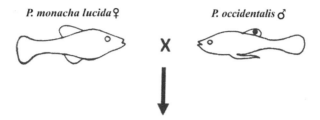

P. monacha occidentalis F₁ all ♀ progenies with dorsal fin spot

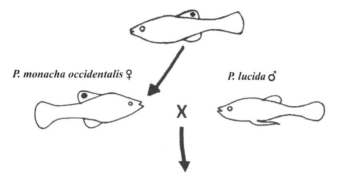

P. monacha lucida F₂ all ♀ progenies without dorsal fin spot

FIGURE 7.2

Hybridogenesis in poecilids. Note the haploid sperm of the related species *Poecilopsis occidentalis* fertilizes the eggs of *P. monacha-lucida* and the trait, dorsal fin spot of *P. occidentalis* is expressed in F₁ progeny but is lost in F₂, when its eggs is fertilized by *P. lucida* (from Pandian, 2011).

clonal diversity by discovering the mosaic forms. In contrast to all these unisexual fish species, which have experimented by modifying oogenesis one way or another, and array of hybrid female forms *Squalius alburnoides* has uniquely chosen to examine the possibility of altering spermatogenesis and the usage of hybrid males. On the whole, *the existence of only 30 unisexual species indicates that unisexualism does not facilitate species diversity.*

8

Clonals and Stem Cells

Introduction

The clonals multiply agametically to produce genetically identical progenies. With no meiosis and fertilization, they may not generate new gene combinations—the raw material for evolution and speciation. Table 8.1 provides some examples from annelids, which solely multiply by cloning for long years. The cloning process may involve (i) fragmentation (e.g. sponges, polychaetes, nemerteans), (ii) fission (e.g. planarians, holothurians), (iii) autotomy (e.g. asteroid starfishes), (iv) budding (e.g. cnidarians, entoprocts), (v) strobilization (e.g. scyphozoans), (vi) sporulation (e.g. myxozoans) or formation of (vii) gemmule (e.g. sponges) and (viii) statoblast (e.g. bryozoans), and the like. On the basis of whether the clonal progeny appears by simple conversion of parental somatic body, or by growth and development of new progeny from the parental body, the clonals may be grouped into two major types: 1. Fragmentation comprising fragmentation, fission, autotomy and the like, and 2. Budding consisting

TABLE 8.1

Obligately clonal reproduction in some annelids (condensed from Pandian, 2019)

Species	Reported observations
Polychaeta	
Cirratulidae Zeppelina monostyla	No sexual reproduction observed for 60 years in Freiburg Aquarium, Germany
Spionidae Polydorella kamakamai	In the field, sexual reproduction is rare or transient in 0.3% population only
Enchytraeidae Cognetia sphargnetorum	Sexually mature specimen is very rare at any season in Denmark. Sexual eggs do not hatch
Aeolosomatidae Aeolosoma viridae	Reproduce only by paratonic fission for > 20 years in the laboratory of Bologna University, Italy
Pristina leidyi	Clonal reproduction for 20 years at the University of College Park, USA and earlier for 4 years at Carolina Biological Supply Co, USA

of budding, strobilization and formation of gemmule and statoblast. Recognizing the scope for limitation by parental somatic resource, Rychel and Swalla (2009) divided fragmentation into two subtypes: (i) unidirectional fragmentation resulting in production of only one functional progeny, the ramet (e.g. earthworms, see Pandian, 2018) and (ii) bi- (or multi-) directional, when a clonal parent, the genet divides to produce two or more functional ramets (e.g. planarians, see Pandian, 2020). Cloning may occur naturally during early embryogenesis as polyembryony (e.g. cyclostomatid bryozoans, see Pandian, 2020, ant-eating *Echidna*), larval (e.g. cnidarians, see Shostak, 1993, echinoderms, see Pandian, 2018) and/or adult stage(s). Being the hub of cnidarian adaptation, cloning seems to have reached a climax in some colonial cnidarians. The following are some examples for the astonishing clonal potency of cnidarians: (i) A translucent and stingy siphonophore in the Australian coast runs to the length of 46 m (Nerida Wilson, Science Foscus, June 09, 2020), (ii) Swarms of *Aurelia* and blooms of *Craspedacusta* are clonally produced, (iii) Hermatypic coral generates monumental colony of > 3 m (in diameter) containing 30 million polyps (Shostak, 1993), as compared to 2.3 million zooids containing bryozoans *Membranipora membranacea*, see Pandian, 2021) and (iv) Sea feathers clonally achieve their magnificent symmetry.

8.1 Taxonomic Distribution

For the first time, Pandian (2020) brought to light that parthenogenesis and clonality mutually eliminate each other among minor phyla (Fig. 8.1). Figure 8.1 (right panel) shows that this is also true of major phyla. The overlap between clonals and parthenogens is limited to oligochaetes. A careful comparison of the reported information listed by Pandian (2019) in Tables 2.5 and 4.2 revealed that the hermaphroditic enchytraeid *Cognettia glandulosa* is the only worm reported as parthenogenic and clonal, which (must be subjected to a more careful study) may be more an exception. *Hence, hermaphroditism and parthenogenesis also mutually eliminate each other in major phyla too (see Chapter 7). These mutual eliminations show that animals do not simultaneously tolerate potential inbreeding depression arising from cloning and parthenogenesis as well as hermaphroditism and parthenogenesis.* However, cloning occurs in gonochores (e.g. echinoderms), and protandric (e.g. annelid *Salmacina australis*) and self-fertilizing (e.g. annelid *Bispira brunnea*) hermaphrodites. It can also occur in diploids and polyploids (e.g. *Dugesia ryukyuensis*).

It has been possible to almost precisely estimate the number of animal species possessing clonal potency. The number of this clonal species ranges from 3 in Placozoa to 10,856 in Cnidaria. From Table 8.2, the following may be inferred: 1. As many as 20,026 species from major phyla and 11,287 species

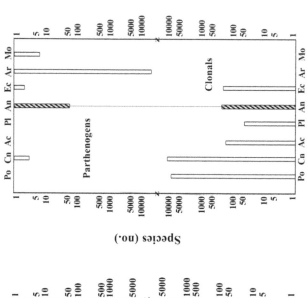

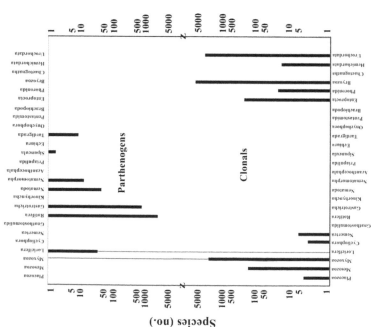

FIGURE 8.1

Parthenogenesis and clonality mutually eliminate each other in minor (left panel) and major (right panel) phyla. Parthenogenic earthworms are functional females following testicular degeneration or reciprocal gonochores. Po = Porifera, Cn = Cnidaria, Ac = Acnidaria, Pl = Platyhelminthes, An = Annelida, Ec = Echinodermata, Ar = Arthropoda, Mo = Mollusca.

TABLE 8.2

Number of clonal species in major and minor phyletic animals. 1. Van Soest et al. (2012), 2. Mapstone (2014), 3. J. Wright, *Animal Diversity Web*, Pandian (4 = 2020, 5 = 2016, 6 = 2017, 7 = 2019, 8 = 2018, 9 = 2021)

Major phylum	Species (no.)	Minor phylum[9]	Species (no.)
Porifera[1]	8553	Placozoa	3
Cnidaria[2]	10,856	Mesozoa	150
Acnidaria[3]	166	Myxozoa	2,200
Platyhelminthes		Loricifera	0
Turbellaria[4]	45	Cycliophora	2
Non-turbellarians	0	Nemertea	4
Arthropoda		Gnathostomulida	0
Crustacea[5]	3[†]	Rotifera	0
Non-crustaceans	0	Gastrotricha	0
Mollusca		Kinorhyncha	0
Gastropoda[6]	1	Nematoda	0
Non-gastropods	0	Nematomorpha	0
Annelida[7]		Acanthocephala	0
Polychaeta	79	Priapulida	0
Oligochaeta	190	Sipuncula	1[†]
Hirudinea	0	Tardigrada	0
Echinodermata[8]	137	Echiura	0
Vertebrata	0	Onychophora	0
Subtotal	20,026	Pentastomida	0
		Brachiopoda	0
Subtotal Major %	20,026/1,496,509	Entoprocta	200
	1.33 %	Phoronida	2
Subtotal Minor %	11,287/46,687	Bryozoa	5700
	24.20%	Chaetognatha	0
Total	31,313/1,543,196	Hemichordata	25
%	2.03%	Urochordata	3000
		Subtotal	11,287

† The incidence within these taxa may be sporadic and exceptional. Hence, they are not included in the present estimate.

from minor phyla are clonals. *Whereas 24.2% of minor phyletic species have clonal potency, only 1.3% major phyletic species have it. This high proportion of clonal potency in minor phyletics has limited species diversity.* 2. Remarkably, none of the pseudocoelomates and hemocoelomates have the potency, although one species each in Sipuncula (*Aspidosiphon brocki*), pteropod Mollusca (*Clio pyramidata*) and three species of colonial parasitic [rhizocephalan] crustaceans (*Polyascus* [= *Sacculina*] *polygenea, Peltogastrella gracilis, Thylacoplethus isoevae*) have been claimed to possess the potency. However, the data assembled in Table 8.2 confirm the hypothesis proposed by Murugesan et al. (2010) and elaborated by Pandian (2016). 3. Remarkably, both sessile colonials (Porifera, Cnidaria, Lophophorata, Ascidiacea), and solitary low motile planarians, annelids and echinoderms have the potency.

8.2 Solitary vs Colonial

An attempt has also been made to quantify the number of solitary and colonial, as well as fragmenting and budding clonal species (Table 8.3). Yet, the following has to be noted prior to making a few generalizations: (i) Whether sponges are individuals or colonials has been debated in detail (e.g. Hartman and Reiswig, 1973). However, they are considered as colonies (see Blackstone and Jasker, 2003), as each osculum with its contributing portion of canal system is usually considered as an individual within the lowest possible grade of a colony (see Hyman, 1940). The class Calcarea comprises small sponges of < 10 × 15 cm size and clonally multiplies mainly by budding. Smaller sponges clonally multiply by budding, but the larger ones may subsequently undergo fragmentation due to wave action, predation and the like. Hence, sponges are considered as colonial budders (Fig. 8.2). For details on construction of the clonal buds in sponges, Sallar (1990) may be consulted. (ii) Among cnidarians, the freshwater *Corymorpha* and *Branchiocerianthus* and marine *Acaulus* and *Myriothela* do not undergo clonal multiplication. Many hydras (e.g. *Protohydra*) are known to undergo transverse or horizontal fission. Among sea anemones *Anthopleura elongantissima* and *Calliactis tricolor* undergo longitudinal fission and actinarians *Gonactinia prolifera* and *A. stellula* transverse fission (see Shostak, 1993). Yet, all the cnidarians are counted as colonial budders. (iii) Dr. S. Sudhakar (M.S. University, Tamil Nadu, India) has shown that the earthworm *Perionyx excavatus* possess clonal potency to generate multiple ramets. But, it is not known whether all the other 55 earthworm species are capable of bidirectional fragmentation. Nevertheless, more or less precise information on the number of clonal species is available for the major phyla but not for minor phyla. In the latter, the number *per se*, in which cloning is experimentally demonstrated is limited to hardly four species in Nemertea and two in Phoronida. But the colonial Entoprocta, Bryozoa and Urochordata are generally recognized as clonals (see Blackstone and Jasker, 2003).

From the values listed in Table 8.3, the following may be inferred: (1) *On the whole, some 1,490 and 29,647 species are fragmenters and budders, respectively. Surprisingly, most budders are colonials but fragmenters are solitaries.* However,

TABLE 8.3

Number and size of solitary and colonial species clonally multiplying by fragmentation and budding. L = length, H = height, D = diameter, 1 and 2 = Hyman (1940), 3 = Pandian (2020), 4 = Pandian (2019), 5 = Pandian (2018), 6 = Pandian (2021), 7 = Pandian (2018)

Taxon	Solitary fragments		Colonial budders	
	Species (no.)	Size	Species (no.)	Size
Porifera[1]				
Calcarea	-	10 cm H	794	-
Non-calcarea	-	-	7,759	-
Cnidaria[1]	-	-	10,856	Up to 65 m L
Acnidaria[2]	166		-	-
Turbellaria[3]	45	2 mm–50 cm L	-	-
Annelida[4]				
Polychaetes	79	2 mm–3 m L	-	-
Earthworms	53	?	-	-
Enchytraeids	29	?	-	
Echinodermata[5]	137	?	-	-
Placozoa[6]	3	5 mm D	-	-
Mesozoa[6]	-	-	150	
Myxozoa[6]	-	-	2,200	
Nemertea[6]	4	-	-	-
Ectoprocta[6]	-	-	200	100 μm–< 5 mm H
Phoronida[6]	2	100 mm H	-	-
Bryozoa[6]	-	-	5,700	0.5 mm–45 cm
Pterobranchia[6]	5		25	
Urochordata[7]	967	-	1,963	
	1,490		29,647	31,137
	4.8%		95.2%	

some 100 aquatic oligochaete species and 75 pelagic tunicate species are budding solitaries. (2) *Of 31,137 clonal species, > 95.2% of them are budders and the fragmenters account for < 4.8% only.* (3) With regard to size, a clear picture has not emerged, due to complications by the presence of different number of individuals in different colonies. *In general, solitary fragmenters are larger in size than the individuals in colonial budders.* Incidentally, Pandian (2018) noted that successive fragmentations progressively reduce the clonal size in echinoderms and urochordates. (4) In clonal multiplication, *fragmentation is a costlier mode than that of budding. Not surprisingly, fragmentation has decisively decelerated species diversity.* (A) For example, the polychaete

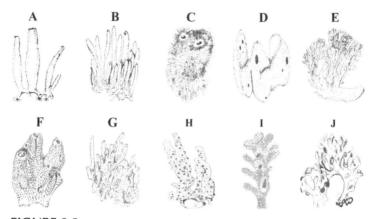

FIGURE 8.2

Calcarea: A. *Leucosolenia* (after Dendy, 1891), B. *Rhabdodermella*, C. *Leuconia*. Desmospongia: D. *Chondrilla*, E. *Microciona prolifera*. F. *Tedania*, G. *Halichondria*, H. *Haliclona*. Hyalospongia/Hexactinellida: I. *Aphrocallistes* and J. *Farrea occa* (all are free hand drawings from Hyman, 1940).

worm *Pseudopotamella reniformis* accumulates resources throughout spring-summer to clonally produce 2–4 progenies during autumn-winter. (B) At the maximum, *Lumbriculus variegatus* can clonally fission 51 progenies (see Pandian, 2019) and (C) Fission frequency in the classical triclad turbellarians ranges from 0.1 time/fluke/d in *Dugesia ryukyuensis* to 0.4 time/fluke/d in *Euplanaria tigrina* (see Pandian, 2019). But the budding generates a larger number of them, for example, a bryozoan colony *Celleporella hyalina* produces 29 zooid/d (see Pandian, 2021, Table 26.2).

8.3 Larval Cloners

Bosch et al. (1989) startled the world of reproductive biologists by the discovery of larval cloning by budding in the asteroid larva of *Luidia* sp. Thus, larval cloning is an intriguing new dimension that confers one or more of the following advantages: (i) Increases the number of offspring arising from the fittest (successfully fertilized and hatched among free-living or parasite that have successfully infected an appropriate host on time); for example, from a single miracidium of digeneans (see below), as many as 150–250 or even up to one million (in strigeids) cercariae are produced (Hyman, 1951a), (ii) Prolongation of the larval duration facilitating the timing of settlement in the free-living cloners or infection of an appropriate host in the parasitic cloners: for example, the duration elapsed between the miracidial entry and cercaria emergence lasts for some weeks and even months (Hyman, 1951a), (iii) enhanced scope for dispersal and settlement in the free-living cloners or dispersal and infection of an appropriate host in the parasitic cloners (Fig. 8.3B) and (iv) recycling of otherwise discarded or resorbed larval tissue (e.g. *Ophiopholis aculeata*, Fig. 8.3C$_4$–C$_5$). Larval cloning occurs in sessile cnidarians in multiple different routes and modes (Fig. 8.3A), low-motile echinoderms (Fig. 8.3C), digeneans (Fig. 8.3B$_2$–B$_5$) and cestodes, in the latter, its incidence is limited to 19 species belonging to four families (e.g. Taeniidae) and occurs sporadically in the mammalian Intermediate Host (IH) (Fig. 8.3D). Among echinoderms, its occurrence is extended to crinoids, which do not clone as adults (see Pandian, 2018). Its incidence among 12,012 speciose digeneans has long been debated to represent either parthenogenesis or polyembryony. From suggestive evidence, Pandian (2020) noted that their osmotrophic sporocysts and macrophagic rediae do possess the germinal balls, which clonally multiply into rediae and cercariae, respectively.

Whereas cnidarian and echinoderm larvae undergo cloning by budding, the germinal balls of sporocysts and rediae of digeneans, cysticerci/ metacercoide of cestodes undergo ontogenetic larval development, recalling a zygote undergoing embryonic development. Following transplantation

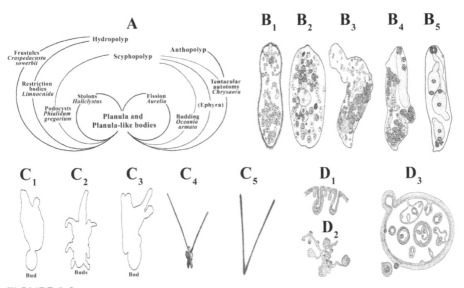

FIGURE 8.3

A. Clonal multiplication and life cycle of cnidarians involving planula and planula-like bodies (modified from Shostak, 1993). B. Digenean larval forms: B$_1$. miracidium, B$_2$. mother sporocyst, B$_3$. daughter sporocyst B$_4$. mother redia and B$_5$. daughter redia (from Phalee et al., 2015). C. Cloning in echinoderm larvae: Budding in C$_1$. auricularia larva of *Parastichopus californicus*, C$_2$. bipinnaria larva of *Luidia* sp and C$_3$ echinopluteus larva of *Dendraster excentricus*. Cloning in ophiopluteus larva of *Ophiopholis aculeata* C$_4$. newly settled juvenile showing posterolateral arms and C$_5$. freshly released posterolateral arms (from different sources, see Pandian, 2018). D. Schematic free-hand drawings to show D$_1$. the inwardly branching coenurus cyst of *Taenia multiceps*, D$_2$. irregular branching coenurus cyst of *Hymenolepis nana* and D$_3$. hydatid cyst of *Echinococcus granulosus* (from Hyman, 1951a).

into naïve *Lymnaea stagnalis*, the rediae of *Echinoparyphium aconiatum* and *Isthmiophora melis* continue to produce rediae up to 27 and 40 successive generations and the fluke continues to retain the clonal potency as well as the cercarial infectivity (see Pandian, 2020). On a intraperitoneal injection of isolated (germ) cells from coenurus cysticerci of *Taenia crassipes* into mice the cells clonally developed into complete cysticerci (Toledo et al., 1997). A climax is that the cultured isolated (germ) cells from the hydrated vesicle of *Echinococcus multilocularis* metacercoide developed *in vitro* into metacercoides (Brehm, 2010). Notably, in both free-living and parasitic larval clonals, cloning occurs by budding or what may be called as 'ontogenetic cloning' to produce the corresponding larval stage but not the adult.

8.4 Stem Cells and Sources

Stem cells divide asymmetrically into two daughter cells, one of which retains the stem cell potency, while the other differentiates into all other cell types including the germline to generate offspring. None of the stem cell type listed in Table 8.4 could qualify, as they undergo symmetric divisions, except the germ balls of the digenean sporocysts and germ cells in the cestode coenurus cysticerci. Based on the level of their potency, the clonal stem cells are classified into (i) pluripotent (e.g. Platyhelminthes), (ii) multipotent (e.g. cnidarians) and (iii) oligopotent (capable of differentiation into a restricted number cell types) (see Rinkevich et al., 2009). Aside from it, this account has chosen to classify the clonal stem cells into three groups on the basis of number of cell types present in the stem cells. Considering the critical mass of stem cell and the resultant 'mass effect', this (Pandian's) classification explains the differences in clonal and regenerative potencies observed among closely related species. Accordingly, Group 1 possesses only a single cell type, as the archaeocytes in sponges, interstitial cells (ic) in cnidarians, agametic cell in mesozoans and neoblasts in platyhelminths (Table 8.4), Group 2 with two stem types, as the Resident Stem Cells (RSCs) and Circulatory Stem Cells (CSCs) arising from ectoderm and mesoderm in echinoderms or arising from the epidermis along with ectodermal or mesodermal or endodermal stem cells, as in urochordates (Manni and Burighel, 2006) and Group 3 with stem cells arising from all three germ layers, namely ectoderm, endoderm and mesoderm, as in polychaeates. The clonals have arisen independently multiple numbers of times and in some cases even within a phylum; for example, oligochaetes with neoblasts and polychaetes with no neoblasts. With involvement of increasing number of stem cell types, the proportion of clonal species decreases from almost 100% in sponges, cnidarians and mesozoans with a single clonal stem cell type (Group 1) to 1.9% in echinoderms in Group 2 (137 species of 7,000, see Pandian, 2018) and to 1.7% in annelids in Group 3 (190 species of 16,911, see Pandian, 2019). The proportion 0.7% (45 species

TABLE 8.4

Stem cells and their source in clonals. 1 = Blackstone and Jasker (2003), 2 = Hyman (1940), 3 = Pandian (2020), 4 = Pandian (2019), 5 = Pandian (2016), 6 = Pandian (2017), 7 = Pandian (2018), 8 = Pandian (2021), 9 = Manni and Burighel (2006)

Major phyla		Minor phyla	
Phylum	**Stem cell source**	**Phylum**	**Stem cell source**
Group 1		**Group 1**	
Porifera[1]	Multipotent archaeocytes	Placozoa[8]	3 cell types act as unipotent stem cells
Gemmules[2]	Cluster of archaeocytes		
Cnidaria[1]	Pluripotent interstitial stem cells	Mesozoa[8] Dicyemida Orthonectida	Agamete in nematogen Agamete in plasmodium
Platyhelminthes[3]	Pluripotent neoblasts		
Digenea Sporocysts + rediae	Asymmetric germinal cells divide to produce germinal and somatic lineages	Myxozoa[8]	Spores from sporoplasm, pansporoblast, plasmodium
Cestoda Cysticerci Metacestoide	Symmetric germinal cells divides to produce germinal blastomere	Cycliophora[8]	Stem cells in buccal funnel of feeding stage
		Nemertea[8]	Pluripotent stem cells from brain and nerve cord
Group 2			
Echinodermata[7]	Pluripotent blastomere Pluripotent Resident Stem Cells (RSCs) form radial nerves and Circulating coelomocytes Stem Cells (CSCs)	Bryozoa[8]	Pluripotent stem cells descending from the top tier cells stored in epithelial cells
		Group 2	
		Entoprocta[8]	Pluripotent stem cells from ecto- and mesoderms
Group 3			
Oligochaeta[4] Polychaeta[4]	Multipotent neoblasts; Dedifferentiated multipotent stem cells arising from 3 germ layers	Phoronida[8]	Pluripotent stem cells vasoperitoneal tissue originated from ectoderm and mesoderm
Arthropoda[5] Rhizocephala	10–15 stem-like cells Cultured in the host; but doubtful cloning	Urochordata[9]	Multipotent stem cells from the epidermis along with ectodermal, or endodermal or mesodermal stem cells
Mollusca[6] *Clio pyramidata*	Stem-like cells from mesosderm and endoderm		

of 6,376 speciose turbellarians, see Pandian, 2020) in turbellarians with a single stem type seems to be an exception. With regard to neoblast, it must be stated that Randolph (1891) discovered it first from the posterior surface adjacent to the ventral nerve cord of *Lumbriculus variegatus*. However, it turned out that neoblasts are multipotent in annelids but pluripotent in turbellarians.

Because of their structural homogeneity and morphological simplicity with not more than six cell types (see Hyman, 1940), even a small fragment of sponges has the minimally required number of archaeocytes and readily reorganizes into an individual clone. This is also true of the parasitic Mesozoa (with only < 50 cells differentiated into 6–7 cell types) and Myxozoa (7–8 cell types, see Pandian, 2021), as well as free-living carnivorous cnidarians. In clonally multiplying planulae, there are seven cell types in hydrozoans and 10 in anthozoans (Shostak, 1993). Though each interstitial cell of *Hydra* is pluripotent, a minimum mass of 2×10^4 interstitial cells is required to clonally generate a whole animal of *H. attenuata* clone with 10^5 cells in it (Gierer et al., 1972). In planarians, the number of cell type is increased to 14 (see Pandian, 2020). The critical mass of stem cells required to clone a whole new individual is estimated to range between 100 and 300 (Rinkevich et al., 2009). The minimal fragment containing 'the critical mass' of neoblasts required to clone a planarian is 1/279th the size of the fluke (Morgan, 1901). The lowest body size that contains the 'critical mass' stem cells is 2 mm in the ophiuroid *Ophiocomella ophiactis* and asteroid *Allostichaster insignis*. Interestingly, to accomplish the same in treble plane fission (to generate 4 ramets), the required size is 68 g in the holothurian *Stichopus herrmanni* (see Pandian, 2018). *The increasing number of cell types, and consequent structural complexity seems to demand stem cells from two (e.g. echinoderms, urochordates) and three sources. In this context, that Hill's (1970) hypothesis is relevant in that each germ layer in eucoelomates like annelids has retained its identity and dedifferentiates (except neoblasts in oligochaetes) into stem cells from the respective germ layer and regenerates their respective organs and systems. In them, the fairly well developed nervous system and brain are able to coordinate the events in morphogenesis. The lower invertebrates with no nerve cells in sponges, mesozoans and myxozoans or no nerve ganglia in cnidaria or with the simplest nerve system in turbellarians have resorted to a single stem cell type. In all these cloners with a single stem cell type, each stem cell is an additive. A 'critical mass' of stem cells and the resultant 'mass effect' can alone accomplish successful cloning, in comparison with what a totipotent single cell, the zygote can do. Arguably, this 'critical mass' of stem cells is present only in the described fragment/body size of 45 species in the 6,376 speciose turbellarains, 190 species of 16,911 annelids (see* Pandian, 2019) *and 137 species of 7,000 echinoderms (see Pandian, 2018), as they are able to successfully clone. For want of this 'critical mass' of stem cells per se, the remaining 6,331 turbellarian, 16,721 annelid and 6,863 echinoderm species are unable to clone.*

At this juncture, it is relevant to consider Pandian's (2016) hypothesis that (i) all the members of pseudocoelomates and hemocoelomates do not clone and (ii) they have not retained the embryonic stem cells. The first part of the hypothesis is valid, except for the very rare incidence of cloning in a sipunculan species *Aspidosiphon brocki* (Rice, 1970). However, the second part of the hypothesis is not correct, as none of the clonal species retain the totipotent embryonic stem cells. In pseudocoelomates and hemocoelomates, the reasons for the lack of clonal potency seem to be related to other features. According to Blackstone and Jasker (2003), the structural body plan of the clade Ecdysozoa, comprising pseudocoelomate phyla and hemocoelomate phylum Arthropoda, has a chitinous cuticular coverage; some of them molt regularly (e.g. Arthropoda, Tardigrada) or occasionally (e.g. Penstatomida). *The presence of the chitinous cuticle and the need for molting has hampered them from cloning. In molluscs, the presence of shell(s) has eliminated the clonal potency. This is also true of brachiopods, a clade within Lophophorata, in which all other phyla Entoprocta, Phoronida and Bryozoa (with no shells) are cloners.*

8.5 Regeneration and Clonals

Regeneration is the potency to repair and replace the voluntary loss of cells, tissues and/or organs of an animal from (i) sub-lethal predation, (ii) self-inflicted spontaneous autotomy and (iii) intolerable shock on encountering frequency (e.g. *Diopatra aciculata*, Safarik et al., 2006), or hypo-osmotic stress (e.g. *Marenzellaria viridis*, David and Williams, 2016) or the like. While it provides immediate benefit of survival, the loss and subsequent regeneration cost a short or long term dysfunction, which deprives competitive ability. In 1889, Morgan classified it into two types: (1) Morphallaxis involving the remodeling of existing tissues into missing ones (by dedifferentiation) without extensive proliferation of cells and (2) Epimorphosis involving massive proliferation of undifferentiated stem cells, e.g. archaeocytes in sponges, interstitial cells in cnidarians and neoblasts in turbellarians as well as formation of blastema. The former may be less energy-intensive than the latter. *However, epimorphosis seems not to have limited species diversity; for example, ophiuroids with regeneration involving epimorphosis are more speciose (2,604) than the 1,000 speciose asteroid characterized by morphallaxic regeneration* (see also Pandian, 2018). Some individuals have an exceptional prodigious potency for regeneration; for example, the head of the oligochaete *Lumbriculus variegatus* can be regenerated as many as 21 times and the tail 42 times and both together 20 times (Muller, 1908).

At the cellular level, almost all animals including eutelic nematodes have the potency to regenerate the lost intestinal mucosal cells lost due to physical and (enzymatic) chemical erosion. This is also true of the replacement of the epidermal cells, except in the 684 speciose hirudineans. Consequently, they are not capable of wound healing (see Pandian, 2019), which may be one reason to deter their species diversity. Hence, *99.5% animals have the potency to regenerate the ectodermal derivative, the epidermis and endodermal derivative and the mucosal cells.* At the organ level, decapod crustaceans (15,000 species, Pandian, 2016) and most insects (1,035,007 species, Zhang, 2011) are known to regenerate a small or larger part of their appendage after a molt. In them, the ectodermal regeneration includes its derivatives epidermis and nerves, and that of mesodermal muscle and connective tissues (see Pandian, 2016). This is also true for molluscs. Their regenerated organs include shell, proboscis, nerves and the brain and rarely the gonad, as in *Arion* and few oeolids in gastropods, and the inhalant and exhalent siphons in bivalves. But the 800 speciose cephalopods do not have the potency for regeneration of any organ (see Pandian, 2017). The vertebrates including humans have the potency to regenerate endodermal derivatives like the liver and mesodermal derivatives like the kidneys and gonads (see Table 8.5). However, the ectodermal derivatives like the brain and nerves and mesodermal derivatives the cardiac and sphinctor muscles cannot be regenerated. In all, of 1,217,490 species, 85.4% have the potency for regeneration of one or more organs. At the animal level, sponges, cnidarians, acnidarians, entoprocts, phoronids, bryozoans, most urochordates and a few hemichordates are capable of

TABLE 8.5

Approximate number of regenerative species in major and minor phyla. 1 = Zhang (2011), 2 = Pandian (2017), 3 = Zhang (2011), 4 = Van Soest et al. (2012), 5 = Shikina and Chang (2018), 6 = Hyman (1940), 7 = Pandian (2020), 8 = Pandian (2019), 9 = Pandian (2018), 10 = Pandian (2021)

Major phyla		Minor phyla[10]	
Taxon	Regenerative species (no.)	Taxon	Regenerative species (no.)
Regenerative potency for cells			
Of 1,543,196 species, 1,542,512, i.e. 99.95% species have the potency for cell regeneration*			
Regenerative potency for organs			
Arthropoda[1]	1,035,007[†]	Nemertea	1,300
Mollusca[2]	117,651	Sipunula	160
Vertebrata[3]	64,832	Subtotal	1,460
Subtotal	1,217,490	%	100
%	85.4		
Regenerative potency for cloning			
Porifera[4]	8553	Entoprocta	200
Cnidaria[5]	10,856	Phoronida	23
Acnidaria[6]	166	Bryozoa	5,700
Platyhelminthes[7]	45	Hemichordata	25
Annelida[8]	235	Urochordata	2,930
Echinodermata[9]	137	Subtotal	8,878
Subtotal	19,992	% of total 9,053 species	98.1
% of total 71,186 species	28.1		

* hirudineans (684) species are unable to regenerate any cells
† appendage regeneration limited to insects and crustaceans

regeneration of all the missing components. However, of 19,992 species, 28.1% have the potency to regenerate all the missing parts of the body. Briefly, 99.9, 85.4 and 28.6% species have the potency to regenerate ectodermal epidermis and endodermal mucosal cells, one or another organ, and all the missing parts of the body, respectively.

Among minor phyletics, the scope for the assessment is limited for want of relevant information. Still, *the eutelism in pseudocoelomatic six phyla (31,504 species) and simpler structural organization in Aorganomorpha + Loricifera and Cycliophora (= 2,359 species) do not let 72.6% minor phyletics (33,893 species) to regenerate. This may be an important factor that has limited their species diversity among minor phyletics.* Interestingly, none of the minor phyletics seems to involve all the three germ layers in regeneration. *Available information suggests that 1,460 species have the potency for regeneration of organs and 8,878 (or 98% of 9,053 species) eucoelomate species have potency to regenerate all missing body parts. In comparison to the clonal potency of major phyletics (28% species), the potency is nearly four-times greater for the minor phyletics* (Table 8.5).

It is usually considered that clonal potency is derived from regeneration potency. Accomplishing an onerous task of annelid species, Zattara and Bely (2016) ascertained that clonal potency is derived from regenerative potency. Incidentally, unlike other radially symmetrical clonal clades sponges, cnidarians and echinoderms, the bilaterally symmetrical body of annelids is composed of metameric segments, each consisting of the same organs and so forth. Hence, amputation at any axial position along the body of annelids results in primarily removal of different quanta of the same organ system rather than removal of a different organ system. With this repetitive segmental system but with the presence of the brain only in anterior part of the body, the regenerative potency of anterior regeneration differs from that of posterior. Hence, it is customary for annelidan experts to separately consider the anterior and posterior as well as anterior cum posterior regenerative potency (Table 8.6). In 149 annelid species, the potency for anterior regeneration includes only 130 species or 87% clonal species. Similarly, only 74% species are clonals with the potency for anterior cum posterior regeneration in annelids. These observations seem to support the proposed view of Zattara and Bely (2016) that clonal potency is derived from regenerative potency.

Considering (i) the occurrence of six clonal species among the aquatic oligochaetes (Table 8.6) and (ii) in the absence of neoblasts, none of the oligochaetes neither regenerate nor clone, this view of Zattara and Bely (2016) is not acceptable to Pandian (2019). Table 8.6 lists the number of regenerative and clonal species in turbellarians, echinoderms and annelids. Of course, regeneration and cloning are related but independent processes that have originated independent of each other. In support of this new proposal, the following could be noted: 1. In turbellarians, the number of species with regenerative and clonal potency is 65 and 45, respectively. But it is 23, 7 and 1 for regeneration in triclads, acoelomorpha and catenulids, respectively, in

TABLE 8.6

The number and percentage (in brackets) of total, regenerative and clonal turbellarians and echinodermates (compiled and modified from Pandian, 2018, 2019, 2020)

Taxon	Total	Regenerative	Clonal
Turbelleria			
Tricladida	1000	23 (2.3)	30 (3.0)
Acoelomorpha	350	7 (2.0)	9 (2.6)
Catenulida	120	1 (0.8)	5 (4.2)
Macrostomorpha	200	7 (3.5)	1 (0.5)
Polycladida	2000	13 (1.3)	0
Proseriata	350	7 (2.0)	0
Lecithoepitheliata	30	1 (2.3)	0
Bothrioplanida	200	1	0
Prolecithophora	?	1	0
Rhabdocoela	?	4	0
Subtotal	6576+	65 (0.99)	45 (0.71)
Echinoderamata			
Crinoidea	700	36 (5.1)	0
Holothuroidea	> 1000	61 (6.1)	37 (3.7)
Ophiuroidea	2604	55 (2.1)	50 (1.9)
Asteroidea	~ 1800	50 (2.8)	47 (2.6)
Echinoidea	> 800	2 (0.25)	3 (0.3)
Subtotal	6904	204 (2.95)	137 (1.98)
Annelida			
Anterior only	149	149 (100)	130 (87.2)
Posterior only	206	209 (100)	19 (9.2)
Anterior + posterior	143	143 (100)	61 (74.3)
Anterior + posterior but no cloning	34	34	0
Anterior + posterior with cloning	19	19	19 (100)
No anterior + posterior but cloning	6	0	6
Total	557	554	235

comparison to 30, 9 and 5 for clonals. 2. In echinoderms, species number with potency for cloning is limited to 137 species, whereas 204 species are reported to have the regenerative potency. It is difficult to comprehend the existence of three clonal species but only two regenerative species in echinoids. 3. In annelids, there are six clonal oligochaete species, which do not have the potency for anterior-posterior regeneration. Of 143 species with anterior-posterior potency, only 61 species are clonals. Additional support also came from the earlier proposed Pandian's 'mass effect' hypothesis. Accordingly, the so called stem cells are not totipotent but are pluripotent or multipotent. *Hence, a critical mass of them is required to accomplish cloning. In the presence of this critical mass of stem cells, an individual has the clonal potency. The critical mass of stem cells required for the cellular regeneration may be a few, but more for regeneration of organ(s) and may be even more for complete regeneration of all the missing parts of the body. Hence, Pandian's 'mass effect' hypothesis can explain the existence of cellular, organ and individual level regeneration as well as clonal potency.*

8.6 Clonality and Diversity

Whereas the incidence of hermaphroditism and parthenogenesis is limited to < 1% each, clonality accounts for ~ 2%. The functional/behavioral gonochoric hermaphrodite can still generate new gene combinations, as both meiosis and cross-fertilization occur in them. Sporadic occurrence of males in mitotic parthenogens may purge deleterious mutations in them. With no scope for elimination of accumulated deleterious mutations, it is not clear how animals can tolerate 2% clonality. It is also not clear how some phyla like the colonial Porifera, Bryozoa and Urochordata can be speciose, despite being hermaphroditic and clonals. Two features namely fusion among the colonial fragments and the regular colonial rejuvenation by degeneration and regeneration may be considered as factors responsible for the speciose nature of these colonial phyla. Notably, not all the degenerated colonies or their members regenerate, but only a few perhaps the fittest ones alone are regenerated. However, not much information is yet available on this aspect.

Many authors have reported valuable information on fission/fragmentation but only a very few have inadvertently noted the fusion among the fragments appearing from different individuals/colonies. For example, Wilson (1907) described the phenomenon of coalescence between fragments of the sponge *Microciona prolifera*. Subsequently, Van de Vyver (1970a, b) confirmed such coalescence in some sponges. Warburton (1958, 1961, 1966) reported the successful reproduction by fusion of larvae in *Cliona celata* and other sponges. In didemid ascidians, colonies fuse indiscriminately (Bishop and Sommerfeldt, 1999), while fusion is regulated by genetic histocompatibility in botryllids (Oka and Watanabe, 1957). Among ascidians, Carlisle (1961) observed the occurrence of true fusion in *Trididemnum tererum* in a wide size range from the smallest of 0.9 mm^2 to the largest 116 mm^2. During their investigation, Stocker and Underwood (1991) found that the fusion rate of *Trididemnum solidum* was twice the times more (0.7 time/colony/d) in the presence of sponge substratum than that in the absence of a sponge (0.4 time/colony/d). In this ascidian, the rate increased from 0.05 time/

colony/d in a small colony of 0.5 mm^2 size to 0.4 time/colony/d in a larger colony of 20 mm^2. Importantly, all these authors have not looked at the fusion from the angle of 'gamete' fusion at fertilization and thereby providing the new gene combinations. *Yes, the fusion of two fragments or colonies or larvae may indeed represent the 'fertilization-like events of two mega-gametes' to produce new gene combinations, as it happens at fertilization of two gametes appearing from two individuals. Hence, the regular colonial rejuvenization and fusion of fragments/ colonies/larvae may have facilitated species diversity in colonial phyla.*

The need to revisit sexuality is obvious, as selfing hermaphroditism and mitotic parthenogensis are limited to 0.5 and 0.6% of animal species and the incidence restricted from Pseudocoelomata to Hemocoelomata, whereas the occurrence of cloning is 2% and is stretched from Aorganomorpha to Eucoelomata (e.g. ascidians). The raw material for evolution and speciation is generated from three sources: (i) random mutation in all organisms leading to horizontal gene transfer (including those prior to the discovery of sex ~ 2 MYA), (ii) recombination during meiosis and at fertilization leading to vertical gene transfer in sexually reproducing organisms. Hence, sex has only introduced an additional source for generation of new gene combination and accelerated the rate of evolution and speciation. As they can gain no new gene combination, the strictly selfing hermaphrodites and mitotic parthenogens need not necessarily suffer inbreeding depression leading to evolutionary dead end. For example, finding a level of genetic diversity unprecedented among strictly selfing hermaphroditic killifish *Kryptolebias marmoratus*, Turner et al. (1992) reported 42 different clones among of 58 individuals. The haploweb and GMYC analyses of 576 individuals of bdelloid rotifer *Adineta vaga* revealed the existence of six species (Debortoli et al., 2016). Not surprisingly, the only parthenogenic bdelloids have not only persisted but also diversified into 461 species during the last 60 million years. It is not equally surprising that the sustenance of experimental clonal multiplication in the cirratulid polychaete over 60 years (see Table 8.1). Whereas the 0.5% selfing and < 0.6% mitotic parthenogens are not supplemented or rarely supplemented by sexual reproduction, the clonal multiplication is always supplemented by sexual reproduction. In fact, cloning is only a supplementary to sexual reproduction in most animals. This may explain the 2% preponderance of clonal mode of multiplication.

B2
Gametogenesis and Fertilization

Gametogenesis is the process through which gametes are produced. In all sexually reproducing animals, only one fertilizable oocyte is generated from an oogonium, whereas four spermatids are produced from each spermatogonium. Being mother cells, oogonia and spermatogonia alone can undergo meiotic reduction division to generate haploid gametes. Fusion of the female and male gametes at fertilization restores diploidy. As indicated earlier, new gene combinations are produced during meiosis and at fertilization. Not surprisingly, gametogenesis can accelerate or decelerate species diversity. For example, animals produce either a large number of smaller eggs or a small number of larger eggs. The scope for genetic variations among the large number of smaller eggs in the former is greater than that in the latter. The mating system has a profound implication to species diversity. Polygamy may increase genetic diversity more than that of monogamy. This may also be true for iteroparity in increasing temporal diversity among progenesis, in comparison to semelparity releasing the entire basket of gametes in a single event.

Incidentally, Indians were perhaps the first to develop the castration technique and application of the nose thread to effectively gain muscular energy from the bulls in an agricultural operation. Chanakya, the guru and minister to King Chandra Gupta (Maurya: 321–297 BC), the grandfather of Emperor Asoka, instructed the Royal Veterinary Superintendent to offer in farm training to stalkholders to castrate less painfully 96 out of 100 bulls leaving one bull for every 25 cows or buffaloes, and 9 out of 10 rams, leaving one ram for every 9 dam goats or sheep (Rangarajan, L.N., 1987, *Arthashastra*: A condensed English translation published by Penguin Books, New Delhi).

9

Mitosis and Meiosis

Introduction

The gamete mother cells, the oogonia and spermatogonia are derived from Primordial Germ Cells (PGCs). The PGCs are roundish, relatively large cells with large nuclei, a clear nuclear envelope and granular chromatin. Being the progenitors of the gonial cells, they are considered 'immortals'. On induction by the Germ Cells Supporting Somatic Cells (GCSSCs), their bisexual potency is reduced to unisexual potency to generate either oogonia or spermatogonia. During ontogenetic development, not only gonads but also somatic tissues and organs are progressively sexualized.

In animals, gametogenesis is a long-drawn process. It involves mitosis in two stages and meiosis in a single stage. Being the parental source, Primordial Germ Cells (PGCs) undergo asymmetric mitosis to produce (i) undifferentiated stem cells, which retain their status as PGCs and (ii) differentiated cells that are committed to become oogonia or spermatogonia. The oogonia and spermatogonia divide mitotically, but symmetrically and produce primary oocytes and primary spermatocytes, respectively. These primary gamete cells undergo meiosis to produce one oocyte from a primary oocyte but four spermatids from a primary spermatocyte (Fig. 9.1). The number of asymmetric mitotic division(s) undertaken by the PGCs and the number of times each oogonium and spermatogonium divides to produce primary oocytes and spermatocytes are crucially important determinants on the number of ova and sperms produced by an animal. The total number of mitotic divisions from the PGC to primary oocytes or spermatocytes and the known meiotic division are all critically important events on two counts: (i) it is during these events, new gene combinations—the raw material for evolution and speciation—are generated and (ii) the number of gametes produced is a decisively important determinant in recruitment and sustenance of population and species. Yet, very little information is available for the number of mitotic divisions undergone by the PGCs, and oogonia and spermatogonia. Hence, the objective of this chapter is to emphasize the need for research in this vital but virgin area than to present a description of it.

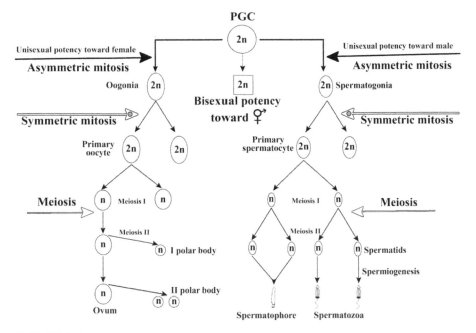

FIGURE 9.1

Gametogenesis commencing from Primordial Germ Cells (PGCs). Note the asymmetric mitosis in PGCs and symmetric mitosis in oogonia/spermatogonia. Information available is very limited in this area.

9.1 The Beginnings

The lineage of PGCs can histochemically be traced from the somatogenic or late blastula stage. With identification of *Vasa* gene, a molecular marker of the PGCs, it is now possible to trace its lineage even from mature oocyte (e.g. zebrafish, Yoon et al., 1997). In the nemertean *Lineus ruber*, the PGCs appear from the mesodermal parenchyma (Table 9.1); they are embedded in the cystid mesothelium in bryozoans. In other minor phyla Rotifera, Nematoda, Sipuncula and Chaetognatha, they could be traced as descendants of 4 D blastomeres. In diploplastic Porifera also, they originate from the archaeocytes located between the endodermal and ectodermal layers (Table 9.2). These observations indicate their mesodermal origin. In crustaceans, molluscs, echinoderms and teleosts, due to the *vasa* gene, it is possible to trace the origin of PGCs from mature eggs. In zebrafish, the PGCs are inherited from maternally supplied mRNAs. On surgical removal of the mRNA, no PGCs are formed in the resultant embryo. Hence, it is now well established that PGCs are the maternally inherited mRNAs. The maternal germplasmic structure called the 'nuage' contains *vasa* and *nanos* genes. The

TABLE 9.1

Source and origin of germline Primordial Germ Cells (PGCs) in minor phyla (from Pandian, 2021a)

Phylum	Reported observations
Mesozoa	Axial cell. Endogonial proliferation and gametogenesis
Myxozoa	Pansporoblast-Plasmodium, in which gametes are formed
Loricifera	Presence of ovary is indicated
Cycliophora	Dwarf male but the number germ cells is not reported
Nemertea *Lineus ruber*	PGCs arise from the mesodermal parenchyma. Ovaries extend from the preformed oviduct
Rotifera	From 4 D blastomere, the germline lineage and PGCs arise. Sex is irrevocably determined at 2nd cleavage in this determinate radial type. The fate of 4 D is not known in the obligate clonal bdelloids
Gastrotricha	Like rofiers
Kinorhyncha	The gonad consists of an anterior apical cell (PGC) from its lineage, all other reproductive cells arise.
Nematoda	From 4 D lineage, each gonad receives a single PGC. It is from the germ cell lineage gonads and gonoduct(s) are developed. Determinate spiral type
Sipuncula	From 4 D blastomere lineage, mesodermal bands arises
Bryozoa	From the PGCs embedded in the cystid mesothelium, spermatogonia and perhaps oogonia arise
Chaetognatha	The PGCs are descendants of 4 D blastomere lineage. The PGCs migrate to the coalescence with gonadal primordium

nanos is required for proper development of the PGCs, their migration to gonadal anlage and maintenance of germline stem cells (see Pandian, 2012).

9.2 Proliferation of PGCs

In fishes, gonadal development commences with the formation of an undifferentiated primordium consisting of the GCSSCs and PGCs. Differentiated at the early blastula, the PGCs migrate dorso-laterally to reach the gonadal anlage (Fig. 9.2B, C). In most teleosts, especially in male heterogametics, the ovarian differentiation precedes and commences with the appearance of proliferating GCSSCs and arrival of PGCs (e.g. chub mackerel *Scomber japonicus*). For example, a leaf-like gonadal primordium is formed at the mid-dorsal mesentery on the 4th day post-fertilization (dpf) in male heterogametic medaka *Oryzias latipes*. Following two to four consecutive rounds of divisions, many cysts are formed in the female embryo on the 5th dpf. Hence, the number of PGCs is more in females (275) than in males

TABLE 9.2

Source of germline Primordial Germ Cells (PGCs) in major phyletics

Phylum/Reference	Reported observations
Porifera (Fell, 1993, Hyman, 1940)	Enlarged archeocyte with large and conspicuous nucleus is transformed into an oocyte, Archaeocyte or choanocyte can be transformed into spermatogonium
Cnidaria (Frank et al., 2009, Shikina and Chang, 2018)	The gametes are developed from interstial cells (i-cells) arising from epidermis in hydroids but from endodermis in other cnidarians. The i-cells contain 2 types of stem cells: (i) Multipotent Stem Cells (MPSCs) with potency for both self-renewal, and differentiation into somatic and germline lineages (GSCs) and (ii) GSCs can self renew but can differentiate only into germline lineage
Platyhelminthes (Rinkevitch et al., 2009, Pandian, 2020)	Neoblasts constitute 3.5% (*Macrostoma lignano*) to 15% (*Convolutriloba longifissura*) of all cells. Among these neoblasts, 3.5% alone can differentiate into somatic and germline lineages. Following gonodectomy in *M. lignano* and *Schmidtea lugubris*, gonads can arise anew from the 3.5% neoblasts. These 3.5% neoblasts contain chromatid body, of which *vasa*-related *DJvlgA* and *DJvlg B* have been isolated from the testicular and ovarian regions, respectively
Annelida (Pandian, 2019)	PGCs emerge earlier and independent of somatic cells of the mesodermal posterior growth zone of the segment addition zone. In clonal species, the PGCs can be transmitted (without expression) over 1,000–3,000 generations lasting for 30–60 years
Crustacea (Feng, 2011)	*Vasa*-like mRNAs are maternally inherited. The PGCs are established as a cluster at the post-larval stage
Mollusca (Pandian, 2017)	*Vasa*-like gene transcript, localized on the vegetal pole of unfertilized oocyte, is maternally inherited
Echinodermata (Pandian, 2018)	The nest of PGCs emerge from the gonad-basis located in the dorsal mesentery
Pisces (Pandian, 2011)	PGCs are inherited through maternal mRNA and subsequently located in the latero-mesoderm at the neural stage. They migrate to reach and coalesce with gonadal primordium

(65) at hatching. Similarly, the PGCs are also more in number in another male heterogametic Nile tilapia *Oreochromis niloticus* and rosy barb *Puntius conchonius*. However, the trend for the number of PGCs is reversed in favor of males (1,159) but 712 in female heterogametic (?) catfish *Silurus meridionalis*. Experimental depletion of PGCs during the embryogenic stage results in the production of all male progenies in medaka, pearl danio and zebrafish, as their GCSSCs is predisposed toward male development. Hence, it is the GCSSCs that determine sex of fishes (Pandian, 2013). Further development of PGCs is elaborated in Chapter 23.

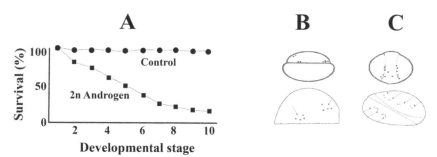

FIGURE 9.2

A. Embryonic developmental sequence in the androgen of the gold rosy barb *Puntinus conchonius* in the genome-inactivated eggs of the tiger barb *P. tetrazona* (from Kirankumar, 2003). B–C. Whole-mount *in situ* hybridization showing dorso-lateral migration of PGCs in goldfish *Carrasisus auratus* embryo and their corresponding schematic figures shown above (modified from Otani et al., 2002).

9.3 Spermatogonium and Oogonium

Spermatogonium: In a striking contrast to that of oogenesis, a key event of spermatogenesis is ~ 85% reduction in cellular volume and nuclear contents. During this period of reduction, the number of germ cells as well as the supporting Sertoli Cells (SCs) increases. Clearly, spermatogonia continue to generate primary spermatocytes through a spermatogonial cycle in successive waves of mitotic divisions. A search for the production of oogonia/spermatogonia from the PGCs reveals that in nematodes, one PGC or two PGCs enter into monodelphic ovary or didelphic ovaries of nematodes (Pandian, 2021). In teleosts, relevant information available may be summarized hereunder: The number of mitotic divisions that a spermatogonium undergoes prior to the formation of primary spermatocyte is 6 or 8 in Salmoniformes, 8 in Characiformes, 8 or 9 in Cypriformes, 7–10 in Perciformes, 10 in Anguilliformes and Beloniformes, 11 in Gadiformes and 14 in Cyprinidontiformes (see Pandian, 2014). In *Gadus morhua*, spermatogonia undergo 11 mitotic divisions (Almeida et al., 2008). Hence, gadiform fishes can generate more number of sperms than solmoniform fishes. Being the resource providing source, a single Sertoli Cell (SC) can support ~ 100 piscine spermatids, which is 10-times more in numbers than that of mammals. The main functions of SCs are to (i) support survival and development of germ cells and (ii) phagocytize by apoptosis the residual bodies discarded by spermatids and residual sperms. Hence, the crucial role played by the SCs in spermatogenesis becomes important. In amniotic vertebrates, they proliferate until puberty; their adult testis contains a fixed number of SCs. Conversely, cyst-forming SCs of fishes and amphibians retain the potency to also proliferate in adults (Schulz et al., 2010). The cystic spermatogensis is more efficient than the non-cystic, as indicated by ~ 70% residual loss of germ cells in rats (Hess and Franca, 2007), in comparison to ~ 30% in fishes (Matta et al., 2002). The duration of spermatogenesis is species-specific and lasts from 12 days in medaka *Oryzias latipes* to 14 days in tilapia *Oreochromis niloticus*, 14.5 days in guppy *Poecilia reticulata* and 21 days in molly *P. sphenops* (Vilela et al., 2003). These durations are relatively shorter than that (e.g. 16 days in human) in mammals. *With the higher potential of Sertoli cells and shorter duration of spermatogenic cycle, the piscine spermatogonial system facilitates greater scope for production of a large number of sperms within a short duration to sustain external fertilization* (see Pandian, 2014).

Oogonium: A computer search revealed the availability of only one publication by Beaumont and Mandl (1962) on mitotic cycles of oogonium in a neonatal rat. In the rat, the number of oogonial mitotic divisions increases from 5.5 times at the age of 15th day covering only 32% of the oogonial germ cell population to the peak covering 96% of the population but subsequently decreases to 0.04 time and covering 1.3% of the population. Hence, the mitotic rate of oogonial cells, following the peak in mid age, decreases to a lowest level prior to entry into menopause.

9.4 Allogenics and Xenogenics

Ruling out any role for endocrines, Schulz and Miura (2002) considered that the number of mitotic spermatogonial cycle is genetically fixed. However, Okutsu et al. (2006) estimated that the spermatogonial cells of the allogenic Atlantic rainbow trout *Oncorhynchus mykiss* undergo 27-times, instead of 6–8 times, typical of Salmoniformes. Animals generated by transplantation of homospecific PGCs or Oogonial Stem Cells (OSCs, e.g. Wong et al., 2010) or Spermatogonial Stem Cells (SSCs, e.g. Yoshizaki et al., 2000) are called allogenics. And those generated by transplantation or grafting from heterospecific donor species to recipient species are called xenogenics (Pandian, 2011b, Chapter 8, Pandian, 2013). Interestingly, the developments in allogenics and xenogenics in a dozen fish species seem to indicate that the number of mitotic spermatogonial cycle is not limited to 6–8 times and the lack of species specificity of PGCs, i.e. the PGCs from a donor are readily accepted and function normally in a recipient species. In recent years, Japanese scientists have successfully transplanted a vector containing PGCs along with the marker *Green Fluorescent Protein* gene (*Gfp*) (e.g. Yoshizaki et al., 2000) or grafted the lower blastula containing PGCs (e.g. Yamaha et al., 2003) of a species into recipient species. From findings of these investigations, the following are relevant to the PGCs: 1. PGCs of a strain can survive, generate their own gametes and progenies through the recipient surrogate strain (e.g. wild black and orange strains of *O. mykiss*, Takeuchi et al., 2003).

2. Similarly, the PGCs from donor *O. mykiss* (Takeuchi et al., 2004) and *Carassius auratus* (Yamaha et al., 2003) to the recipients of *O. masou* and *Cyprinus carpio*, respectively, survive, generate the respective donor's gametes. 3. In 'hybrid gonads', in which both the endogenous recipient's PGCs and exogenous donor's PGCs peacefully coexist and produce their own respective gametes. 4. On transplantation of the PGCs of the diploid donor *O. mykiss* to the triploid sterile surrogate recipient *O. masou*, the recipient *O. masou* commences to generate the donor's gamete and progenies only. 5. The transplantation of PGCs can also be advanced to late the blastula stage instead of the alevin stage. 6. The transplantation of 'W' bearing PGCs into a male chick embryo led to colonization and the conversion of the female gonadal primordium into a male gonad and produce sperm (Tagami et al., 2007). *Hence, the PGCs are neither a strain- nor species-specific trait, at least within these closely related investigated fish species.* It must also be noted that mortality rate is high in some of these investigated fishes. Incidentally, fertilization by heterospecific sperm, a product of the PGCs from the rosy barb *Puntius conchonius* into the genome-inactivated eggs of the tiger barb *P. tetrazona* led to progressively increased mortality (Fig. 12.2A) suggesting that the cytoplasmic RNAs are unable to accept the genetic code arising from the nucleus of the alien sperm (Pandian and Kirankumar, 2003).

9.5 Minor Phyla

The simple structural organization and reported lifetime fecundity of some minor phyla provides an opportunity to estimate the possible number of times the PGCs undertake asymmetric mitotic divisions as well as oogonia symmetric mitotic division. The number of PGCs reported is only one (i) for Placozoa, (ii) Cycliophora and (iii) Kinorhyncha (Hyman, 1951b) and two for the didelphic nematode *Caenorhabditis elegans* (Sommer, 2015). Incidentally, the digenean fluke *Haplipegus accentricus* is reported to produce only 1 egg and 32 sperms (Pandian, 2020). Similarly, the rotifer *Asplanchna sieboldi* male has only 50 sperms (Pandian, 2021). Hence, their PGC(s) are not likely to have undertaken any asymmetric mitotic divisions. The figure provided by Otani et al. (2002) suggests that four PGCs at late blastula stage undergo the first two asynchronous mitotic divisions and produce eight PGCs and the third and fourth prior to the dorso-lateral migration to the gonadal primordium and produce 16 PGCs (Fig. 13.2B–C). Therefore, with increasing structural complexity, the PGCs increase the number of asymmetric mitotic divisions from zero in Placozoa to four times in goldfish. It seems that symmetric mitotic division by the oogonia/spermatogonia, which doubles the number of cells, is preferred over the asymmetric mitotic division by the PGCs. Hence,

TABLE 9.3

Approximate number of PGCs, oogonia and primary oocyte in some minor phyletics (compiled from Pandian, 2021)

Phylum	Lifetime fecundity	Primary oocyte	Oogonia	PGC
Placozoa	1	0	0	1
Cycliophora	1	0	0	1
Loricifera	4–12	3–6	1*	1
Nemertea	22	11	1*	1
Gnathostomulida	4–6	3	1*	1
Rotifera				1
Monogononta	14–40	7–20	1*	
Bdelloidea	2-6	1–3		
Gastrotricha	6	3	1*	1
Kinorhyncha	3	2	1	1
Nematoda	1000?	500?	2?	2

* Number of cells.

it is likely that the oogonia of Nemertea to Tardigrada undertake symmetric mitotic divisions only, as they may not have more than one PGC, in the light of the fluke holding only one PGC. The number of mitotic divisions undertaken by Rotifera to Tardigrada (Table 9.3) is tentative and is yet to be confirmed by future research. Accordingly, the primary oocyte may undertake mitotic divisions from 2-times in Kinorhyncha to ~ 100-times in Nematoda.

10

Oogenesis and Vitellogenesis

Introduction

Oogenesis is the process, through which a single fertilizable oocyte is generated from an oogonium. It is a complex process and involves folliculogenesis, steroidogenesis, vitellogenesis, meiotic maturation, ovulation, hydration and spawning (e.g. fishes, Pandian, 2014, Chapter 2). Figure 10.1 summarizes the major cytological events that occur during oogenesis in teleost fishes. Notably, meiosis takes place over a prolonged period of time to facilitate vitellogenesis. The process is arrested at the diplotene stage of prophase I by a nuclear structure, the Germinal Vesicle (GV); following GV break down, the process is resumed and continued up to the expulsion of the first polar body, when meiosis I is completed. But meiosis II is completed only at the entry of the sperm, when the second polar body is released. To this common time-schedule, the 900 speciose echinoid echinoderms alone are an exception. In them, the entire process of maturation is completed within the ovary and haploid ova are released (see Walker et al., 2001).

Nurse eggs are reported from polychaetes (e.g. Rasmussen, 1973), nemerteans (e.g. Schmidt, 1932) and branchiopod crustaceans (e.g. Munusamy and Subramoniam, 1985). The bryozoan oocytes develop in pairs, one cell of the pair acting as a nurse cell, which fuses with the oocyte proper at their entry into the coelomic cavity (see Pandian, 2021). Incidentally, two types of oogenesis are distinguished—panoistic and meroistic. In panoistic oogenesis, all the germ cells have the potency to become the oogonia. But in meroistics, only a few germ cells become an oogonia, while the remaining cells differentiate into nurse cells, trophocytes and the like that support the oocyte development. Being interconnected by stable cytoplasmic bridges, clusers of germ cells are formed (Ereskovsky, 2018). As a result, germ cells are released as a bunch called cysts (e.g. oligochaetes, Ferraguti, 1984, Boi et al., 2001).

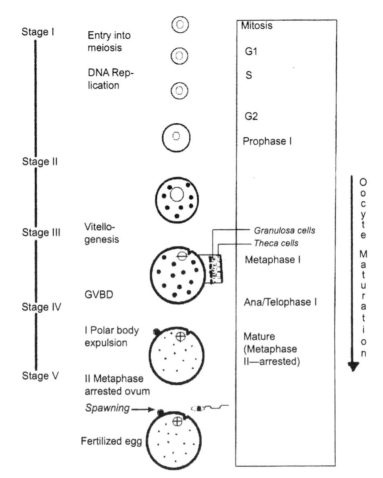

FIGURE 10.1

Oocyte development and differentiation in fishes. Stage I = primary growth, Stage II = cortical alveoli growth, Stage III = early vitellogenic oocytes, Stage IV = late oogenesis and Stage V = mature/ovulated oocyte. Note the formation of micropyle, through which sperm enters the egg (from Pandian, 2014).

10.1 Vitellogenesis

In vertebrates, vitellogenesis is more a generic name for incorporation of (i) vitellogenins, (ii) yolk protein, (iii) maternal mRNA, (iv) lipids, (v) glycogen, (vi) vitamins, (vii) hormones and (viii) hatching enzymes. Vitellogenesis constitutes 60–90% of the soluble yolk protein within an egg (see Eckelbarger, 1983). They are phospho-glycoproteins synthesized by hepatocytes and are released into the blood circulation. Prior to their entry into the ooplasm, they are cleaved into lipoprotein, phosphorin and others.

The hepatocytes also synthesize yolk precursor proteins and release them into circulation. Recruited from peripheral tissues and muscles, vitamins A and E are transported by vitellogenesis. Table 10.1 summarizes the maternal contribution of selected steroids into the sexual and all female eggs of the European perch *Perca fluviatilis* and their utilization during embryonic and larval stages. For example, the decreasing trend in testosterone shows that it is converted into estrogen for female differentiation.

TABLE 10.1

Levels (pg/g) of sex steroids in unfertilized 'sexual' and 'all female' eggs, embryos and larvae of *Perca fluviatilis* (compiled from description of Rougeot et al., 2007)

Developmental stage	'Sexual' (XX♀/XY♂)	All-female (XX♀)
Testosterone		
Egg	1634	1513
Embryo	1331	1281
Larva	542	727
Estrogen		
Egg	554	550
Embryo	494	254
Larva	726	156
11-Ketotestosterone		
Egg	1513	629
Embryo	760	718
Larva	432	0

10.2 Patterns of Oogenesis

Through an impressive series of publications, Eckelbarger (1983, 2005) recognized two patterns, namely the intra-ovarian and extra-ovarian oogenesis in polychaetes. Classifying the oogenesis in Porifera into two patterns, Ereskovsky (2018) named them autosynthetics (= intra-ovarian) and heterosynthetics (= extra-ovarian). Hence, Eckelbarger's classification can be extended to all animals. Accordingly, in the intra-ovarian pattern, the entire process of oogenesis including vitellogenesis occurs within the ovary. But in the extra-ovarian pattern, the required nutrients for vitellogenesis are supplied by special cells like chloragogue in oligochaetes or from organs like the hepatopancreas or liver (cf Fig. 18.1A–B). In the intra-ovarian oogenesis, follicular cells in polychaetes and Nutritive Phagocytes (NPs) in echinoids facilitate vitellogenesis (Table 10.2). Nutrients from nurse cells and chloragogue

TABLE 10.2

Approximate number of species characterized by intra-ovarian and extra-ovarian pattern of vitellogenesis in major and minor phyla

Minor phyla (Species no.)

Phylum	Intra-ovarian	Extra-ovarian	Cells/Organ
Porifera	5	8,548	Ameobocytes, trophocytes
Cnidaria	10,856	0	
Acnidaria			
Platyhelminthes	2,400	25,300	Vitelline glands
Polychaeta	2,980	10,390	Coelomic fluids
Oligochaeta	0	3,850	Chloragogue
Arthropoda	0	1,242,040	Hepatopancreas
Mollusca	0	118,451	
Echinodermata			
Astroidea	0	1,800	Pyloric caeca
Echinoidea	900	0	Nutritive
Others	4,300	0	phagocytes
Vertebrata	0	64,832	Liver, muscle
Subtotal	21,441	1,475,211	
%	1.4	98.6	

Minor phyla (species no.)

Intra-ovarian	Extra-ovarian	Cells/Organ	Phylum
2,353	0		Aorganomorpha
136	0		Acoelomorpha
0	1,300	Nurse cells	Nemertea
31,504	0		Pseudocoelomata
19	0		Priapulida
0	160	Hemoelomic fluid	Sipuncula
0	200		Echiura
0	1,391	Hepatopancreas?	Panarthropoda
0	6,314	Coelomic fluid	Lophophorata
150	0		Chaetogntha
0	130	Hepatic caeca	Hemichordata
0	3,000		Urochordata
34,162	12,495		Subtotal
73.2	26.8		%
55,603	1,487,706		Grand total
3.6	96.4		%

cells or from a special storage organ like the pyloric caeca, hepatopancreas or liver support vitellogenesis in the extra-ovarian oogenesis. Keeping this in mind, a survey was made to identify and quantify the number of species, in which the intra- or extra-ovarian pattern of oogenesis occurs. In sponges, the presumptive oogenic archaeocyte either engulfs another archaetocyte or receives nutrients from the trophocytes (Hyman, 1940). Hence, the 8,553 speciose Porifera are assigned to the extra-ovarian pattern, except in *Suberiteo massa, Aplysina cavernicola, Tetilla serica, Stelleta grubi, Raspaciona aculeata* (see Ereskovsky, 2018). Throughout the lengthy description of Cnidaria, Hyman (1940) has not mentioned the word 'ovary' but hints that the potential oogenic cell is surrounded by an invagination of epidermal (in hydrozoans) cells. Apparently, these cells have adequate nutrients. Hence, all the cnidarians are tentatively assigned to the intra-ovarian pattern. In Platyhelminthes, the female gonad is separated into two structures: (i) the ovary proper and (ii) the vitelline yolk glands to supply yolk to the eggs. Considering this unique feature, Laumer and Giribet (2014) recognized the 2,400 speciose endolecithal Archoophora and 25,300 speciose ectolecithal Neoophora. Hence, the former belongs to the intra-ovarian pattern and the latter to the extra-ovarian pattern. In the 3,850 speciose Clitellata, the chloragogues meet the requirements of vitellogenesis (see Pandian, 2019). Eckelbarger (1983) recognized the intra-ovarian oogenesis in 43 species belonging to 8 polychaete families and extra-ovarian oogenesis in 527 species in 19 families. Considering 13,002 species (Pandian, 2019) in 85 families (Benbow, 2009) in polychaetes, separate values for the intra- (7.5%) and extra- (92.5%) ovarian patterns were assessed (Table 10.2). Among echinoderms, the Nutritive Phagocytes (NPs) of echinoids grow in number and size during the pre-vitellogeneic phase. With the growth of oocytes, the size and number of NPs diminish, clearly indicating that the NPs within the ovary meet the requirement of vitellogenesis. With no definite storage organ, the crinoids, echiuroids and holothuroids are also included into the intra-ovarian pattern. In carnivorous asteroids, the pyloric caeca serve as a storage organ and ensure oocyte growth. Negative correlation between indices of the gonad and caeca has been reported for many asteroids (e.g. *Coscinasteria scalamaria*, Crump and Barker, 1985). In arthropods and molluscs, the hepatopancreas serve as a storage organ to ensure oocyte growth. In minor phyletics, three phyla within Aorganomorpha belong to the intra-ovarian type, as their structure is too simple to have a separate storage organ. In all six phyla of Pseudocoelomata and Priapulida, the eutelism has not relegated any cell or organ to store nutrients to ensure vitellogenesis (see also Table 13.1). Hence, they are included within the intra-ovarian pattern. In Lophophorata, the pre-vitellogenic oocytes are shed into the fluid-filled coelom, where vitellogenesis is accomplished by coelomocytes. This is also true of Sipunculan and Echiura (Table 10.2). *Loxosomella ovipares* seems to be of intra-ovarian type, as their intraovairan oocytes are fed by follicular cells. However, it may be more an exception.

Many gastropods produce a large number of small oligolecithal benthic eggs; in them, the low level of vitellogenesis is compensated by (i) ingestion of nurse eggs, (ii) acquisition of capsular nutrients, and/or (iii) adelphophagy in prosobranchs and (iv) by the Extra Capsular Yolk (ECY) in opisthobranchs. With the need to deposit benthic eggs left with almost no parental care, the majority of gastropods and cephalopods cover their eggs in enveloping structures. These structures range from a multi-layered capsule to fragile gelatinous mass, strands or strings. The tubular string of the West Indian *Strombus* reaches a length of 50 m and contains 460,000 eggs (Hyman, 1967). Clark and Jensen (1981) considered that the oligolecithality and the consequent reduction in vitellogenesis may be an adaptive mechanism to accelerate cleavage by reducing the egg size and allowing the embryo to grow faster. The uncleaved nurse eggs are ingested as a whole (e.g. *Crepidula dilatata*) or sucked by a developing embryo (e.g. *C. capensis*). The vermetids *Vermetus triquerrus* and *Thylaeodus regulosus* not only ingest the nurse eggs but also cannibalistically predate their own smaller siblings. This adelphophagy is also reported for the polychaete *Boccardia proboscida*. The number of eggs consumed by a gastropod embryo ranges from 1.7 nurse eggs in the Pacific muricid *Acanthinucella spirata* to 50,000–100,000 nurse eggs in the North Atlantic deep sea buccinid *Volutopsius norwegicus*. The number of nurse eggs available for an embryo within a capsule of *Searlesia dira* decreases from 172 in a capsule with one embryo to seven in a capsule with 72 embryos (see Pandian, 2017). The capsular wall of prosobranchs is permeable to water and salts. An egg of *C. furnicata* absorbs 0.52 µg salts during embryogenesis for shell construction (Pandian, 1969). The internal spongy layer of the capsule serves as an additional source of nutrients. For example, the thickness of the spongy wall of *C. furnicata* is reduced from 12 µm at the zygote stage to 3 µm at the veliger stage. Hence, the transferred nutrients account for 15 µg/capsule equivalent to 0.015 J/capsule. In lieu of nurse eggs and capsular fluids in prosobranchs, the strategy of opisthobranchs is to provide ECY. In them, the egg mass contains ECY connecting every egg capsule, within which the embryo develops. The ECY granules enter the capsule through minute tears at the point of contact. The ECY are akin to nurse eggs.

The provision of nutrients required for vitellogenesis in growing oocytes within the ovarian sources like the follicular cells or nutritive phagocytes may be cheaper. On the other hand, the provision through motile cells within the coelom or through a circulating system from a storage organ to the 'distantly located' ovary may certainly be costlier, especially, as the larger molecules like the vitellogenins have to be cleaved prior to their entry into the ooplasm. Nevertheless, the extra-ovarian pattern offers adequate scope for the provision of different nutrients like vitellogenins, maternal mRNAs, steroids and so forth appearing from different sources. Not surprisingly, evolution has chosen the extra-ovarian in preference over the intra-ovarian pattern. *Among major phyla, 98.6% are characterized by the extra-ovarian pattern.*

This value for minor phyletics is 73.2% only. In all, of 1,543,196 animal species, 96.4 and 3.6% belong to the extra- and intra-ovarian pattern, respectively (Table 10.2). Though not precise and exhaustive, the analysis represents the very first attempt to identify and quantify these patterns in animals. The proportion in this estimate may change a little, but the conclusion arrived in this analysis would remain valid.

10.3 Micropyle and Hybridization

The fully developed fertilizable ova are shed, as such or covered by a jelly coat (e.g. asteroids: *Dendraster excentricus*, see Pandian, 2018), or one or more fertilized eggs are encapsulated within a capsule (e.g. platyhelminths, gastropods) or cocoons (e.g. clitellates). In general, an acrosomal sperm enters the ovum by penetration. The entry is mainly decided by the species-specific acrosomal reaction. In all teleostean fishes, it is, however, through a hole called micropyle of the ova (Fig. 10.2) and is determined by the compatibility between the diameter of the micropyle and head of the sperm bearing no acrosome. This system of entry through the micropyle by an acrosomeless sperm has led to hybridization in many fishes.

For a long time, hybridization is known to occur in nature (e.g. Cyprinids: *Rutilus rutilus* x *Scardinius erythrophthalmus*); it is an enigmatic phenomenon that defies the very definition of species. For, hybrids carry an amalgamation of diverse genomes. They occur widely across the entire spectrum of animal taxa. With acrosomeless spermatozoa, hybridization occurs more often in teleosts, which are regarded as a representative example. Of 16,050 records on hybrids, fishes have a lion share of 21%, and some 200 and odd records on hybridization in fishes are reported every year. The incidence of hybrids ranges from 2% in clupeids (e.g. *Drosoma capedianum* x *D. petenense*) to 39% in

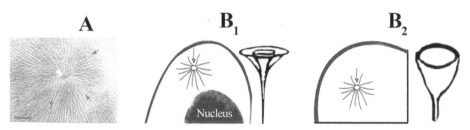

FIGURE 10.2

A. *Synbranchus marmoratus*: magnified view of animal pole showing the furrows converging into a micropyle. White arrow indicates the location of the micropyle and dark arrows indicate the furrows leading to the micropyle (from Ravaglia and Maggese, 2002). Schematic views of the respective micropyle in eggs of B_1. *Hemigrammus caudovittatus* and B_2. *Gymnocorymbus ternetzi*. Arrows indicate the micropylar locations (from David, 2004).

salmonids (e.g. *Salvelinus alpinus* x *S. fontinalis*) (see Pandian, 2011). Drawing information from 158 publications covering 168 species and 139 species pair, of which 47 are intergeneric hybrids, Scribner et al. (2001) concluded that *anthropogenic interventions have been historically an important contributor to hybridization in fishes.*

Hybrids usually do not survive due to incompatibility between parental species with regard to (i) egg yolk content, (ii) karyotypes and (iii) nucleo-cytoplasm. When they survive, they may be sterile. The sterility may be overcome by adopting escape routes like (a) triploidization (e.g. *Clarias macrocephalus* ♀ [2n = 54] x *Pangasius sutchi* ♂ [2n = 60] producing triploid with 54 + 30 = 84 chromosomes) or (b) parthenogenesis, in which the hybrid female generates unreduced diploid eggs (see Pandian, 2011). Hybrid survival decreases from 80% in interspecific hybridization between *Danio rerio* and *D. frankei* (Kavumpurath and Pandian, 1992) to 52% for the intergeneric hybridization between *Heteropneustes longifilis* ♀ x *C. gariepinus* ♂ (Legendre et al., 1992) and to < 2% for the interfamily hybridization between *C. macrocephalus* (Clariidae) x *Pangasius sutchi* (Pangasiidae, NaNakorn et al., 1993). Among the relatively compatible hybrids, progeny production is progressively reduced with successive generation. For example, it is reduced from 2–50% in F_2 generation of hybrid fish belonging to 10 families (see Pandian, 2011). In them too, it breaks down by the F_3 generation. Rarely, it proceeds up to F_7 generation but is broken down at F_8 generation in the spake, the hybrid between *Salvelinus namayacush* x *S. fontinalis* (Argue and Dunham, 1999). Briefly, the process of progeny production by hybrids is progressively reduced through successive generations and is completely broken down at the F_8 generation.

Amazingly, Dr. Liu and his colleagues at the Hunan National University, Changsha, China, created a new species by 'synthesizing' bisexual tetraploids. They thereby compressed the time scale within one generation, while nature may require many years to create a new species (cf bdelloid rotifers, see Pandian, 2020). Liu and his colleagues established a hybrid line using 47.0% fertile males and 44.3% fertile females produced by crossing *Carassius auratus gibelio* ♀ with *Cyprinus carpio* ♂. Unusually, some F_3 hybrid females simultaneously spawned haploid and eudiploid eggs. Likewise, some F_3 hybrid males also milted twice a larger volume of semen, which consisted of 40, 49 and 11% haploid, diploid and others, respectively. Fertilization of diploid eggs with diploid sperm generated bisexual (0.5 ♀ : 0.5 ♂) hybrid tetraploids. As of 2001, progenies up to F_{16} were mass produced, showing that these hybrid tetraploids are stably inherited from one generation to other (Liu et al., 2001, 2004, Sun et al., 2003). As a 'new species' with 200 chromosomes, the allotetraploid became an academically and economically important source of research. The allotetraploid males (F_{16} generation) were successfully mated with the diploid Japanese crucial carp *C. auratus cuvieri* and generated an allotriploid consisting of equal number of genomes of *C. auratus gibelio*, *Cy. carpio* and *C. auratus cuvieri*. This allotriploid is characterized by high survival and faster growth to the extent that > 100,000 tetraploids and > 300 million triploids are annually produced in China. Hence, hybridization can be a source of species diversification. Indeed, it has been the major source for the origin of unisexual complexes in *Poecilia*, *Poeciliopsis* and *Gobitis* (see Pandian, 2011). Incidentally, introgressive hybridization may also contribute to extinction of some fish species (McMillan and Wilcove, 1994).

11

Spermatogenesis and Spermiogenesis

Introduction

Spermatogenesis is a highly organized and coordinated developmental process, during which a small number of diploid Spermatogonial Stem Cells (SSCs) proliferate and produce a large number 'spermatids' carrying a haploid recombined genome (Schulz et al., 2010). Unlike oogenesis, the entire process of spermatogenesis is completed within the gonad. Spermiogenesis, which follows spermatogenesis, produces four spermatozoa. It involves (a) nuclear shaping, (b) chromatin condensation, (c) acrosome formation, (d) development of a short (e.g. Bivalvia, Fig. 11.1F) or a long flagellum (Fig. 11.1E) and one or more tightly packed mitochondria. As a result, a typical sperm consists of an apical acrosome, condensed thin nucleus, mid-piece, one or more mitochondria and a mid, thin long or short flagellum.

11.1 Sperm Types and Quantification

From a survey, the following sperm types are recognized: (i) typical spermatozoa, (ii) typical spermatozoa with no acrosome (e.g. teleosts, Fig. $11.1K_1-K_2$), (iii) typical spermatozoa with no flagellum (e.g. nematodes Fig. $11.1L_1-L_4$), (iv) typical spermatozoa with a thick flagellum (e.g. cephalopods Fig. 11.1P) or with no flagellum (e.g. crustaceans) packed in a spermatophore (Fig. 11.1N), (v) typical spermatozoa, bundled together as spermatozeugma (Fig. 11.1S) or (vi) parasperm with no fertilizing ability (Fig. $11.1R_1-R_2$). It must, however, be indicated that the sporadic occurrence of spermatozeugma and parasperms is limited to some species, although production of two types of spermatozoa is widespread in meso- and neo-gastropods (Pandian, 2017). For example, sperms are bundled to form the spermatozeugma in a few species belonging to the polychaete families Maldanidae, Syllidae and Terebellidae. In the sipunculan *Thysanocordia nigra* also, the incidence of spermatozeugma is reported (Adrianova and

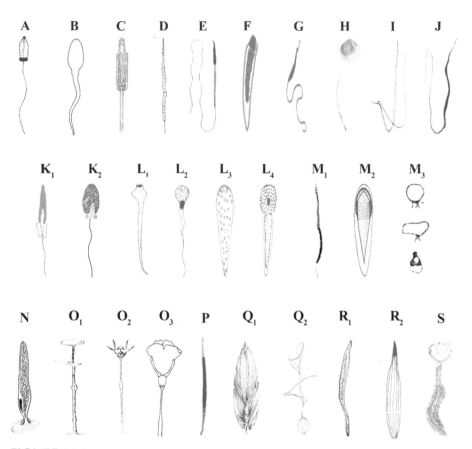

FIGURE 11.1

Spermatozoa: A. Porifera, B. Cnidaria, C. Nemertea (*Malacobdella grossa*), D. Turbellaria (*Maera stichopi*), E. Mollusca, F. Bivalvia (*Nutricola tantilla*), G. Cephalopoda (*Nautilus*), H. Echinodermata (*Psolus chitonoides*), I. Phoronida (*Phoronis australis*), J. Onychophora (*Peripatus*). K_1. viviparous and K_2. oviparous teleost fishes. L_1–L_4. Nematoda, M_1–M_3. Gnathostomulida. Spermatophores: N. Crustacea (*Eupagurus bernhardus*), O_1–O_3. Chelicera, P. Cephalopoda (*Loligo bleekeri*), Q_1–Q_2. Phoronida, R_1–R_2. parasperm and its structure of a mollusc, S. spermatozuegma (All are free-hand drawings from different sources).

Maiorova, 2010). The parasperm is reported from *Tubifex tubifex* (Boi et al., 2001), a marine snail *Fusitriton oregonensis* (Buckland-Nicks et al., 1982), an abyssal echinoid *Phrissocystis mullispira* (Eckelbarger et al., 1989) and a sculpin *Hemilepidotus gilberti* (Hayakawa et al., 2002). Thanks to Boi et al. (2001), it is known that during meiosis, mitotic spindle is not formed in *T. tubifex*. As a result, the parasperms undergo a peculiar nuclear fragmentation with the uneven distribution of a variable amount of DNA among spermatids. In the cottid fish too, the flagellumless non-motile parasperm is large (~ 10 times). On collision with each other, the parasperms of *H. gilberti* lump and obstruct the late arriving sperm and thereby minimize competition from sneakers. In *T. tubifex*, the clumping of parasperms reduces diffusion of euspermatozoa and enhances fertilization success.

Though relevant information for sperm types is available, it is widely scattered (e.g. Adiyodi and Adiyodi, 1990). For the first time, an attempt has been made to present a comparative account on these sperm types (Fig. 11.1) and their taxonomic distribution (Table 11.1). The typical sperm with acrosome and flagellum is present in all the members of 17 taxonomic groups from Porifera to Platyhelminthes and to Echinodermata, as well as Insecta to tetrapod vertebrates. It is also true for nine minor phyletics. In the motile typical spermatozoa, flagellum size ranges widely from the almost

TABLE 11.1

The number of species and sperm types in major and minor phyla

Phylum	Species (no.)	Phylum	Species (no.)
Typical Spermatozoa		Spermatophore	
Porifera	8,553	Myriopoda[1]	11,880
Cnidaria	10,856	Chelicera[2]	155,760
Acnidaria	166	Cephalopoda	800
Platyhelminthes	27,700	Gnathostomulida	100
Insecta	1,020,007	Gastrotricha	813
Bivalvia	9,856	Phoronida	23
Echinodermata	7,000	Onychophora	200
Vertebrata: Tetrapoda	32,322	Subtotal	169,578, 11.0%
Nemertea	1,300	Typical sperm or spermatophore	
Priapulida	19	Polychaeta 12,254	748
Echiura	230	Clitellata 1,225	2,684?
Tardigrada	1,047	Gastropoda 52,000	26,000
Entoprocta	200	Acanthocephala 550	550
Bryozoa	5,700	Branchiopoda 185	185
Chaetognatha	150	Subtotal 66,214, 4.3%	30,167, 2.0%
Hemichordata	130	Flagellumless sperm or spermatophore	
Urochordata	3,000	Rotifera 785	785
Subtotal	1,128,236, 73.1%	Crustacea 10,910	43,474
+ polychaete & others	66,214, 77.4%	Subtotal	45,505, 3.0%
Typical sperm with no acrosome		Spermatophore = 243,219 (16.0%)	
Pisces	32,510, 2.1%	Spermatozoa = 1,254,135 (81.3%)	
Sperm with no flagellum			
Nematoda group	27,360, 1.8%	1 = Wright (2012), 2 = Weygoldt (1990)	
Total	1,254,135, 81.3%		

stout sperm of bivalves (Fig. 11.1F) to the long ones (Fig. 11.1A–E, G–J). Incidentally, the classification of two orders in Kinorhyncha is based on the spermatozoa. In *Echinoderes*, the cigar-shaped mature sperm measures up to 25% of the worm length (Higgins, 1974). The crustacean ostracods are also known to produce long sperms. For example, *Propontocypris monstrosa* of 0.6 mm body length produces sperms, each measuring 6.0 mm in length, i.e. 10 times longer than its own body length (see Pandian, 2016). Interestingly, Wang et al. (2020) collected 50 μm long sperm of *Myanmarcypris hui* entombed in a 100 million years-old Cretaceous amber. Incidentally, the sperm of *Drosophila bifurca* measures 5.8 cm, which is 20 times longer than the fruit fly's body length. The 32,510 speciose teleost fishes have the typical spermatozoa but with no acrosome (Fig. 11.1K_1–K_2). The typical spermatozoa lack flagellum in the 27,360 speciose Nematoda and Nematomorpha. Some of these sperms are shown in Fig. 11.1L_1–L_4.

In all the members of seven taxonomic groups from Myriapoda to Onychophora have their motile (e.g. Cephalopoda) or non-motile (e.g. Crustacea) sperms packed into a spermatophore (Table 11.1), whose shape (and size) varies widely (Fig. 11.1N–Q_2). However, there are five taxonomic groups, in which some members have the typical motile spermatozoa, while the others carry spermatophore. Approximate estimation of these two types within these taxa posed a problem. However, Brahmachary (1989) hinted that typical motile spermatozoa are present in prosobranchs, however opisthobranchs and pulmonata carry spermatophores. For polychaetes, Schroder (1989) listed 15 genera in 9 families for the presence of spermatophore: From the list, the number of spermatophore-carrying polychaete species was estimated from WoRM. Among polychaetes, Jamieson and Rouse (1989) recognized three spermatozoon groups namely the ecto-aquasperm (38 species), ento-aquasperm (10 species) and interosperm (54 species); this group has implications for internal and external fertilization. For Crustacea, the shown values were estimated using the grouping of flagellated and non-flagellated sperm by Pandian (2016, Table 2.8). For species level identification of spermatophore carrying 71 crustaceans, Subramoniam (1993) may be consulted.

From Table 11.1, the following may be noted: 1. Though the occurrence of the spermatogenesis is hinted by molecular markers in Placozoa, neither the spermatozoa nor their equivalents are found in structurally simple Aorganomorpha (2,353 species) and Loricifera and Cycliophora (34 species). *The 461 speciose bdelloids have no males and do not produce spermatozoa or spermatophore. Hence, ~ 0.2% (2,848 species of 1,543,196) of animals are not capable of spermatogenesis.* 2. The appearance of structurally complex spermatophore commences with its appearance from the pseudocoelomate Rotifera among minor phyla and Annelida among major phyla. It ceases from the enterocoelomates and Echinodermata onward among minor and major phyla. 3 (a) *More than 73.1% of animals engage the typical motile spermatozoa with an apical acrosome. Another 2.1 and 1.8% of animals have sperm with no acrosome and sperm with no flagellum, respectively.* From the group possessing *both spermatozoa and spermatophore ~ 4.3% of animals have typical spersmatozoa. On the whole, 81.3% of animals engage typical spermatozoa to transfer the sperm to females. The corresponding value for spermatophore is 16.0%.* The economics and efficiency of sperm transfer by these two types of sperm are elaborated in Chapters 16 and 17.

12

Female- vs Male-Heterogamety

Introduction

The majority (99%) of animals are gonochores. In them, sex is primarily determined by gene(s) at fertilization. Gene(s) involved in sex determination are rarely distributed over the genome (e.g. mitochondria, Zouros et al., 1994a,b, B-chromosomes, Beladjal et al., 2002) but mostly located on a single chromosome and restricted to a single gene locus (e.g. *Sry* gene on morphologically distinguishable sex chromosome) in mammals conserved over 166 Million Years [MY] (Graves, 2013) or *Dmrt1by* in a couple of medaka fishes (Nanda et al., 2002) conserved over 100 MY (Ramanov et al., 2014). The single locus control restricts sex to be determined by a single sex chromosome namely Y in the male heterogametic system (XX♀ : XY♂) or W in the female heterogametic system (ZW♀ : ZZ♂). Beladjal et al. (2002) reported that the paternally transmitted B-chromosomes determine the male sex in the fairy shrimp *Branchipus schaefferi*, which contains 10 autosomes and one to three B-chromosomes. The incidence of B-chromosomes is also recorded from Platyhelminthes (Spakulova and Cassanova, 2004) and fishes; in the latter, they are stably inherited over eight generations in *Poecilia formosa* (Nanda et al., 2007). However, Carvalho et al. (2008) considered that they are an unstable, dispensable constituent of the genome in many fishes. Hence, the sex determining role played by B-chromosome in *B. schaefferi* may be an exceptional case. At best, B-chromosomes may increase the diversity of cytotypic races. The study of mitochondrial genome, which has become a 'hot area' of research during the last two decades, has shown that bivalves are the only known clade to comprise some species, in which two mitochondrial genomes stably coexist. It is, however, not known whether they are involved in sex determination independently or jointly with sex chromosomes (see Section 23.1).

12.1 Chromosomal Mechanism

Sex chromosomes are carriers of sex determining gene(s), and chromosomal mechanism of sex determination is well established. The number of 2n chromosomes ranges from 3 to 8 in turbellarians, 6 to 14 in digeneans and 3 to 14 in cestodes (Birstein, 1991, Spakulova and Cassanova, 2004), from 12 in *Pristina aequiseta* to 54 in *Aporrectodea rosea* in annelids (Pandian, 2019), from 4 in *Drosophila* to 191 in a lepidopteran among insects (Blackmon et al., 2017), from 18 to 44 in molluscs (Pandian, 2018) and from 12 to 500 in fishes (Pandian, 2012). The chromosome size in fishes, for example, ranges from 2 to 5 μm (Pandian, 2012). Among the chromosomes, the sex chromosome can microscopically be identified by size (e.g. W chromosome in fish *Parodon hilarii*, Fig. 12.1A) or by shape (Y chromosome in *Hoplias malabaricus*, Fig. 12.1B). Examination of chromosome pairing in the synaptonemal complex during the pachytene stage of the spermatocyte or oocyte can reveal the presence of sex chromosome regions, which are otherwise not visible (see Pandian, 2012). However, most sex chromosomes remain cytologically indistinguishable, as they are perhaps at a low level of differentiation. In some animals, the 'existence' of chromosomal mechanism of sex determination is genetically identified by tracing sex-linked color genes (e.g. eye color in *Drosophila*, Morgan and Bridges, 1916) or by estimation of the sex ratio among progenies in experimental crossing involving sex reversed female (XY) with normal male (XY) (e.g. *Oreochromis mossambicus*, Varadaraj and Pandian, 1989), or using a marker chromosome (Fig. 12.1B$_1$–B$_2$).

Apart from these well established female-male heterogametic systems, crustaceans have as many as five other systems. They carry XX/Y, Xn/Y, XX/XO, Xn/O or ZZ/W$_1$W$_2$ (Table 12.1). A survey of sex determination systems of Crustacea (Table 12.1), Insecta and Teleostei (Table 12.2) reveals that apart from a simple pair of the sex chromosomal system, there are others, in which sex is determined by multiple numbers of chromosomes; for example, some crustaceans carry XX/Y, Xn/Y, XX/XO, XN/O or ZZ/W$_1$W$_2$ (Table 12.1). Interestingly, ostracods have explored the use of five systems

FIGURE 12.1

Karyotype of (A$_1$). *Parodon hilarii* showing W chromosome (from Moreira-Filho et al., 1993) and (A$_2$). *Hoplias malabaricus* showing Y chromosome (from Rosa et al., 2009). Arrows indicate marker chromosome in (B$_1$). diploid and (B$_2$). triploid *Oreochromis mossambicus* (from Varadaraj and Pandian, 1989).

TABLE 12.1

Heterogametism and its complexity in crustaceans (from Pandian, 2016)

Type	Clada/Species name
	Anostraca
XY/XO ZW/ZZ	*Cheirocephalus nankinensis* *Branchipus vernalis* *Artemia salina*
	Spinicaudata
ZW/ZZ	*Eulimnadia texana*
	Ostracoda
XX/XY Xn/Y XX/XO XX/XnO	*Cyclocypris laevis* *Cypria compacta, C. deitzei, C. exsculpta, C. fordiens, C. whitei, C. ophtalmica,* *Cyc ovum, Heterocypris incongruens, Cypria* sp, *Cyprinotus incongruens* *Cyc. globosa* *Notodromas monacha, Physocypris kliei, Platycypris baueri, Scottia browniana*
	Copepoda
XY/XO ZW/ZZ	*Acartiella gravelyi, Acartia keralensis, Tortanus barbatus, T. forcipatus,* *T. gracilis* *Acartia centura, A. negligens, A. plumosa, A. spinicauda, Centropages typicus,* *Ectocyclops strenzki* *Acanthocyclops vernalis, Eucyclops serrulatus, Megacyclops viridis, Cyclops* *viridis*
	Isopoda
XX/XY XX/XO ZW/ZZ ZZ/ZW$_1$W$_2$	*Anisogammarus anandalei* *Tecticeps japonicus* *Paracerceis sculpta, Idotea balthica* *Jaera albifrons, J. albifrons forsmani, J. ischiosetosa, J. praehirsuta, J. syei*
	Amphipoda
ZZ/ZW XX/XO XX/XY	*Gammarus anandalei* *G. pulex, Marinogammarus marinus, M. pirloti* *Armadillidium vulgare, Naesa bidentata, Orchestia cavimana, O. gammarellus*
	Decapoda
XX/XO XX/XY XX/X$_2$Y ZW/ZZ	*Eriocheir japonicus, Gaetice depressus, Hemigrapsus penicilatus,* *H. sanguieneus, Pachygrapsus crassipes, Plagusia dentipes* *Ovalipes punctatus* *Cervimunida priniceps* *Macrobrachium rosenbergii, Penaeus monodon, Fenneropenaeus chinensis,* *P. japonicus, Cherax quadricarinatus, P. vannamei*

within the said male heterogamety: They are X_2/O, X_3/Y, X_4/Y, X_6/Y and X_{4-7}/Y in *Notodromas monacha, Cypria compacta, C. exsculpta, Heterocypris incongruens* and *Cyclocypris ovum*, respectively (see Pandian, 2016). Among crustaceans, ostracods remain a goldmine for cytogenetic studies. Insects (X_3Y_3) and fishes (three male heterogametic X_4/X_2Y, X_4/X_2Y_2, X_3/Y_3; female

TABLE 12.2

Number of heterogametic types in Teleostei, Crustacea and Insecta (compiled from Pandian, 2012, 2016, Blackmon et al., 2017)

Heterogametic type	Teleostei	Crustacea	Insecta
Male Heterogamety			
XX/XY	144	15	5,577
XX/YY	0	0	51
XX/XO	12	13	1,612
XX/Y	18	1	40
X_4/X_2Y	6	0	0
X_3Y_3	0	0	2
X_4/X_2Y_2	1	0	0
Xn/Y	0	10	0
Xn/O	0	4	0
Type: Subtotal	5	5	5
Female Heterogamety			
ZZ/ZW	80	18	30
ZZ/ZO	3	0	10
Z_4/Z_2W_1	4	0	0
Z_4/Z_1W_2	1	0	0
ZZ/W_1W_2	0	5	0
Type: Subtotal	4	2	2
Type: Total	9	7	7

heterogametic Z_4/Z_1W_2, Z_4/Z_2W_1) use not only different combinations of heterogamety but also more number of sex chromosomes (Table 12.2). The latter is named multiple (Pandian, 2011b) or complex (Blackmon et al., 2017) sex determining system. On the whole, teleosts, crustaceans and insects use each of five systems in male heterogamety but in female heterogamety, four in teleosts and two each of insects and crustaceans (Table 12.2). The males have diversified the multiple non-conventional sex determining mechanisms more than that of females. *Still, it must be stated that the single pair of chromosomal system is in operation in 80–90% teleosts and > 80% insects. Hence, the single locus carrying sex chromosomes Y and W has fostered more species diversity within male- and female-heterogametics than those carrying sex determining loci in multiple number of chromosomes.*

12.2 Taxonomic Distribution

For following reasons, quantification of male and female gametic species has indeed been an onerous task. For example, the chromosomal mechanism of sex determination in fishes (e.g. Pandian, 2011b) and amphibians (e.g. Schmid, 1983) has involved the repeated emergence of new sex chromosome systems from autosomes/female heterogametic system (e.g. Bachtrog et al., 2014), bringing the sex determination cascade under new master regulators (Volff et al., 2007). In amphibians, Hillis and Green (1990) indicated that the XX/XY system has independently evolved at least seven times and in one case reversed back to the ZW/ZZ system. Hence, the dynamic unstable heterogametic system of fishes and amphibians renders quantification as difficult. As a result, the heterogametic system differs within a genus and in some within a species; for example, within the genus *Poecilia*, *P. reticulata* is male heterogametic (Kavumpurath and Pandian, 1993a, b) but female heterogametic in *P. sphenops* (George and Pandian, 1995). Within a single species *Gambusia*, the strain *G. affinis holbrooki* is male heterogametic but *G. affinis affinis* is female heterogametic (Black and Howell, 1979). The platyfish *Xiphophorus maculatus* is male heterogametic in Mexico but a female heterogametic in Honduras (Gordon, 1952). Nevertheless, an attempt has been made to quantify the gametics in different taxa.

Karyotyped	(no.)	Karyotyped	(no.)
Chromosome no. only	2,205	Homomorphic chromosomes	159
Parthenogens	1,215	Haplo-diploidy + paternal genome exclusion	1,891
Total no. of non-gametic karyotyped species			5,470

Arthropoda is a speciose phylum; in it, Insecta is the most speciose. Hence, an attempt was made to assign them into one or the other gamety. Fortunately, Blackmon et al. (2017) summarized a valuable account on the gamety of insects. Of 1,020,007 species, karyotype is reported for 13,473 or 1.3% of insect species. Of them, 5,470 karyotypes (see box above) cannot be included in a gametic estimate. The remaining values related to gamety under male- and female-heterogamety are listed in Table 12.3. Of 30 orders, male heterogamety occurs in 28 orders (including Diptera) and female heterogamety in three orders namely Trichoptera, Lepidoptera and Diptera. In 1,958 dipteran species, for which gamety is reported, 1,951 species (99.64%) are male heterogametics. Only 7 species (0.36%) are female heterogametics, which may be more an exception. In all these 30 orders, chromosomal sex determination systems remain virtually stable and conversed. Whereas mammals and birds are male- or female-heterogametics at Class level, insects display it at Order level. Hence, it is reasonable to transform and

TABLE 12.3

Approximate male- and female-heterogametic insect species (calculated from Blackmon et al., 2017)

Order	Species (no.)	XY (no.)	XO/XYC (no.)	Karyotyped (no.)
		Male heterogamety		
Diplura	1,000	-	-	-
Protura	700	3	0	3
Archaeognatha	500	-	-	-
Zygentoma	300	0	3	3
Collembola	8,000	0	17	17
Ephemeroptera	3,000	6	2	8
Odonata	5,500	20	403	423
Psocoptera	3,000	2	91	93
Phthiraptera	5,000	1	-	1
Thysanoptera	5,000	-	-	-
Hemiptera	90,000	284	156	440
Blattodea	4,500	0	108	108
Isoptera	2,600	2	62	64
Mantodea	2,300	1	100	101
Zoraptera	30	1	0	1
Orthoptera	20,000	49	231	280
Phasmatodea	3,000	14	69	83
Embiioptera	300	0	8	8
Notoptera	40	2	0	2
Plecoptera	2,000	1	11	12

Order	Species (no.)	XY (no.)	XO/XYC (no.)	ZW (no.)	ZWC (no.)	Karyotyped (no.)
		Male heterogamety				
Siphonaptera	2,500	2	4			6
Dermaptera	2,000	22	30			52
Hymenoptera	100,000	-	-			0
Coleoptera	350,000	3198	989			4187
Strepsiptera	500	1	0			1
Neuroptera	5,000	70	4			74
Megaloptera	300	4	0			4
Raphidioptera	300	6	0			6
Mecoptera	550	0	14			14
Diptera	125,000	1,893	58			1951
Subtotal	742,920	5582	2359			7940
				Female Heterogamety		
Trichoptera	11,000			-	15	0
Lepidoptera	160,000			-	18	22
Diptera	125,000			1893	7	0
Subtotal	171,007			1893		62
Grand total	913,920					6972

apply the reported values to the total number of species in each of the 30 orders. Accordingly, as all the reported values for 280 species are male heterogametics, the 20,000 speciose order Orthoptera is considered as male heterogametics. Following this procedure, 7,42,920 and 171,007 insect species are assigned to male- and female-heterogamety, respectively, i.e. whereas 81.3% of insects are male heterogametics, 18.7% are female heterogametics.

The assignments of insects have led to divide the animals into two groups: Group 1, for which hardly a few data (< 1% of the respective taxa species number) are available and for Group 2, with availability for > 1% *per se*. In Group 1, sex chromosome number is known for 7 species or 0.59% of the 13,002 speciose polychaetes (see Pandian, 2019). Of these seven species, six are male heterogametics and one is female heterogametic (Table 12.4). In the 118,451 speciose Mollusca, 4 are male heterogametics and 3 female heterogametics. In two minor phyla namely Nematoda and Acanthocephala, for which information is available, all 13 species are male heterogametics.

TABLE 12.4

Number of species characterized by male or female heterogamety systems (compiled from 1 = Pandian, 2020, 2 = Pandian, 2019, 3 = Pandian, 2017, 4 = Pandian, 2016, 5 = Blackmon et al., 2017, 6 = Chowdaiah, 1965, 7 = Oliver, 1989, 8 = Tsurusaki and Cokendolpher, 1990, 9 = Pandian, 2011b, 10 = Schmid, 1983, 11 = Hillis and Green, 1990, 12 = Miura, 2017, 13 = Alam et al., 2018)

Taxon	XX/XY	XX/XO	ZW/ZZ	ZO/ZZ
Group 1				
Platyhelminthes[1]	0	0	1	0
Polychaeta[2]	5	1	1	0
Mollusca[3]	3	1	1	2
Nematoda	7	4	0	0
Acanthocephala	1	1	0	0
Crustacea[4]	30	13	18	5
Myriapoda[6]	4	3	0	0
Chelicera[7,8]	12	12	14	1
Group 2				
Insecta[5]	745,920		171,007	
Teleostei[9]	21,684		10,842	
Amphibia[10,11,12]	4,005		1,223	
Reptilia[13]	6,386		3,159	
Aves	0		10,038	
Mammalia	5,513		0	
Subtotal	783,508 (99%)		196,269 (43%)	
Grand total	783,605 (80%)		196,312 (20%)	

For crustaceans, the available values are for 43 and 23 species male- and female-heterogametics, respectively. Reporting male heterogamety (4 XX/XY species, 3 XX/XO species) in 7 species, Chowdaiah (1965) hinted that all myriapods (11,000 species), especially diplopods are male heterogametics. A computer search revealed the availability of only two publications for spiders by Tsurusaki and Cokendolpher (1990) and one for ticks by Oliver (1989). The reported values are: (i) spiders – XX/XY = 3 and (ii) ticks – XX/XY = 9, XX/XO = 12, X_4X_3/Y_1 = 1, ZZ/ZW = 14, ZZ/ZO = 1. In molluscs, the only archaeogastropod *Theodoxus meridionalis*, for which information is available, is male heterogametic (Vitturi and Catalano, 1988). Male heterogamety is also reported in three gastropods. Hence, as data available for Group 1 is scarce, it may be premature to transform and apply the values to the respective taxa. Whereas all other clades share the presence of both gametics, nematodes and mammals operate only with male heterogametism.

On the other hand, for Group 2 available data are fairly large. With regard to vertebrates, for the 32,510 speciose teleosts, gametic data are available for 1,700 species, i.e. for 11%. In them, male- and female-heterogamety occurs in 66.7 and 33.3% or 21,684 and 10,842 species, respectively (see Pandian, 2012). Of 175 values available for the 5,228 speciose Amphibia also, 134 (76.6%) and 41 (23.4%) species are male- and female-heterogametics. Hence, 4,005 and 1,223 species are assigned to the former and latter, respectively. For Reptilia, 6,386, (121 species or 66.9% of identified male heterogametics) and 3,159 (60 species or 33.1% *per se*) are also assigned to male- and female-heterogamety (Table 12.4).

Three publications have listed gamety for amphibians. The details are listed below:

Type	Schmid (1983)	Hillis and Green (1990)	Miura (2017)	Total
XX/XY	30	49	55	133
ZW/ZZ	9	14	17	40
WO/ZZ	0	1	0	1

An idea proposed by many authors, who have limited their observations only to a single clade (e.g. amphibians, Schmid, 1983) is that the female heterogamety is primitive, from which the male heterogamety has evolved independently a number of times. To prove their claim, the authors have also provided histological and molecular biological evidences. However, the holistic view of the existening gametics provides a different picture.

On the whole, gamety is recognized for 979,917 species; of them, 783,605 or 80.0% are male heterogametic, while the others 196,312 (20.0%) are female heterogametics. The reasons attributed for the dominance of male heterogamety are: (a) For every primary spermatocyte, four spermatids are produced, whereas only one ovum is produced from a primary oocyte. (b) Females produce far less number of eggs than the number of sperms generated. (c) Females enter menopause earlier than the males, i.e. while females have ceased to produce eggs, males continue to generate sperms. Hence, males produce far more new gene combinations than the females—the raw material for evolution and speciation. This is why many biologists consider the evolution as a male driven process.

12.3 Survival of Heterogametics

The restricted ability to recombine with its respective counterpart namely the X or Z, the Y or W chromosome may degenerate sooner or later to a small heterochromatic element with a few functional loci (Charlesworth and Charlesworth, 2000). For example, the human X chromosome carries ~ 1,000 genes but the Y harbor only a few dozen genes (Bachtrog et al., 2014). Even among these dozens, if the Y chromosome carries a recessive gene like that of hemophilia, the scope for survival as a function of age in Y carrying male or W carrying female may become increasingly reduced (see p 9). As may be expected, survivorship trends for the male heterogametics is relatively lower for the Y carrying males of the chimpanzee (Fig. 12.2A), spotted hyaena

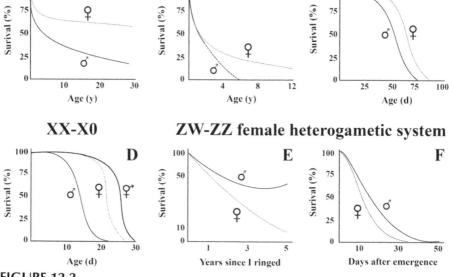

FIGURE 12.2

Survival of the female, male and hermaphrodites in A. chimpanzee *Pan troglodytes* (from Littleton, 2005), B. spotted hyaena *Crocuta crocuta* (from Watts and Holekamp, 2009), C. fruit fly *Drosophila melanogaster* (from Lashmanova et al., 2017), D. nematode *Ceanorhabditis elegans* (from Hotzi et al., 2017), E. eastern kingbird *Tyrannus tyrannus* (from Murphy, 1996) and F. diamondback moth *Plutella xylostella* (from Shirai, 1995). All are simplified free hand drawings.

(Fig. 12.2B) and fruitfly (Fig. 12.2C) than that of their respective females. Similarly, the survival trends for the W carrying female heterogametic eastern kingbird (Fig. 12.2E) and diamondback moth (Fig. 12.2F) are lower than those of their respective males. Interestingly, sex in the male heterogametic nematode *Caenorhabditis elegans* includes the male, female and hermaphrodite. Their survival decreases in the following order: male (XY) > female (XX) > hermaphroditic (XX/XY) (Fig. 12.2D).

13

Eutelism and Parasitism

Introduction

For the first time, Pandian (2021) recognized and highlighted the occurrence of eutelism in all six phyla of Pseudocoelomata. In eutelics, mitosis is ceased in all somatic cells after hatching. For example, the number of cells, as counted by the number of nuclei in the rotifer *Epiphanes senta* is ~ 888 (Table 13.1); even after the inclusion of testis and others, the number may not exceed 900 nuclei. For the monogonont rotifers, the number may range between 900 in *E. senta* and 959 *Asplanchna priodonta* but is limited to 428 only in entirely parthenogenic bdelloids (Hyman, 1951b). An important consequence of eutelism is the limitation of fecundity to < 30 eggs. Apart from rotifers, limited information is available on the eutelic cell number for a few nematodes (e.g. *Turbatrix oceti*, Pai, 1927). In nematodes, cell division is ceased in all somatic cells during late embryonic development, except in the midgut and epidermis. In fact, the number of cells increases from 35 in Juvenile 1 (J_1) to 200 in the adult *Camallanus sweeti* (Moorthy, 1938). Rusin and Malakhov (1998) found that the hypodermal cells are not eutelic in six free-living nematode species. Cunha et al. (1999) also discovered increases in the epidermal cell number of 13 nematode species. More detailed information is reported by Sulston and Horvitz (1977) for the number of cells in *Caenorhabditis elegans*. There are 887 and 940 cells present in hermaphrodite and male worm, respectively (Table 13.1). More importantly, the authors found that the number of cell increased from ~ 550 in J_1 to 970 in an adult, i.e. 887 somatic cells + 83 germ cells in a hermaphrodite and from ~ 496 in J_1 to 971, i.e. 940 somatic cells + 31 germ cells in a male. Remarkably, two features should be noted: *1. Unlike rotifers, the nematodes have postponed eutelism from hatching to adulthood. 2. Unlike rotifers with only a one celled ovary, the nematodes have 83 germ cells in a hermaphrodite and 31 germ cells in a male. This difference in the number of germ cells between rotifers and nematodes seems to have liberated the latter to be more fecund.*

TABLE 13.1

Number of cells present in the eutelic rotifer *Epiphanes senta* (from Hyman, 1959a) and nematode *Caenorhabditis elegans* (from Sulston and Horvitz, 1977)

Ectodermal derivatives	(no.)	Endodermal derivatives	(no.)	Mesodermal derivatives	(no.)
Rotifera : *Epiphanes senta*					
Coronal epidermis	172	Esophagus	15	Circular muscle	22
Trunk foot epidermis	108	Stomach	39	Retractor muscle	40
Mastax epidermis	91	Gastric gland	6 +	Mastax muscle	42
Brain	183	Intestine	14	Protonephridium	28
Peripheral nerves	63			Ovary	1 +
Mastax nerve cells	34			Vitellarium	8
Pedal glands	19			Oviduct	3
Subtotal	670			Subtotal	144 +
Total cell types = 18, cell (no.) = 888					
Caenorhabditis elegans ♀					
Epidermal cells	186	Intestine	34	Muscles and others	122
Nerve cord & others	123			Head	311
				Tail	31
Subtotal	309	Subtotal	34		464
Total cells (no.) = 887					
C. elegans ♂					
Epidermal cells	250		34	Muscle and others	146
Nerve cord & others	129			Head	315
				Tail	66
Subtotal	379	Subtotal	34	Subtotal	527
Total cells (no.) = 940					

13.1 Parasitism and Fecundity

In eutelic pseudocoelomates, the life cycle is simple and direct in Rotifera, Gastrotricha and Kinorhyncha. For example, more or less fully developed neonates are released from their eggs. Contrastingly, the cycle is complicated and indirect in parasitic Nematomorpha and Acanthocephala. In nematodes, the hatchlings pass through four juvenile stages. About 16,000 species or 59% of nematodes are parasitic; expectedly, the passage of their juvenile stages is complicated with the need to infect a suitable host at an appropriate time. Arguably, *Nematoda, Nematomorpha and Acanthocephala have progressively postponed the manifestation of eutelism from the hatchling to the adult stage. They seem to have also increased the number of germ cells.* For example, one cell out of 888 cells is an ovarian cell in rotifer (Table 13.1). But there are 31–83 germ cells out of ~ 900 cells in nematodes. This has enabled the free-living nematode like *Caenorhabditis elegans* to produce > 1,000 eggs (see Klass, 1977), in comparison to < 30 eggs in rotifers. In this context, a point of interest is that the ratio between somatic and genital cell number is 1 : 900 in rotifers but 31–83 : 900 cells in nematodes. This has also led to more and more fecundity in plant (500–1,000 eggs), invertebrate (25,000 eggs) and vertebrate (10,000–12,000 eggs) parasitic nematodes (Fig. 13.1). With assured food supply, the parasites allocate more of their resources for egg production. A consequence of the presence of relatively more number of eggs and the consequent generation of a new gene combination is that the parasitic taxa are more prone to species diversity than the free-living species. *The trend for the relation between fecundity and parasitic minor phyla is positive and linear, but its slope level is so high that at a given level, the fecundity of parasitic minor phyletics is nearly 10–1,000 times more than that of free-living minor phyletics.* The parasites are more fecund not only in minor phyletics but also in major phyletics. For example, Batch Fecundity (BF) of a free-living isopod *Idotea pelagica* is in the range of 17–37 eggs (see Johnson et al., 2001). It is 250–750 eggs in the cymothoid parasitic isopod *Ichthyoxenus fushanensis* with a direct life cycle and 100,000–250,000 eggs in

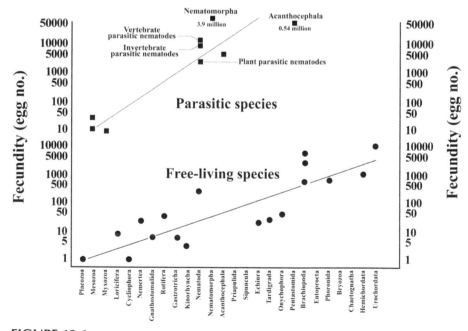

FIGURE 13.1

Fecundity in free-living (lower panel) and parasitic minor phyla (upper panel).

the bopyrid parasite *Argeia pugatenensis* involving an indirect life cycle with two intermediate hosts (Pandian, 2016, Fig. 7.7). Further, with assured food supply, the fecundity is increased to a few thousands in parasitic nematodes and a couple of millions in Nematomorpha.

13.2 Sperm Limitation and Sterility

In fact, resource allocation is more for egg production (with 83 germ cells in ♀, i.e. 83 − 31 = 53 for ovarian germ cells) than for spermatogenesis (with only 31 germ cells) in the lower minor phyletics. As a result, sperm production is more severely limited than that of egg production. In all the examined dicyemid species, the number of sperms produced is fewer than the number of eggs (see Furuya and Tsuneki, 2003). In mictic rotifers, 5 mictic eggs alone are fertilized but 22 amictic eggs remain unfertilized (see Pandian, 2021, Table 9.3). In free-living nematodes, only a few sperms are produced so that the first few eggs are fertilized and those appearing subsequently are not fertilized (Potts, 1910). Klass (1977) found that of 1,185 eggs produced by *Caenorhabditis elegans*, only 66% eggs are fertilized. Hyman (1951b) listed other nametodes like *Rahbditis elegans*, *Diplogaster robustus* and *Cephalobus dubius* producing a high percentage of sterile eggs. Due to protandry, the selfing *C. elegans* produce only 280 sperms, but ~ 1,000 eggs are produced following irrevocable switching to the female phase. As a result, there is only 0.3 sperm to fertilize an egg (Cutter, 2004). Table 13.2 summarizes a comparative account on selected reproductive features of free-living and parasitic nematodes. Understandably, with assured food supply, the parasitic *Nippostrongylus brasiliensis* produce 1,000 or 40-times more number

TABLE 13.2

Effect of eutelism on some reproductive features in free-living and parasitic nematodes (compiled from Phillipson, 1969, 1970, Duggal, 1978)

Reproductive feature	Free-living *Panagrellus redivivus*	Parasitic *Nippostrongylus brasiliensis*
Potential sperm output (no.)	246	15,000
Insemination duration (min)	1	1
Male's ability copulate (time/d)	9.6	12
Spermatozoa (no./insemination)	28	1,250
Eggs produced (no./♀/h)	25	1,000
Sperm availability (no./egg)	0.89	0.80

of eggs than its free-living counterpart *Panagrellus redivivus*. But the ability of the parasite to produce the number of sperms per insemination is 1,250, which is 44-times more than that of *P. redivivus*. Despite its ability to copulate 12 time/d, 20% eggs of *N. brasiliensis* remain unfertilized, in comparison to ~ 11% unfertilized eggs in *P. redivivus*. *The sperm deficiency seems to be a common feature of both free-living and parasitic gonochoric nematodes.*

14

Monogamy and Polygamy

Introduction

In animals, group formation is a common phenomenon. It has led to a social structure and mating system. Four patterns of mating systems are recognized: (i) Monogamy and Polygamy, the latter is divided into (ii) Polygyny, (iii) Polyandry, and (iv) Promiscuity. Of these, polygyny and promiscuity are the most common patterns. Monogamy is defined as a pair bond between two conspecific individuals of opposite sexes. It may lead to inbreeding and decrease in genetic diversity. In polygamy, polygyny involves an individual male mating with multiple numbers of females and polyandry refers to a female mating with one or a few males. The polygynic males are often 1.5–2.0 times larger in size than females. In promiscuis pattern, it is free for all, i.e. any male individual can mate with any number of females and any number of times and the reverse holds true for the females. Table 14.1 provides examples for these patterns regarding the number of matings gained and the sex subjected to intense selection in fishes. Polygyny may be facilitated by skewing a sex differentiation process in favor of females, as in parasitic nematodes and polyandry by haplodiploidy or

TABLE 14.1

Mating systems leading to intense selection (modified from Searcy and Yasukawa, 1995)

Mating system	Successful mate(s)		Sex at intense selection	Example
	no./♀ mating	no./♂ mating		
Monogamy	One	One	Neither	*Typhogobius californicus*
Polygamy				
Polygyny	One	Two	Male	*Lamprologus ocellatus*
Polyandry	One	> Two	Female	*Gambusia holbrooki*
Promiscuity	Many	Many	Either or neither	*Gasterosteus aculeatus*

paternal genome elimination. In almost all echinoderms, sex ratio oscillates between 0.45 and 0.55, clearly indicating that the ratio is altered by sex-specific mortality but not by altering the sex differentiation process. Not surprisingly, neither polygyny nor polyandry is reported in them. Most animals are promiscuous.

14.1 Taxonomic Survey

In the hermaphroditic turbellarian *Pseudoceros bifurcatus*, reciprocal, i.e. monogamic insemination occurs between pairs (Michiels and Newman, 1998, Table 14.2). Natural monogamy is reported in *Nereis acuminata* and enforced monogamy in *Ophryotrocha* spp. A microsatellite study has revealed the occurrence of exclusive monogamy in the earthworm *Homogaster elisae*, in which all the four spermatotheca stored sperms are from the same male (see Diaz-Cosin et al., 2011). In arthropods, monogamy is a rarity among myriapods and insects. However, it does occur in a few caridean shrimp and the like (Table 14.2). Schneider and Fromhage (2010) described a peculiar monogamy or bigamy in spiders. In *Argiope bruennichi* female, there are two gonopores, into one of which the first male inserts his spermatophores. But if he lingers long to ensure his paternity, the cannibalistic female eats him. If he escapes by leaving the female earlier, a second male may attempt to court her but in most cases, he is unable to locate the free female gonopore. Hence,

TABLE 14.2

Taxonomic survey of mating patterns in animals

Taxon	Reported observations
Turbellarians	Monogamy: *Pseudoceros bifurcatus*, 3.8 insemination/pair
Annelids	Monogamy: *Homogaster elisae, Nereis acuminata* Enforced monogamy: *Ophyrotrocha* spp
Molluscs	Polygyny: 1.9 sire/brood to 5.1 sire/brood (see Table 14.4)
Crustaceans	Monogamy: *Alpheus angulosus, A. armatus, A. heterochaelis, A. roquensis, Synalpheus* sp, *Lysmata amboinensis, Stenopus hispidus* (Subramoniam, 2017)
Spiders	*Argiope bruennichi* (Schneider and Fromhoge, 2010)
Nematodes	Polygyny: Altering sex differentiation to skew ratio toward female
Sharks	Monogamy in *Sphyrna tiburo* (Klug, 2018)
Teleosts	Mostly polygynics (Table 14.5)
Reptiles	Monogamy in Sleeping lizard *Tiliqua rugosa* (Bull, 2000)
Birds	90% mostly social monogamy (Black, 1996a, b)
Mammals	3–5% monogamy (Kleiman, 1977)

A. bruennichi can be monogamous or bigamous. In many parasitic nematodes, monogamy occurs (e.g. *Blatticola monandros*). With the incidence of two females in a cockroach host, fecundity is reduced due to competition for the resources (Zervos, 1988). However, polygyny occurs, when a single male is present for many females in *Wetanema hula, Thelastoma collare, T. moko* and *Leidynema appendiculata*. In them, sex differentiation skews the ratio in favor of females. Or the presence of a male inhibits the differentiation of any other presumptive juvenile males. However, the reverse occurs in mermithids; the sex ratio of *Mermis subnigrenscens* remains equal until the grasshopper host is infected with three worms but the ratio is skewed toward the male, when the worm number increases to 13–15. With more than 25 worm/host, only males are developed; so much so, for every 24 males, there is only a single polyandric female. Hence, polygyny or polyandry is achieved in parasitic nematodes by altering the sex differentiation process. Monogamy is reported to occur in the shark *Splyrna tiburo* (Klug, 2018). Among amphibians and reptiles, monogamy is rare. But Bull (2000) reported the occurrence of it in the Australian sleeping lizard *Tiliqua rugosa*. Incidentally, it may be difficult to quantify the different mating patterns, as available information is limited. Further, the pattern changes from race to race and place to place in fishes. In the Nigerian cichlid *Pelvichromis pulcher*, the red morph is a haremic polygynic but the yellow morph is not. Some fishes like *Halichoeres maculipinna* is haremic in Florida but not in Panama. Conversely, *H. garnoti* is haremic in Panama but not in Florida (see Pandian, 2011).

14.2 Monogamy

The common features of monogamy are (i) site fidelity, (ii) territoriality and (iii) size equality. A fairly large volume of information is available for monogamy in fishes, birds and mammals (e.g. Pandian, 2011, 2012, 2015, Black, 1996a, b, Kleiman, 1977, Lukas and Clutton-Brock, 2013). Monogamy is divided into genetic and social, albeit no publication is available to show the existence of the former in any animal. Social monogamy refers to the co-habitation of a male and a female to collect and store resources in a shelter, protect eggs and young ones. It occurs in a population, which is small and dispersed, and allows the consistent availability of a partner and the partners need not waste time and energy on searching for a suitable mate. A second reason that has driven the animals to monogamy is the parental care. In fact, increased parental investment enhances fidelity and extends monogamy duration (Kleiman, 1977). The social monogamy is further divided into three groups: (1) exclusive monogamy, in which copulations occur between the same partners. In them, reciprocal benefit renders each partner to be fidelous to the other. (2) Manipulated monogamy, in which one partner aggressively and unilaterally monopolizes the other; the former defends and excludes an intruder, as in haremic species and (3) Biparental monogamy, in which a pair of male and female remains together as partners after fertilization until the offspring(s) no longer require their care.

Exclusive monogamy: Among birds, the most common examples cited for exclusive monogamy are black vultures, swans, emperor penguins and hornbills. A black vulture attacks others that participate in extra-pair copulation and thereby reduces promiscuity. Hornbills are fidelous for their lifetime. The emperor penguins are fidelous, even when one of the partners is far away. For mammals, Kleiman (1977) listed 105 species (Marsupialia 1, Insectivora, 5, Chiroptera 11, Primates 11, Rodentia 28, Mysticerti 1, Carnivora 26, Pinnipedia 4 and Astiodactyla 7). The prairie vole is extremely fidelous and attacks females that approach him. This type of behavior has been linked to vasopressin, which is released, when the male mates and cares for young. Due to the hormone effect, the males experience a positive feeling, when they maintain monogamy.

Conversely, many monogamic birds and fishes may not be as fidelous as the prairie vole and black vulture are. A widowed/deserted mate shall soon find another mate. More often, the females desert the males in fishes (e.g. *Gobiosoma evelynae*, Harding et al., 2003). The occurrence of a size mis-matched pair and overlapping home ranges facilitate mate desertion. For example, with increasing size mis-match, the pairing duration is reduced from 120 days to 50 days in *Valenciennea longipinnis* (Fig. 14.1A). In this coral reef goby, the female spawns once in 13 days throughout the year; a male tends the eggs in the nest for 1–4 days and can mate a second female. However, most pairs mate monogamously over multiple numbers of cycles, although the number of such pairs decreases with successive cycles (Fig. 14.1B). Yet, exclusive monogamy is common among coral fish species belonging to the ghost pipefishes, seahorses Syngnathidae, seabasses

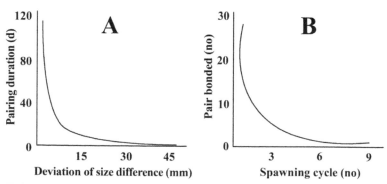

FIGURE 14.1

Velanciennea longipinnis: A. relationship between pairing duration and deviation of size difference between mates and B. number of pairs remaining together as a function of spawning cycle (modified from Takegaki and Nakazono, 1999).

Serranidae, Solenostomidae, tilefishes Branchiostegidae, butterflyfishes Chaetodontidae, angelfishes Pomacentridae, damselfishes Pomacentridae, gobies Gobiidae, surgeonfishes Acanthuridae, triggler fishes Balistidae, filefishes Monacanthidae and sharphose puffers Canthogasteridae.

Enforced monogamy is represented by harems that reduce the diversity level in the polygynic pattern. In the gonochoric cichlids, a harem master may hold from two females, as in *Lamprologus ocellatus* or up to seven brooding females and nine non-brooding sub-adults in *Neolamprologus furcifer* (Kuwamura, 1997). In marine protogynics, the master usually holds three-four females but may also hold up to 10 females, as in *Paraperceis snyderi*. Rarely, the Mediterranean wrasse *Xyrichthys novacula* holds four or six female/territory and also simultaneously holds two territories, each with six females. In all, the number of polygynics holding harem is 21 species from 6 families (Pandian, 2011). For the Tankanyikan cichlids also, the number is 20 (Kuwamura, 1986, 1997). *Obviously, the haremic system deters species diversity.* Even within some of these harems, a few or more females 'steal' fertilization from other male morphs. With increasing number of females, the cumulative fecundity of a harem increases. For example, it increases in the wrasse from 24,482 to 36,864 and 56,777 egg/d, respectively. In the harem, the master is able to increase sperm release from 5×10^6 for 0.5×10^4 eggs to 15×10^6 for 60×10^4 eggs. Yet, Fertilization Success (FS) decreases from 100% in the harem holding four females to 87% to that holding a dozen females (Fig. 14.2).

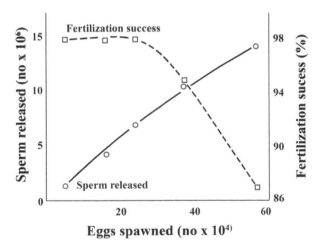

FIGURE 14.2

With increasing number of eggs spawned by all the females in the harem, the number of sperm released is also increased but fertilization success decreases in *Xyrichthys novacula* (drawn from data reported by Marconato et al., 1995).

Biparental monogamy is limited to the egg stage alone in marine fishes like the gobies (Gobiidae), gunnels (Pholidae), wolffishes (Anarhichanidae) and clingfishes (Gobierocidae). In the most speciose (> 1,000 species) Cichlidae, two methods of parental care are recognized: (i) 29% guarding substrate-brooders and (ii) 70% mouth-brooders. In the 131 Tanganyikan cichlid species, Kuwamura (1997) recognized five types of parental care: (i) Maternal guarding (25 species, 19%), (ii) Biparental guarding (21 species, 16%), (iii) Maternal mouth brooding (60 species, 46%), (iv) Female to male shift in mouth-brooding (15 species, 11%) and (v) Maternal mouth-brooding + biparental guarding (10 species, 8%). Hence, it is mostly the female's responsibility to protect eggs and young ones.

14.3 Polygamy

14.3.1 Polygyny

In a large number of animals, polygyny occurs but this description is limited to molluscs, fishes and crustaceans. For them, one or a few matings are adequate for a female to fertilize all her eggs. For example, a single mating provides enough sperms to fertilize 1,000 to 4,400 eggs of the snail *Bulinus globosus*. Hence, monogamy is expected of it. Contrary to it, polygyny is ubiquitous not only in *B. globosus* but also in many molluscs and even within the Simultaneous Hermaphroditic (SH) molluscs. Sexual selection can also occur in hermaphrodites, which can adjust their allocation to mating in female and male roles. Though mating can be costly in terms of time and energy, multiple matings, leading to multiple paternity and polygyny, is a significant adaptive strategy in many animal taxa. In molluscs, multiple paternity has been estimated initially using a biochemical allozyme marker and subsequently with a molecular marker, the microsatellites. Pandian has (2017) summarized paternity estimates in 17 molluscan species that includes all the major three classes of molluscs, gonochores and SH, sedentary/slow motiles and motiles, as well as representation from freshwater and marine habitats, internal and external fertilization (Table 14.3). 1. Surprisingly, SHs have not limited their option to a single paternity; they have opted at least for two sire/brood. Multiple paternity ranges from 2.7 sire/brood in externally fertilizing freshwater *Hyriopsis cumingii* to 4.0 sire/brood in *Crepidula* spp with internal fertilization. In the latter, paternity increases from 2.2 to 3.1 sire/brood with increasing population density from 100 g/m^2 to 600 g/m^2 (not shown in Table 14.3). The highest paternity of 3.5 to 7.6 sire/brood occurs in the slow motile gonochoric snail *Littorina* spp, which have multiple sperm storage systems, and the stored sperm remain fertilizable for 3 months to 1 year. *Remarkably, the presence of sperm storage organs in the female*

TABLE 14.3

Paternity of selected groups in Mollusca (from Pandian, 2017). SH = Simultaneous hermaphrodites

Species	Paternity	
	Sires/Brood	Mean
Gonochores with Sperm Removing Males		
Neptunea arthritica	1.7	1.9
Octopus vulgaris	2.0	
Low Motile Unilateral/Facultative SH		
Aplysia californica	2.1	
Bulinus globosus	2.0	
B. africanus	2.0	2.0
Biomphalaria obstructa	2.0	
Physa fontinalis	2.0	
Motile Gonochoric with Consort/Sneaker males		
Loligo bleekeri	2.0	2.6
L. pealeii	3.2	
Sedentary Gonochore		
Hyriopsis cumingii	2.7	2.7
Sedentary Protandrics		
Crepidula furnicata	4.0	
C. coquimbensis	4.0	4.0
Ostrea edulis	4.0	
Low Motile Gonochores		
Littorina saxatilis	7.6	
L. obtusata	5.0	5.1
Busycon carica	4.3	
Rapana venosa	3.5	

reproductive system and fertilizability of the stored sperms are the most important traits that increase multiple paternity.

The Alternative Mating Strategy (AMS) also increases the diversity. In the polygynous mating system, competition for the female mate becomes more and more intense. Male of a lower competitive ability may adopt AMS "to make use of a bad situation" (Taborsky, 2001) and pave ways for the evolution of two or more mating morphs (Gadgil, 1972). However, "the struggle between the male 'morphs' for possession of females .... result not in death of unsuccessful competitors but in a few or no offspring" (Darwin,

1859). Oliveira (2006) classified 140 species belonging to 28 piscine families that displayed AMS into three patterns and these patterns may be adopted to crustaceans and molluscan cephalopods, as well. The patterns are: (1) Plastic reversibles, in which they switch reversibly in either direction back and forth from one morph to another. (2) Plastic transformants irreversibly switch from one morph to another in a single direction. (3) Fixed sex-linked non-reversibles, in which the non-reversible morphs have a genetic basis (Table 14.4). In the first, the submissive and dominant morphs switch reversibly in either direction in fishes like *Astatotilapia burtoni*, in crustaceans like *Orconectes immunis*, the copulatory morph (during spring and autumn) switches to the non-copulatory morph (during summer and winter) (Table 14.4). For plastic transformants, an example for fish is *Symphodus roissali*, which shifts from sneaker → satellite → territorial → pirate morph. In the prawn *Macrobrachium rosenbergii*, the switch is from the submissive, non-courting pink morph to an aggressive but non-courting orange morph, and then to courting and guarding the blue morph. In the cephalopod *Loligo bleekeri*, the shift is from the externally fertilizing sneaker morph to internally fertilizing consort morph. Examples for the fixed sex-linked non-reversibles are (a) the hooknoses and jacks in *Oncorhynchus tshawystcha* and (b) abraded

TABLE 14.4

Characteristics in patterns of morphotypes that display Alternating Mating Strategy (AMS) (compiled from Pandian, 2014, 2016, 2017)

Pattern	Fishes	Crustaceans	Cephalopod
Plastic reversibles	*Astatotilapia burtoni* *Neolamprologus pulcher* Submissive ↔ Dominant	*Orconectes immunis*: Copulatory morph during spring + autumn ↔ non-copulatory morph during summer + winter	
Plastic transformants	*Symphodus roissali*, *Porichthys notatus*, *Xiphoporus nigrensis*, *Telmatochromis vittatus*, *Parablannius parvicornis*, Sneaker → satellite → territorial → pirate	*Macrobrachium rosenbergii*: submissive pink → aggressive orange → courting blue morphs *Dynamella perforata* Penis absent → present	*Dorytheuthis plei* & *Loligo bleekeri* Sneaker vs consort
Fixed sex-liked non-reversibles	*Onchorhynchus tshawystcha* Hooknoses vs Jacks *Lepomis macrochirus* Junior- & senior-sneakers vs Nest holders *Lamprologus callipterus* Senior sneaker vs Nest holder	*Euterpina acutifrons* Small warm adapted vs large cold adapted morphs, *Paracerceis sculpta* non-haremic satellite vs territorial haremic, *Libinea emarginata*, small gonad + abraded claw vs large gonad + unabraded claw	

clawed morph with a small gonad and aggressive morph with large gonad in the crab *Libinea emarginata*.

14.3.2 Polyandry

It is relatively rare in animals. In fishes, for example, it may occur in gonochoric *Gambusia holbrooki*, haremic gonochore *Julidochromis marlieri* and protogynic haremic *Amphiprion alkalopsis* and *A. bicinctus*. The harem of *J. marlieri* consists of the harem mistress, one small male and one or two helpers. Similarly, that of *Amphiprion* spp also consists of the harem mistress, one small male and many (up to 10) sub adults and juveniles. Essentially, these harems are more of monogamic than polyandric. But true polyandry occurs in nest-holding fishes, in which the male also guards eggs that are not sired by him (Table 14.5). It also occurs in egg guarding *Pseudorasbora parva* and egg gestating *Syngnathus typhle* females. In the androdiocious mating

TABLE 14.5

Reports from natural and experimental observations on the effect of different combinations sex ratio (from Muthukrishnan, 1994, Pandian, 2011, 2012)

Groups	Species name	No. of ♂/♀	Fecundity (egg no.)
Polyandry			
Pelagic spawners	*Gadus morhua*	3♂/1♀	-
Egg guarders	*Ditrema temmincki*	1♂/2♀	-
Gestators	*Syngnathus typhle*	2♂/1♀	85
		1♂/2♀	58
	Pseudorasbora parva	1♂/1♀	1996
		3♂/1♀	2053
		10♂/1♀	1717
		1♂/3♀	1170
		1♂/10♀	632
Coeloptera	*Anthononous grandis*	1♂+1♀	388
		10♂+1♀	238
		20♂+1♀	88
		40♂+1♀	56
Coleoptera	*Tribolium confusum*	1♂+1♀	1010
		2♂+1♀	988
		3♂+1♀	644
Orthoptera	*Gryllodes sigillathus*	3♂+1♀	2979
		8♂+1♀	2300

system, each of the hermaphroditic cirripede *Scalpellum scalpellum* carries 3.5 dwarf males (Pandian, 2016). Lifelong polyandry occurs in colonial isopterans and hymenopterans; in them, each queen may hold up to 12–14 drone males (e.g. *Apis mellifera*, Rangel and Fisher, 2019). For details on social insects, see next.

Interesting field and experimental observations have been reported on the effects of different sex combination over fecundity (Table 14.5). At least three males succeed in fertilizing millions of eggs spawned in a single pulse by pelagic spawning *Gadus morhua*, suggesting natural polyandry. The imminent spawning *G. morhua* is encircled by at least three competitive males. DNA marker analysis has revealed the most dominant male fertilized 84% of her eggs, the next dominant 11% of eggs and the least dominant fertilized the remaining 4% eggs. Simple but meaningful experimental alteration of operational sex ratio has brought to light interesting information. Fecundity is the highest in the combination of 1 ♀ : 1 ♂ in tested coleopteran and orthopteran (monogamy?) insects. But it is the heighest in oviparous fish *Danio rerio*, with a combination of 2 female/male. Contrastingly, combination of 1 ♀ : 2 ♂ is required in the male gestating *Syngnathus typhle* and 1 ♀ : 3 ♂ in *Pseudorasbora parva*, in which males guard the eggs. Among nest holding fishes, maximum promiscuity occurs. For example, the nest holding male guards eggs stealthily brought by females from 5 (e.g. *Lepomis punctatus*) to 15 (e.g. *Gobiusculus flavescens*) nests.

15

Semelparity vs Iteroparity

Introduction

A vast majority of animals are iteroparous; in them, the reproductive episode recurs and is distributed over space and time. Contrastingly, semelpares reproduce only once in a 'big bang' during their entire lifetime. Their oocyte maturation requires longer durations. To spawn only once, some semelparous like the polychaete *Glycera dibranchiata* require 4.0 years (see Pandian, 2019), 13 to 17 years in the cicada *Magicicada septendecula* (Kot, 2001), 7 to 20 years in the lampreys, 10–21 years in the salmoniforman gallaxidids and 8–88 y in anguillids (Finch, 1990). *Further, by choosing to dump the entire basket of lifetime fecundity in a single event at a single location, the semelpares are not as speciose as iteropares.* In them, fecundity ranges from 129 in *Baetis macani* to 2,792 in *Siphlonurus lacustris* among ephemeropterans and from 100 in *Capnia atra* to 900 in *Nemoura avicularis* among plecopterans (Brittain, 1990). But it is 20.8×10^4 in the iteroparous dipteran melonfly *Dacus dorsalis* (Vargas and Chang, 1991). The Lifetime Fecundity (LF) of the semelparous cephalopod *Ocythoe tuberculata* is 21,000 eggs, in comparison to 2 million veligers annually produced by iteroparous *Ostrea edulis*; the life span of *O. edulis* is 5–10 years; assuming it spawns 5-times in its life, the LF of *O. edulis* may be in the range of 10 millions (see Pandian, 2017). In the semelparous Pacific salmon *Oncorhynchus mykiss*, the fecundity is 12,700 (Serezli et al., 2010), whereas that the iteroparous Atlantic cod *Gadus morhua* is 4.3×10^5 egg/kg body weight (see Pandian, 2011a). These reproductive features of the semelpares significantly reduce the progeny number and consequently new gene combinations—the raw material for evolution and speciation.

15.1 Taxonomic Distribution

Not surprisingly, the incidence of semelparity is limited to a very few species. The number of iteroparous species was arrived by subtracting the number of semelparous species from the total species number. The taxonomic distribution and quantification of semelpares are summarized in Table 15.1. Repeated computer searches have revealed that there are no semelparous species in Myriapoda and Chelicera among Arthropoda, as

TABLE 15.1

Taxonomic distribution of semelparity in animals

Taxa	Species (no.)	Reported species name
Turbellaria	3+470	*Dendrocoelum lacteum, Planaria torva, Bdellocephala punctata* (Pandian, 2020). In the 350 speciose, Acoelomorpha and 120 speciose Catenulida gametes exit at death of the parent
Aelostomitidae	29	As above
Hirudinea	1	*Erpobdella obscura* (Govedich et al., 2010)
Oligochaeta	1	*Stylaria lacustris* (Pandian, 2019)
Isopoda	2	*Paracerceis sculpta* (Shuster, 1991), *Schizidium tiberianum* (Warburg and Cohen, 1991)
Decapoda	3	*Palaemonetes kadiakansis, P. paludosus, Eriocheir sinensis* (Cumberlidge et al., 2015, Sewell, 2016)
Insecta		
Hemiptera	4	*Magicicada cassinii, M. septendecim, M. septendecula* (Kot, 2001), aquatic *Parastriches japonensis*
Megaloptera	1	*Corydalus cornutus* (Pennuto and Stewart, 2001)
Diptera		
Chironomidae	2	*Chironomus tentans, C. circumdatus* (Section 24.1), *Polypedilum vanderplenki* (McLachlan and Yonow, 1989)
Simuliidae	2	*Simulus damnosum, S. venustum* (*Anim Div Web*)
Lepidoptera	6	*Hapalonotus* (=*Malacosoma*) *pinnotheroides, H. reticulatus, Thyridopteryx ephemeraeformis, Lymantria dispar, Alsophila pometaria, Operophtera brumata* (Fritz et al., 1982)
Ephemeroptera	3,000	Brittain (1990), Blackmon et al. (2017)
Plecoptera	2,000	
Polychaeta	62	*Nereis diversicola, Nicolea zostericola, Platynereis dumerilii* (Eckelbarger, 1983), 62 epigamic epitokous species in 12 families (Pandian, 2019)
Mollusca		Pandian (2017)
Opisthobranchia	2,000	
Cephalopoda	800	
Urochordata	145	Pandian (2018, Table 10.2), WoRMS
Pisces	~ 100	Finch (1990) includes cyclostomes
Mammalia	23	23 (*Antechinus*-12, *Phascogale*-3, *Dasykaluta*-1, *Mermosops*-7) (Fischer et al., 2013)
Total	8,655	
Of 1,543,196 major phyletic species, semelpares constitute 0.56%		

well as in Echinodermata. Sporadic occurrence of one to five semelparous species is reported for other phyla (Table 15.1). In 470 species of Turbellaria, gametes are exited through pores in the parent's body; hence, they all must be semelpares. For the semelparous polychaetes, the number was reached from several reports summarized by Pandian (2019) indicating the incidence of 62 epigamic epitokous species in 12 families. The number of species for Ephemeroptera (3,000 species) and Plecoptera (2,000 species) were taken from Blackmon et al. (2017). Fritz et al. (1982) confirmed semelparity in a dozen lepidopterans: *Malacosoma* spp, *Thyridopteryx ephemeraeformis*, *Lymantria dispar*, *Alsophila pometaria* and *Operophtera brumata*. Many other authors hinted at a semelparous egg shedding crustaceans and (unisvoltinism) insects. But no one has confirmed the number of semelparous species in arthropods.

For fishes, Finch (1990) reported the occurrence of semelparity in 1% of 20,000 fish species, perhaps, including those, which have become extinct. However, an estimate revealed a maximum of 100 species for living semelparous lampreys and teleosts. From an analysis on shelled and non-shelled molluscs, Pandian (2017) found that the presence of shell(s) afforded iteroparity and a relatively longer life in prosobranch, pulmonates and bivalves but in its absence, semelparity occurs in opisthobranchs and cephalopods. This observation, supported by adequate evidence, led to the estimate of semelparity in 2,000 speciose opisthobranchs and 800 speciose cephalopods. There are exceptions: some nautiloid cephalopods are shelled; some terrestrial pulmonates have no shell. In minor phyletics too, no incidence of semelparity is known, albeit male semelparity is reported for a few Tardigrada. Table 15.1 shows that among the major phyletics, a maximum of 8,655 species are semelpares. Of 1,543,196 species including major and minor phyletics (with no semelparos species), only 0.56% is semelparous.

15.2 Semelparity and Migration

Added to the burden of a single spawning and relatively less fecundity, semelparous species undertake catadromous (61 species) and anadromous (11 species) migration, respectively prior to spawning and subsequent death. The number of these semelpares is listed here under:

Clade	Semelpares	Catadromous	Anadromous
Petromyzontiformes	42	42	0
Anguilliformes	18	18	0
Salmoniformes	23	1*	10
Clupeiformes	5	0	1
Atheriniformes	2**	0	0
Gasteriformes	2	0	0
Perciformes	8	0	0
Total	100	61	11

* Imanga famous for spawning in tidal banks, ** gruneons riskly spawn in beaches

Added to the list, the crab *Eriocheir sinensis* is known to undertake catadromous migration. Apart from these horizontal migrations, the epigamic epitokous 62 polychaete species undertake vertical migration prior to the 'big bang' and subsequent death. Hence, of 8,655 semelpares, at least 135 species (72 fish species + 1 crab species + 62 polychaete species), i.e. 1.6% of them undertake horizontal or vertical migration prior to the 'big bang' spawning.

Thankfully, Quinn and Myers (2004) estimated the distance travelled by the semelparous migrant salmonids; the estimated distances are 463 km for the chinook *Oncorhynchus tshawytscha*, 503 km for the pink *O. gorbuscha*, 543 km for the coho *O. kistuch*, 935 km for the sockcye *O. nerka*, 1,870 km for the chum *O. keta* and 2,013 km for the rainbow trout *O. mykiss*. With a cylindrical long body possessing no powerful lateral fins, the angullids travel 4,000 to 7,000 km for the mega event of spawning once in their lifetime (Yokouchi et al., 2012, 2018). *Eriocheir sinensis* also travels a minimum distance of 65 km and maximum of 1,500 km (Sewell, 2016). The epitokous polychaetes climb a vertical distance ranging from 6 m by *Nereis falcaria* to 4,000 m (4 km) by the ctenodrillid *Raricirrus variabilis*. The climbed distance decreases from the longest 4,000 m in 2 ctenodrillid species to 6–52 m in nereidids (Fig. 15.1). The same holds true for the non-semelparous schizogamic epitokous syllids; in them, there are 8 species climbing distances of 2–75 m, 4 species

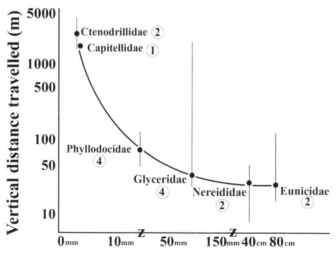

Body length

FIGURE 15.1

Effect of body length on vertical distance climbed by epigamic epitokous polychaetes. Values in circles indicate the number of species at the vertical level. Vertical lines indicate the range of distance climbed (modified from Pandian, 2019).

189–320 m and only one species climbing 4,000 m. It was suggested that all the polychaete species that climb 6–100/150 m distance, use muscle energy and the others may be lifted by reducing specific gravity with absorption of water (Pandian, 2019). During the migration period, the salmonids do not feed, as their digestive system is degenerated to produce energy for the ordeal. They are reported to invest 80–85% of their body (other than gametes) energy on migration. *Alosa sapidissima* also incurs 42% weight loss during the migration. The degeneration of the gut and body wall provides the required energy for vertical migration. In this process, the body weight loss incurred ranges from 75–79% (e.g. *Perinereis cultrifera* and *Nereis pelagica*) (see Pandian, 2019).

15.3 Escape Routes

For the reasons described above, it may not be a surprise to note that of 68 salmonid species, only 11 are semelparous migrants; it may also not be surprising that of 2,070 speciose Cypriniformes, barely five are semelpares. Within the described semelpares, at least four routes are recognized, through which semelpares escape into iteroparity: (1a) Among Anguilliformes, *Nemichthys scolopaceus* and *Conger vulgaris*, not all the members are semelpares. (b) Subsequent to the 'big bang' spawning, 20% of the gobiid *Pomatoschistus microps* and 53% of *Alosa pseudoharengus* survive to spawn again. Among polychaetes also, a few partially milted males of *Glycera dibranchiata* return to the burrows to milt again. The anterior fragment of the epigamic epitokous *Eunice siciliensis*—when reared under optimal conditions—survives and regenerates the posterior genital segments and undertakes another epitoky. These observations may explain how the schizogamic epitokous 45 syllid species have returned to iteroparity. (2) The populations inhabiting above 32°N in *A. sapidissima* and those of *O. mykiss* inhabiting the northern latitudes alone are semelpares, while those in the southernly habitats are iteropares. (3) In osmoroid Salmoniformes, there are two distinct races in *Hypomesu solidus*, *Plecoglossus altevis* and *Thalaeichthys pacificus*: of the two, one race undertakes migration, while the other remains land locked. (4) In the perciformid *Callionymus lyra*, females are iteroparous but the males are not. That leads one to look at the dichotomic male semelparity and female iteroparity in a number of eusocial colonial insects and others (Table 15.2).

15.4 Semelparous Males

As maintenance of males in a population/species is a luxury, animals may choose to reduce the (i) male ratio or (ii) male size. Another alternative is to

TABLE 15.2

The number of semelparous males in eusocial insects (compiled from Seeley, 1985)

Order	Eusocial species (no.)
Insecta	
Hymenoptera	15,250
Isoptera	2,600
Hemiptera	50
Thysanoptera	6
Coleoptera	1
Crustacea	
Synalpheus	1
Mammalia	
Bathyergidae	12
Total	17,908

reduce the male's life span, which results in semelparity, i.e. by either natural death or cannibalistically devouring the mated male partner. Following the exit of sperms and insemination, natural death of males is known in some aelostomatids (e.g. *Stylaria lacustris*) and limnic tardigrades (Bertolani, 2001) and archiannelids. The dwarf male of *Dinophilus gyrociliatus* with no gut inoculates into his many immature sisters and dies (see Pandian, 2019). In Acanthocephala, semelparous males of *Reighordia sternae* copulate with the still juvenile females and die (see Pandian, 2021a). Many spiders (e.g. *Argiope bruennichi*) and scorpions (e.g. *Bothriurus bonariensis*) avidly devour the male following insemination. As an exception in the polychaete *Neanthes arenaceodentata* and *Platynereis massiliensis*, females lay eggs and die; but the males fertilize the eggs, protect and ventilate the fertilized eggs (Premoli and Sella, 1995). Aside from these sporadic incidences of male semelparity, it occurs in all the social insects. Literature on social insects is voluminous. However, this account is limited to the semelparity alone. In isopteran termites and hymenopterans, the eudiploid queen is a sexually reproducing female. Through haplodiploidy, she produces diploid females, which become workers and haploid drones or males. Unmated female workers can also produce haploid males, and thereby create a competition for production of males and competition for mating among these males, as well (Olejarz et al., 2015). Eusociality is a form of social organization, where female workers reduce their own lifetime reproductive potential to raise the offspring of others. In comparison to the mean Life Span (LS) of 0.1 ± 0.2 years for 87 solitary insect species from eight families, LS of the queen is 10.0 ± 6.6 years., 11.5 ± 5.4 years and 5.6 years for ants, termites and honeybees, respectively, with an overall mean of 10.1 ± 6.4 years for the eusocial queens (Keller and Genoud, 1997). For the female workers, the LS range from 0.83 months in wasps to 1.33 months for ants (Kramer and Schaible, 2013). The value for drone males ranges from 60 to 70 days only. With LS of 10 years, the queens seem to increase genetic diversity by mating with a large number of males, and thereby try and test a new genome with every mating male. In *Cardiocoudyla abscurior*, the frequency of mating seems to determine the LS of males; when allowed to more frequently, it is 60 days but as long as 120 days, when allowed to mate less frequently. The males seem to opt for "live fast, die young" strategy (Metzler et al., 2016).

16

Spawning and Oviposition

Introduction

In the life history of animals, spawning/oviposition is a fitness-deciding, crucially important event that lets the exit of time-long gametic investment. Restricted to aquatic habitats, spawning and milting are events of releasing female and male gametes, respectively. However, many authors consider both together as spawning. In general, many aquatic animals spawn both eggs and sperms and ensure External Fertilization (EF) (e.g. sponge *Tetilla serica*, Fell, 1989, abalone *Haliotis iris*, see Pandian, 2017). In the others involving oogamic fertilization, sperms alone are released (e.g. sponge *Halisarca duyardini*, Fell, 1989). Thereby, they accomplish (i) cross fertilization and (ii) dispersal. However, these fertilization events are achieved at the cost of huge investment on one or both gametes; for example, Fertilization Success (FS) is ensured at sperm density of 100,000 to 190,000/ml (Shepherd, 1986). Of numerous sperms released by the bryozoan *Lophopus crystallianus* characterized by OF, only 115 sperms reach the ovary to fertilize 18 oocytes (see Ostrovsky, 2013). Contrastingly, terrestrial animals are characterized Internal Fertilization (IF), which economizes investment on gamete production. But they may have to invest on synthesis of structures like capsule, cocoon and the kind to protect the eggs. The chances of their dispersal are limited by wind and/or rain water or they have to depend on inadvertent ingestion of their indigestible eggs/cocoon by animals.

The fertilized eggs of many turbellarians are continuously released (e.g. *Stylochus ellipticus*, see Pandian, 2020). In some nematodes, spawning is measured as a daily output (see Pandian, 2021a). The spawning frequency ranges from once every 19 minutes in Owens pupfish *Cyprinodon radiosus* to 15 days in the fightingfish *Betta splendens* and to 19–61 days in *Tilapia zilli*. However, most aquatic animals select a favorable season, during which they spawn in a single bout (e.g. *Gadus morhua*) or multiple pulses (e.g. 41 pulses in *Serranus atricauda*, Pandian, 2011). In terrestrial habitats, many snails oviposit on a daily basis (e.g. *Biomphalaria glabrata*, Vianey-Liaud, 1995). But the duration of the oestrus cycle is prolonged to once a month in humans and once a year or so in cattle.

16.1 Taxonomic Survey

A survey of structural and/or behavioral modes of gamete transfer reveals that (1) with no distinct reproductive system, the Porifera, Cnidaria and Acnidaria are obligated to external fertilization alone (Table 16.1). (2) Irrespective of possessing a simple (e.g. echinoderm) or complicated (e.g. pulmonates) reproductive system, polychaetes, molluscs and echinoderms lack a functional intromittent structure, which has led to external fertilization in them. (3) Even without a penis, the behavior-mediated clitellar juxtaposition in earthworms and juxtaposition in nemerteans, possibly in myriapods, and reciprocal/unilateral insemination in opisthobranchs, pulmonates and gnathostomulids have led to external cum internal epizoic fertilization or internal fertilization. (4) Interestingly, cloacal apposition in birds has achieved as effective internal fertilization, as mammals have it with a penis. Hence, the lack of an intromittent organ can be neutralized by behavioral insemination. (5a) The less efficient spermatophore-mediated sperm transfer ensures epizoic fertilization in some polychaetes, rotifers, acanthocephalans and phoronids (Table 16.2). (5b) Contrastingly, the effective spermatophore transfer by structurally more adapted appendage(s) in chelicerans (except in Araneae) and heterocotyle in cephalopods results in internal fertilization. (6) Even with a scelerotized or long protrusible penis, ostracods and cirripedes are unable to accomplish internal fertilization. Clearly, there are animals, which can achieve internal fertilization by behavioral 'copulation', while those possessing an intromittent organ may not ensure internal fertilization. Yet, in the absence of an intromittent organ, 44% females in the holothuroid *Leptosynapta clarki* remain unfertilized (see Pandian, 2018). (7) A hypodermic injection is reported from a dozen polychaetes (see Schroder, 1989) and a few gnathostomulids. Dermal penetration is reported in a turbellarian *Pseudoceros bifurgatus* (Pandian, 2020) and a couple of cephalopods (Equardo and Marian, 2012). Incidentally, a hypodermic injection releases the sperms into the parenchyma in turbellarians or into the body but not inside the reproductive system. Similarly, a dermal injection is followed by penetration through the body to reach the reproductive system. However, these are sporadic and exceptional. These two costly routes of sperm transfer have almost deterred species diversity.

TABLE 16.1

Structural and behavioral modes of gamete transfer in major invertebrate phyla (compiled from Mann, 1984, Weygoldt, 1990, Schroder, 1989, Pandian, 2016, 2017, 2018, 2019, 2020)

Phylum/Clade	Reported observations
Porifera, Cnidaria Acnidaria	No distinct reproductive system. Obligate transfer of both eggs and sperms or sperms only
Platyhelminthes	Penis, Penis papilla, Cirrus; mostly internal fertilization
Polychaeta	Nephridia serve as exits for gametes. Hypodermic injection in a ~ dozen species. Spematophore-mediated epizoic transfer
Aquatic oligochaetes	Nephridia serve as exits for gametes
Aeolostomatids	Dermal pores serve as exits for gametes
Earthworms	Clitellar apposition effects epizoic fertilization (e.g. *Disenia*). Dermal piercing in *Lumbricus terrestris*
Hirudinea	Penis: but spermatophore-mediated epizoic fertilization
Crustacea	Sceletized penis in ostracods, long protrusible penis in cirripedes, penis papilla in amphipods with flagellated sperms. In others, thoracic appendage or pleopod, and in some penis-like structure on appendage with non-flagelled sperms held in spermatophore(s)
Myriapoda	Spermatophore-mediated epizoic fertilization?
Insecta	Internal fertilization effected by intromittent organ
Chelicera	Spermatophore(s) transferred by appendage to gain internal fertlization except in Araneae
Archaeogastropoda	Eggs and sperms are spawned
Other prosobranchs	Only sperms are released
Opisthobranchia, Pulmonata	Reciprocal or unilateral insemination suggests internal fertilization
Bivalvia	Eggs and sperms are spawned except in a dozen with release of sperms only
Scaphopoda	Eggs and sperms are spawned
Polyplacophora	Eggs and sperms are spawned
Cephalopoda	Heterocotyle arm facilitates internal fertilization Dermal penetration in a couple of species
Apolyplacophora	Internal fertilization
Echinoderamta	Eggs and sperms are spawned

16.2 Gamete Transfer Routes

In sexually reproducing animals, this account recognizes the following four routes of gamete transfer: (1) External fertilization (EF) in water involving the transfer of eggs and sperms in a synchronized (e.g. *Xestospongia muta*,

TABLE 16.2

Structural and behavioral modes of gamete transfer in vertebrates and minor phyla (compiled from Hyman, 1951b, Seale, 1987, Pandian, 2011a, 2021)

Phylum/Clade	Reported observations
Teleostei	Eggs and sperms are spawned, except in 577 viviparous species, in which penial setae insert sperms. Internal fertilization
Amphibia	Eggs and sperms are spawned, except in 72 caecilian species with internal fertilization
Reptilia	Penes. Internal fertilization
Aves	Cloacal apposition leads to internal fertilization
Mammalia	Penis. Internal fertilization
Nemertea	Eggs and sperms spawned
Ganthostomulida	Reciprocal insemination
Rotifera	Protrusible penis: Spermatophore-mediated epizoic sperm transfer. Hypodermic injection in some
Gastrotricha	Penial spicule: Internal fertilization
Kinorhyncha	Internal fertilization
Nematoda	Spicule dialates gonopore but copulatory muscles push the spermatopore into oviduct
Nematomorpha	Spciule absent. Spermatophore-mediated epizoic fertilization
Acanthocephala	Copulatory transfer of sperms/spermatophores
Priapulids, Sipuncula, Echiura	Eggs and sperms are spawned
Tardigrada, Onycho-phora, Pentastomida	Internal fertilization
Phoronida	Spermatophore-mediated epizoic fertilization
Bryozoa	Only sperms are released
Brachiopoda	Eggs & sperms spawned; a third of them are selfing hermaphrodites
Chaegognatha	Only sperms are released
Hemichordata	Eggs and sperms are spawned
Urochordata	Eggs and sperms spawned except in internally fertilizing salpids and doliolids

see Fell, 1989) or asynchronized (e.g. *Tetilla japonica*, see Fell, 1989) episodes, (2) Oogamous Fertilization (OF) involves the release of sperms only and retention of eggs within parent body. Interestingly, Thorson (1950) listed a number of marine invertebrates, in which chemical cues synchronize spawning in marine benthics (Table 16.3). (3) Spermatophore-Mediated Transfer (SMF) occurs by implanting it in and around the female gonopore in

TABLE 16.3

List of animals, in which synchronized spawning is induced by chemical cues (compiled from Thorson, 1950)

Phylum	Synchronizingly spawning species
Anthozoa	*Cerianthus lloydii, Pachycerianthus multiplicatus, Sagartia troglodytes*
Nemertea	*Cerebratulus* sp
Priapulida	*Priapulus caudatus*
Sipuncula	*Phascolosoma vulgare*
Brachiopoda	*Lingula* sp
Polychaeta	*Eunice harassi, Eusyllis blomstrandi, Hydroides dianthus, Nereis cultrifera, N. dumerili, N. limbata, N. vexillosa, Pionosyllis lamelligera*
Mollusca	*Chiton cinereus, C. marmoratus, C. squamosus, Ischnochiton magdalenensis, Chaetopleura apiculata, Mopalia lignosa, Gibbula cinerarius, G. magus, Haliotis gigantea, H. tuberculata, Crassostrea gigas, C. virginica*
Echinodermata	*Antedon bifida, Cucumaria frondosa, Holothuria tubulosa, H. marmorata, Thyone briareus, Leptosynapta inhaerens, Asterias rubens, Strongylocentrotus lividus, Amphiura filiformis, Ophiopholis aculeata*

Crustacea and the route is named epizoic (see Pandian, 2016). The difference between OF and SMF is that in the latter, the mating partners are in close contact with each other but not in the former and (4) Intromittent organ-mediated sperm transfer into the female's body (by a hypodermic injection or dermal penetration) or reproductive system results in internal fertilization.

The incidence of external transfer of eggs and sperms or sperms alone is limited to aquatic habitats. The spermatophore- or intromittent organ-mediated sperm transfer occurs in both aquatic and terrestrial habitats; for instance, spermatophore-mediated transfer occurs in aquatic crustaceans and terrestrial chelicerans. The taxonomic distribution is summarized for the gamete transfer through the identified four routes for major phyla (Table 16.4) and for minor phyla (Table 16.5). Whereas the quantification of each of this transfer route was relatively easier for minor phyla, it was indeed a task for major phyla (for lack of adequate information). In his account, Fell (1989) named 6 and 12 species, in which eggs and sperms, or only sperms were emitted, respectively. Hence, the ratio of 1 : 2 for EF and OF was taken to assign the gamete transfer routes in sponges. Ereskovsky (2018) listed at least a dozen orders of Porifera, in which hermaphroditism and viviparity occur, and suggests internal fertilization in them. However, the list did not indicate the number of species for each of the identified orders. An estimate from WoRMS suggests 5,365 of 9,342 species of Porifera as potential vivipares. Ostrovsky et al. (2016) limited the number to 34 matrotrophic species. Considering all these observations, 2,837 and 5,680 sponge species are assigned to EF and OF, respectively. Shikina and Chang (2018) considered that external fertilization is typical of cnidarians; however,

TABLE 16.4

Number of major phyletic species transferring gametes in identified routes. 1 = Ereskovsky (2018), 2 = Ostrovsky et al. (2016), 3 = Shikina and Chang (2018), 4 = Sasson and Ryan (2016), 5 = Pandian (2020), 6 = Pandian (2019), 7 = Pandian (2017), 8 = Brahmachary (1989), 9 = Steiner (1993), 10 = Kocot et al. (2019), 11 = Abadia-Chanona et al. (2018), 12 = Pandian (2016), 13 = Wright (2012), 14 = Mann (1984), 15 = Pandian (2017), 16 = Pandian (2014), 17 = Hillis and Green (1990)

Phylum	External		Spermatophore, Epizoic	Internal
	♂, ♀ Gametes	♂ Gametes only		
Porifera[1,2]	2,840	5680	0	34
Hydrozoa[3,2]	0	3,163	0	1*
Others[4]	3,845	3,844	0	3*
Acnidaria	0	166	0	0
Turbellaria[5]	2050	0	0	3,450
Other flukes[5]	0	0	0	22,200
Polychaeta[6]	5225	7,000	787	19
Aeolosomatids[6]	29	0	0	0
Aquatic oligochaetes	2,000	0	0	0
Other clitellates[6]	0	0	0	1,850
Gastropoda[7,8]	~ 4,900	75,100	0	26,000
Bivalvia[7]	974	12	0	0
Cephalopoda[7]	0	0	0	800
Scaphopoda[9]	570	0	0	0
Aplacophora[10]	0	0	0	263
Other molluscs[11]	960	0	0	0
Crustacea[12]	3,600	26,000	24,713	71*
Insecta	0	0	0	1,020,007
Myriapoda[13]	0	0	11,800	0
Chelicera[14]	0	0	140	155,620
Echinodermata[15,2]	6,941	0	0	59*
Pisces[16,**]	31,933	0	0	1,857
Amphibia	5,009	0	0	0
Caecilians[17]	67	0	0	72
Reptiles	0	0	0	9,545
Birds	0	0		10,038
Mammals	0	0	0	5,513
Total	70,943	120,965	37,440	1,257,402
% for 1,489,750	4.8%	8.1%	2.5%	84.6%

[†] by Clocal apposition, ** = inclusive of 1,280 elasmobranch species

TABLE 16.5

Number of minor phyletic species transferring gametes through the identified routes (from Pandian, 2021). * = Ostrovsky et al. (2016), Adiyodi and Adiyodi (1989, 1990), Harzsch et al. (2015), Temereva and Malakhov (2016), Cantell (1989)

Phylum	External		Spermatophore, Epizoic	Internal
	♂, ♀ Gametes	♂ Gametes only		
Mesozoa	0	0	0	150
Loricifera	0	0	0	1
Nemertea	-	0	1	1,300
Gnathostomulida	0	0	0	100
Rotifera	0	0	785	785
Gastrotricha	0	0	0	813
Kinorhyncha	0	0	0	200
Nematoda	0	0	0	27,000
Nematomorpha	0	0	360	0
Acanthocephala	0	0	0	1,047
Priapulida	0	0	0	19
Sipuncula	147	0	0	0
Echiura	230	0	0	0
Tardigrada	0	0	0	100
Onychophora	0	0	0	200
Pentastomida	0	0	0	144
Entoprocta	200	0	0	0
Phoronida	0	0	20	3
Bryozoa	0	4,856	0	844*
Brachiopoda	0	261	0	130
Chaetognatha	3	150	0	0
Hemichordata	125	0	0	5?
Urochordata	1,967	2,033	0	-
Subtotal	2,672	7,300	1,165	32,841
% for 43,978	6.0	16.6	2.7	74.7
Major Phyla	70,943	120,965	37,440	1,257,402
% for 1,486,750	4.8	8.1	2.5	84.6
Total	73,615	128,265	38,605	1,290,243
% for 1,530,728	4.8	8.4	2.5	84.3

internal fertilization does occur in brooding cnidarians. Ostrovsky et al. (2016) indicated that occurrence of matrotrophic viviparity in four cnidarian species only. Notably, the indications by Hyman (1940) and others suggested the retention of eggs and release of sperms alone in most hydrozoans. Fautin et al. (1989) indicated the almost equal occurrence of EF or OF in other cnidarians. For acnidarians, a couple of publications hint internal fertilization and release of larvae (e.g. *Mnemiopsis leidyi*, Sasson and Ryan, 2016).

In Platyhelminthes, the 2,050 speciose turbellarian orders Prolecithophora, Lecithoepitheliata, Typloplanoida, Dalyelloida, Catenulida, Acoelomorpha, Macrostomorpha and Proseriata emit both eggs and sperms. Interestingly, polychaetes have representations for all the four modes of gamete transfer (Table 16.4). (i) Covering 307 species in 36 families and 10 orders, Wilson's (1991) analysis found that of 307 polychaetes, 40%, or of 13,062 speciose polychaetes, 5,225 species spawn both eggs and sperms (EF). With regard to the remaining 60%, neither Wilson (1991) nor Carson and Hentschel (2006) named them as brooders but have not defined whether they can be grouped under EF or OF. In polychaetes and most aquatic oligochaetes, the nephridia serve as exits for the gametes. Clearly, these two groups retain the eggs in the parent body and release the sperms alone. In fact, the tentacles of *Nicolea zostericola* and *Neolepa septochaeta* are reported to collect conspecific sperms and ensure OF on the tentacles (see Pandian, 2019). A calculation of Schroder's (1989) data revealed that SMF occurs in 727 polychaetes species. Aquatic oligochaetes are also OF. Except for 29 speciose aeolostomatids, in which the gametes are exited through the dermal pores resulting in EF, internal fertilization occurs in the remaining ~ 1,850 clitellate species. Brahmachary (1989) hinted that complete external fertilization occurs in archaegastropods and other primitive molluscs, but internal fertilization in 2,000 speciose opisthobranchs and 24,000 speciose pulmonates. Devoid of an intromittent organ, the mesogastropods and neogastropods retain their eggs but spawn only sperms. Hence, they are obligate oogamous fertilizers. In fact, the eggs retained in the body of *Fissurella nubecula* are fertilized by sperms brought by the incoming current of water (see Brahmachary, 1989). Except for about a dozen brooding species, EF seems to be the rule for bivalves. EF is reported for two scaphopod species (e.g. *Caudulus subfasiforensis*, *Pulsellum lofotensis*, Steiner, 1993) and a few polyplacophore species (e.g. *Chiton articulata*, Abadia-Chanona et al., 2018). But internal fertilization occurs in aplacophorans (e.g. *Epimenia australis*, *Encylopedia*). Spermatophore-mediated internal fertilization occurs in the 800 speciose cephalopods. Except for 59 matrotrophics, all the remaining 6,941 echinoderm species are obligate EFs. Similarly, except for 844 matrotrophics, all the remaining 4,856 bryozoan species are OFs (Table 16.5). In them, the conspecific sperms are captured by the lophophore to facilitate the oogamous fertilization. Incidentally, Ostovsky et al. (2016) reported the number of matrotrophics in different phylum, but have not indicated the route(s), through which the sperms reach the eggs.

In almost all insects including those transferring spermatophore(s) (Mann, 1984), internal fertilization is the rule. With an effective spermatophore-mediated sperm transfer by one or other appendage, the chelicerans, except for the 140 speciose Araneae, are able to inseminate the spermatophore(s) into the female's gonopore or reproductive system and thereby achieve successful internal fertilization. In the absence of such an effective spermatophore-transferring appendage, the juxtaposition in some annelids and others may be less effective to transfer the spermatophore. According to Table 2.8 of Pandian (2016), ~ 26,000 species comprising ostracods, cirripedes, peracarids including isopods and amphipods, ensure oogamous fertilization (OF) using flagellated motile sperms. About 6.4% crustaceans, i.e. ~ 3,600 species are broadcast spawners (Pandian, 1994). In the remaining 24,713 species ensure epizoic fertilization using one or other modified appendage for spermatophore transfer (Table 16.4). This is also true for the Nematomorpha (Table 16.5). Also in rotifers, the eggs are retained in the parent body. In the Chaetognatha, the spermatophore-like sperm bundle is inseminated and fertilization is mediated by specialized somatic accessory cells in the ovary prior to ovulation.

Though exhaustive, the present estimates may not be precise. However, it claims that the proportions arrived for the four routes of fertilization shall remain valid. (1) For the first time, it has accounted for the routes of fertilization in 94.2 and 99.3% in minor and major phyletic species, respectively (Table 16.4, Table 16.5). On the whole, of 1,530,728 species, the assembled data amount to 99.2% of major and minor phyletics. (2) Internal fertilization occurs in 84.6% major phyletics but only in 74.7% minor phyletics. Being more 'aquatic', the minor phyletics have shares of 6.0 and 16.0% for the complete (EF) and Oogamous Fertilization (OF), respectively; the corresponding values for the major phyletics are 4.8 and 8.1%. (3) *Though efficient in accomplishing internal fertilization by chelicerans and cephalopods, and economic epizoic fertilization by crustaceans, the spermatophore-mediated sperm transfer is not the preferably availed route for fertilization.* Only 2.5% of animals (2.7% for minor phyla, 2.5% for major phyla) have chosen this route for fertilization. (4) Most (~ 206,898 species) aquatic animals have selected either fully (EF) or the Oogamous Fertilization (OF) route. Despite incurring no cost on manifestation and maintenance of reproductive system, mate searching and selection, *the fully external fertilization (PF) is costlier, as it is chosen by only 4.8% major phyletics and 6.0% minor phyletics. To reduce the cost, the Oogamous Fertilization (OF) has almost doubled the proportion to 8.1% (8.2% major phyletics).* (5) Irrespective the cost of manifestation and maintenance of distinct reproductive system, mate searching and selection and so forth, internal fertilization is the most preferred route by 84.3% animals.

16.3 Broadcast Spawners

External Fertilization (EF) involves the transfer of both eggs and sperms. It requires a huge investment on gamete production. For example, the broadcast spawning female abalone *Haliotis iris* spawns > 11.5 million eggs at > 65% cost of its body weight—measured by Ovarian Somatic Index (OSI). Another broadcast spawning snail *Patella vulgaris* invests 33% on sperm production (Testicular Somatic Index, TSI) and 40% on egg production (OSI) (see Pandian, 2017). Some fishes invest as much as 1.5- (e.g. *Canthigaster valentine*) and 2.0- (e.g. *Corynopoma riisei*) times of their respective body weight on OSI (see Pandian, 2011a). To reduce the high cost *per se*, the motile broadcasters adopt one or more morphological, physiological and behavioral strategies, which distinctly differ from those adopted by sessile/sedentary animals. In sessile sponges and cnidarians photoperiodism is an important factor to synchronize the spawning episodes. In the hydroid *Hydractina echinata*, both sexes spawn simultaneously, immediately following sunrise. The most spectacular instance is that the 112 anthozoan species in the Great Barrier Reef spawn only a few nights each year following the spring full moon. In hydroids like *Orthopyxis caliculata* and *Hydra carnea*, Fautin et al. (1989) reported examples for the mature eggs from the parent body releasing attractants to guide sperms. On the other hand, the sponge *Neofibularia nolitangere* expels sperms in a turbid cloud that lasts for a few minutes; by traveling with the current, the sperms induce spawning in others; apparently, they release a pheromone like substance to synchronize spawning. This is also true of many sponges (e.g. *Verongia archeri*, *Geodia* sp, Fell, 1989). Similar epidemic spawning covering a whole population is also known for polychaetes: *Podarke* sp, *Diaptra* sp and *Branchiomma vesicular*, nemertean: *Cerebratulus lacteus*, bivalves: *Mytilus edulis*, *Pecten irradians* and *Tivela sultorum* and sea urchin: *Diadema setosum* (Thorson, 1950).

The motile broadcasters adopt structural (stacking) and behavioral (aggregation) strategies to aggregate the mating partners closer together and thereby reduce the cost of gametic production. Reducing the distance between mating partners to distances less than 20 cm at the time of spawning dramatically increases FS (Levitan, 1991). The antarctic limpet *Nucella concinna* deliberately stacks and each stack consists of up to eight adults, one resting over the shell of the one beneath and thereby reduces the distance between mating partners. But the most spectacular swarming is that of the vestigastropod *Scissurella spinosa*, in which the epidemic nocturnal synchronized spawning and milting occur during November in shallow back-roof of Moorea, French Polynesia (Hickmann and Porter, 2007). Following changes in structural, functional and behavioral, the schizogamic epitokous *Odontosyllis enopla* synchronizes swarming and spawning with precise timing guided by the lunar cycle. The swarming in epigamic epitokes leads to high density, for example, 14,800/m² in *Dendronereis aestuarina* (see Pandian, 2019). Another feature that attracts aggregation is the luminescence. It is the luminescent glow of *O. luminosa* in the Caribbean that is reported to have provided hope for the nearby island mass to the totally exhausted Christopher Columbus in 1492. For other chemicals that facilitate aggregation in epitokous polychaetes, and molluscs, Pandian (2017, 2019) may be consulted. The second set of the adaptive strategy is to retain the eggs within the parent body and let the sperms travel alone to reach the eggs. This oogamous fertilization, considered as primitive, occurs in sponges, cnidarians and bryozoans. Here too, chemical cues arising from the eggs are reported to attract and guide the sperms (see Fell, 1989). The third, perhaps the rarest strategy is known from one third of brachiopods, which have switched to hermaphroditism and self fertilization to reduce the cost of gamete transfer (see Pandian, 2021).

When many broadcasting species simultaneously spawn their eggs and sperms in a single habitat—as it occurs in 112 anthozoan species on the Great Barrier Reef—it becomes necessary for gametes of one species to identify and recognize their respective counterparts. The bindin present in the sperm acrosome serves as an adhesive to attach a sperm to the vitelline layer of a conspecific egg. Hence, it plays an important role in determining compatibility between conspecific gametes and the level of compatibility between closely related (may allow hybridization) and distantly related (and not allow hybridization) species (see Pandian, 2018). Then a question may be raised how conspecific gametes are recognized in species like teleost fishes, which lack acrosome in their sperms. For want of information, the one available for abalones is described, which may help to explain how acrosomeless species identify and recognize its own counterpart. In 60 abalone species, the breeding season and habitats overlap. In them, the amino acid L-tryptophane serves as an attractant to bring the sperms close to the eggs (Riffell et al., 2002, 2004). However, the final recognition between conspecific gametes is brought by species-specific protein present in the gamete membranes (Vacquier, 1998).

16.4 Spermatophore and Economy

In polychaetes, crustaceans, earthworms, some snails and a few other minor phyla, the sperms are transferred in the form of a spermatophore. In them, the spermatophore is placed on the female's body, often adjacent to her gonopore resulting in epizoic fertilization. Within epizoics, spermatophore placement is not precise in polychaetes and molluscs. For example, the spionid *Polydora ligni* male sheds it into water, which is subsequently acquired by the ciliated palps of the female (Rice, 1978). In *Hesionides gonari*, Westheide (1969) reports indiscriminate placement on another male. On the other hand, the placement by crustaceans is discriminate, precise and economic. Their appropriate placement ranges from 68% in *Penaeus japonicus* to 85% in euphausiid *Thysanoessa raschii*. In the copepod *Pareuchaeta norvegica*, it is 72, 20 and 7% for a single, double and treble spermatophore, respectively. With 1.8% TSI and placement of only two spermatophores, *Calanus helagolandicus* fertilizes 100 eggs. *With 50% TSI, the semen transferred by the long protrusible penis fertilizes 300 eggs in cirripedes. Despite being epizoics, the crustacean strategy of spermatophore transfer is more precise and economic. Not surprisingly, many crustaceans have opted for the pathway of spermatophorogenesis, instead of the usual spermiogenesis pathway* (see Pandian, 2016).

17

Fertilization Success

Introduction

The egg is the most valuable single cell. It is highly specialized and contributes not only a half of the genome but also the entire mitochondrial genome. Fertilization is a crucially important event in the reproductive life history of all animals. Internal fertilization is the safest route to transfer male gametes into the female's gonopore. In aquatic animals, the eggs and sperms are spawned in External Fertilization (EF) or only sperms/spermatohores are released (Oogamous Fertilization, OF). In EF, synchronization of spawning is tuned by external factors like the lunar cycle, time of the day or photoperiodism, or internal factors like chemical cues. In some sessile sponges, cnidarians, bryozoans and others, in which OF occurs, the sperms are attracted by chemical cues released by mature eggs from the parent body. In motile aquatic and terrestrial animals, external factors like the lunar cycle or internal factors like luminescence and other chemicals aggregate the mates to a specific space and time to reduce the distance between conspecific mates and ensure Fertilization Success (FS). These fascinating aspects have attracted a large number of interesting studies.

17.1 Aquatic Habitats

As described earlier, fertilization is ensured by adopting one of the four routes: (i) External Fertilization (EF), (ii) Oogamous Fertilization (OF) by emission of eggs and sperms into water in the former and releasing sperms alone in the latter, (iii) spermatophore-mediated partial external fertilization (SMF) and (iv) copulation-, gonopore/cloacal apposition- or spermatophore-mediated internal fertilization. Most terrestrial animals avail the fourth route. In aquatic habitats, the sperms remain immotile until they are released and come in contact with water. Experimental studies have revealed that: (i) FS increases up to 100% at the density of 10^4 sperm/ml in the sea urchin

Strongylocentrotus droebachiensis (Fig. 17.1A) and in many ascidians (Fukumoto et al., 1996) but ~ 50–60% in the serpulid polychaete *Galeolaria caespitosa* (Fig. 17.1A). (ii) Swimming velocity of *G. caespitosa* sperm decreases as a function of the gamete age (Fig. 17.1B). (iii) As a result, fertilizable lifetime of a sperm decreases to 5% within < 50 minutes in *S. droebachinensis* (Fig. 17.1C) and to < 10% in < 30 seconds in the surgeonfish *Acanthurus nigrofuscus* (Fig. 17.1D). Conversely, the eggs remain fertilizable up to 100 minutes in *S. droebachiensis* and up to 300 seconds in the surgeonfish (Fig. 17.1D). Notably the fertilizability of both eggs and sperms of *G. caespitosa* decreases as a function of gamete age (Fig. 17.1E). (iv) Amazingly, the fertilizable lifetime of both eggs and sperms of the lugworm *Arenicola marina* lasts for 3 days. Incidentally, it is also 3 days for the echiuran *Urechis caupo* (see Thorson, 1950). The female lugworm lays the eggs within her burrow. The male releases his sperm bundles on the sediment surface, which are carried to the burrow by tidal waters at the density of 10^6 sperm/ml and the lugworms's eggs and sperms remain fertilizable (40–60%) up to 3 days (see Pandian, 2019).

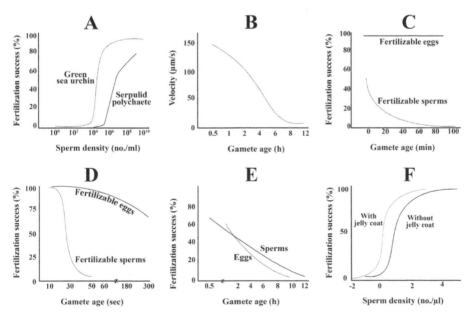

FIGURE 17.1

A. Sperm density as a function of fertilization success in green sea urchin *Strongylocentrotus droebachiensis* (redrawn from Pennington, 1985) and serpulid polychaete *Galeolaria caespitosa*. B. Swimming velocity of *G. caespitosa* sperms as a function of gamete age (from Kupriyanova, 2006). C. The age of fertilizable eggs and sperms in the urchin *S. droebachiensis* (from Pennington, 1985), D. in surgeonfish *Acanthurus nigrofuscus* (compiled from Kiflawi et al., 1998) and E. *G. caespitosa* (from Kurpriyanova, 2006). F. Fertilization success of eggs with and without jelly coat as a function of sperm density of sand dollar *Dendraster excentricus* (simplified from Podolsky, 2002). All are simplified free hand drawings.

Notably, other factors like increasing surface area of eggs by jelly coat or water egg hydration are also important to increase the target surface area in small eggs. In the sand dollar *Dendraster excentricus*, the cost-effective jelly coat substantially increases not only the target area for sperm collision but also reduces the sinking rate and thereby retains the egg in the pelagics for longer duration. For example, an ovum without the jelly coat sinks 12-times faster than that with the jelly coat. Podolsky (2002) showed that FS of the jelly coated eggs of *D. excentricus* is greater at all the tested sperm density than that without the jelly coat (Fig. 17.1F). Incidentally, jelly coat is also formed following fertilization in polychaetes (e.g. *Nereis falcaria, Platynereis dumerilii*); the dejellied eggs of *Lumbrinereis latellieli* lose fertilizability but it can be restored on addition of jelly to the dejellied eggs, as the jelly ensures sperm binding and induction of acrosomal reaction (Pandian, 2019).

Oocyte hydration is another important factor that facilitates floatation of eggs. Interestingly, there is a succint difference in the timing of the hydration event. In pelagic spawning fishes, oocyte hydration occurs prior to spawning (Fig. 17.2A). But, it occurs immediately after spawning in penaeid shrimp, during embryogenesis in benthic spawning fishes (e.g. herring), toward hatching in brooding crustaceans. Devoid of pelagic larval stage, lobsters and semi-terrestrial *Ligia oceanica* release their young ones at an advanced stage as neonates. Hence, the level of hydration during embryogenesis is low in lobsters or almost nil in *L. oceanica*. In some pelagic spawning fishes, spawning occurs in multiple numbers of pulses but in a single bout in others. With multiple numbers of pulses, the pelagic spawning fishes have increased

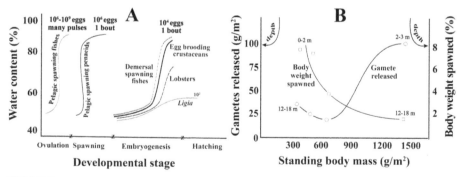

FIGURE 17.2

A. Hydration of oocytes/eggs: Pelagic spawning fishes (e.g. *Solea solea*, Pandian and Fluchter, 1968, panaeids (e.g. *Penaeus monodon, Metapenaeus monoceros*), benthic spawning fishes (e.g. *Salmo fario* [41% water], *S. irrideus* [66% water], Pandian and Fluchter, 1968), egg brooding crustaceans producing pelagic larvae (e.g. *Crangon crangon*, Pandian, 1967) and egg brooding lobsters (e.g. *Homarus* spp, Pandian, 1970a,b) and *Ligia* producing non-pelagic larvae, neonates (Pandian, 1972). B. Gametes released as a function of standing body biomass of the urchin *Strongylocentrotus droebachiensis* at different depths (drawn from data reported by Keats et al., 1984).

their fecundity to millions and billions, while in others, fecundity is reduced to a few millions, thousands or hundreds of eggs.

Two examples are described to show how adult density and man-made changes in sex ratio alter FS. In the field, biomass of the sedentary green sea urchin *S. droebachiensis*, changes at different depths (Fig. 17.2B). The quantum of gametes spawned—measured as of biomass—decreases from 8.0% at 0–2 m depth to 2.9% at 12–18 m depth. Correspondingly, the body weight of gametes spawned is also decreased from 101.5 g/m² at 2–3 m depth with 1,410 g/m² to 18.1 g/m² at 12–18 m depth with biomass of 522 g/m² (Fig. 17.2B). With increasing gametic biomass (density), up to 75 and 100% FS may occur at ~ 600 and 1,400 g biomass/m², respectively. Fishing pressure on large sized males of protogynic fishes leaves more females for every male. For example, in the black grouper *Mycteroperca bonaci*, for every male, there are 15.4 females in Florida, 33.3 in Cuba and 77.6 in Mexico (see Pandian, 2011). Large *Thalassoma bifasciatum* male milts 1,000 times more number of sperms than that of eggs spawned with the ratio of 1 egg : 1,000 sperms. (cf *Clarias gariepinus*, Runangwa et al., 1998). Alonzo and Mangel (2004) developed a model to show the effect of selective fishing of larger males and their consequent mortality (Fm) on the stock dynamics and sperm limitation in protogynic fishes. Accordingly, with availability of 5 billion sperms, 95–98% eggs spawned by 60 females can be fertilized. But only 85 and 75% eggs of 300 and 700 females can be fertilized, respectively. With increasing Fishing mortality (Fm) from 0.5 to 3.0, the number of eggs spawned remains constant; however, the FS/recruitment is reduced. A comparison of the scenario between gonochoric and protogynic fishes reveals that for every unit increase in FM, FS decreases by 8.5% in gonochores but only 7.6% in protogynics. Hence, the protogynics seem to have economized their sperm requirement.

17.2 External Fertilization—Surprises

Available FS values for external fertilizers are listed in Table 17.1; therein, the values are also grouped into (i) fully, and (ii) partially or (iii) spermatophore-mediated external transfers: The latter is further divided into crustaceans, which implant the spermatophore(s) precisely and those (polychaetes, gastropods), which implant them indiscriminately. Being costlier, the external route of gamete transfer—irrespective of EF or OF—is limited to sessile animals (12.5% of aquatic animals or 2.9% of all animals, see Table 24.2) and pelagic spawners. *Surprisingly, the external fertilization ensures > 80 to 100% FS in a wide range of aquatic animals from polychaetes to fishes.* Arguably, the route allows scope for a high level of genetic diversity. It is also true for the OF involving the transfer of sperms only; in them also, the FS ranges from ~ 60 to 100%. *The external and partially external routes provide considerable scope for*

TABLE 17.1

Fertilization success (FS) in some animals (from Rice et al., 2008, Pandian, 2011a, 2015, 2016, 2017, 2018, 2019, 2021, Robertson, 1990, Gould et al., 2001 and others)

Species	FS (%)	Species	FS (%)
Complete externals		**Partial externals**	
Polychaeta		**Polychaeta**	
Nereis virens (epitokous)	100	*Arenicola marina***	40–60
Riftia pachyptila (thermal vent)	86	*Galeolaria caespitosa*	60
Gastropoda		**Bryozoa**	
Placopecten magellanicus	93	*Hippodiplosia insculpta* (gonochore)	100
		Gymnolaemates (♀)	83–100
Bivalvia		*Crisrina corbicula* (brooder)	100
*Crassostrea gigas**	98		
*C. virginica**	77	**Partial Externals by spermatophore**	
Sipuncula[†]	-	**Crutacea:** Precise placement	
		Acartia tonsa	84
Echiura		*Euterpina acutifrons*	83
Lottia gigantea[††]	> 90	*Calanus helgolandicus*	81
Urochordata (solitary)		*Centropages typicus*	80
		Pareuchaeta norvegica	79
Pisces		*Thysanoessa raschii*	85
Acanthurus nigrofuscus	98	*Penaeus merguiensis*	78
Xyrichthys novacula	~ 97**	*P. japonicus***	68
Thalassoma bifasciatum	95**		
Oryzias latipes	100		
Fundulus heteroclitus	95	**Polycheata**	
Hippocampus kudo	92		
Danio rario	85	*Polydora cornuta*	63–99
Gobiocypris rarus	79	**Annelida:** Indiscriminate by placing[†]	
Amphibia		**Gastropoda:** Indiscriminate by placing[†]	
Uperoleia laevigata	> 90		
Echinodermata		* = Cryopreserved sperm	
Dendraster excentricus	100	** = in natural field	
Luida clathrata	100	†† = ammonia induced	
Strongylocentrotus droebachiensis	100	† = data not available	

not only diversity but also selection of the fittest gametes, especially the fertilizing sperms have to travel distances to reach the eggs retained in the parent body. More surprisingly, the spermatophore-mediated partial external route also ensures high FS. In crustaceans, the FS ranges from the field observed 68% in *Penaeus japonicus* to 84% in the experimentally observed copepod *Acartia tonsa*. It must be noted that (i) the percentage of hatching of *Calanus helagolandicus* was more or less equal in the laboratory-spematophored and ocean-spermatophored famales. (ii) The 84% FS value was arrived from *A. tonsa* maintained over seven years and extended for 32 filial generations (see Pandian, 1994).

17.3 Terrestrial Habitats

Terrestrial animals are characterized by internal fertilization. A fairly large volume of literature is available for terrestrial vertebrates and insects. Due to page limitation, the description is limited only to some insects. After successful completion of copulation, males tend to guard their mates against resemination by other males and ensure that their sperms alone are used to sire the eggs. Yet, females of many species welcome repeated insemination to (i) produce more genetic variations and (ii) acquire nuptial gift of nutrients. However, a single male can mate the same or other female(s) many times. The damselfly *Calopteryx maculata* male can mate once every 10 minutes without affecting his potency. With much of its storage, *Aedes aegypti* can mate six females in succession. The duration of copulation in *Helicornius erato* lasts for > 48 hours, that the male repeatedly reseminates his mate. In the fly *Scatophaga sterocoraria*, with increasing duration of copulation, 100% eggs are fertilized, only when the copulation lasts for 100 minutes.

The honey bee *Apis* maintains the sperms of each male separately. Of 23 females investigated, the sperms of the first male predominate in 11 females, and the last male in 5 females; in others, an equal volume of sperms from each male are used. In another extreme, sperms of the first mated males are displaced by the second male in *Calopteryx maculata*. About 82% of the offspring are sired by the second male in the bruchid *Callasobruchus maculatus*. But *Bombyx mori*, the first mated male has the priority of siring 76% offspring. By increasing mating frequency, many insects have explored to increase mating frequency and mating partners to acquire as much sperms as possible and nuptial gifts and to select the fittest sperms for fertilization (see Muthukrishnan, 1994).

18

Fecundity

Introduction

The fertilized eggs, collectively called fecundity, are a decisively important factor in recruitment and sustenance of population size. Fecundity ranges from one egg in the microscopic placozoan *Trichoplax adharens* (Pandian, 2021) to one billion eggs in the long living snookfish *Centropomus indecinalis* (Pandian, 2014). In this regard, the following may be noted: 1. *Barring parasitic clades, fecundity is limited to < 100 eggs in minor phyletics from Placozoa to Onychophora. In them, it depends more on structural complexity to support vitellogenesis rather than body size.* 2. Within the structurally organized fishes, it ranges from 0.15–2.00 egg/g fish/d in *Tilapia zilli*, 12–22 egg/g/d in *Betta splendens* and 24–56 egg/g/d in *Danio rerio*. But the large long living (175 years) *Hoplostethus atlanticus* spawn 0.00031–0.00038 egg/g/d. *Small annuals allocate more of their nutrient reserves to the reproductive output, whereas the long living perennials allocate more for somatic growth.* 3. Life history traits of large animals can broadly be divided into two types: (i) Type 1 Microphages inclusive of fluid suckers and (ii) Type 2 Macrophages—The reproductive characteristic traits of these are listed below:

Trait	Type 1	Type 2
Food	Macrophages	Microphages
Reserve	Liver, Hepatopancreas	Muscle, Viscera
Breeding pattern	Income breeder	Capital breeder
Fecundity	Indeterminate	Determinate
Behavior	Solitary	Gregarious, colonial

As > 80% of animals are macrophages (Table 4.5), the prevalence of income breeders is obvious. Aside from it, fecundity is also complicated by the existence of semelparity and iteroparity, brooding and viviparity, nurse eggs and adelphophagy and poecilogony, i.e. simultaneous release of lecithotrophic larger and planktotrophic small eggs in a few polychaetes and gastropods. It is the diversity that has attracted numerous studies on fecundity. *An important finding from most of these publications is that any life history trait or environmental factor that increases fecundity increases genetic diversity among progenies and accelerates species diversity.* Conversely, a feature like parthenogenesis may reduce genetic diversity and thereby decelerate species diversity.

18.1 Terms and Definitions

With regard to Fecundity (F), the terms used by fishery biologists have to be defined, and they can be adopted to other taxa as far as possible. Batch or clutch or Brood Fecundity (BF) is the number of eggs shed per spawning/clutch/brood. It is a function of body size (in Length, L or Weight, W) and is related to the volume of space available within the (intra-ovarian) ovary or (extra-ovarian) coelomic cavity. To accommodate maturing oocytes, geometry suggests ($F = {}_aL^b$ or ${}_aW^b$) that the length or weight exponent would be 3.0. However, the exponent may deviate from 3.0 due to changes in allometry; for example, the body shape is cylindrical in worms but dorso-ventrally flat in platyhelminths and many leeches. Within fishes alone, the exponent ranges from 1.0 to 5.0 but mostly falls between 3.25 and 3.75. In some animals, BF is better correlated with weight than length (e.g. *Oreochromis niloticus*). In iteroparous species, seasonal/annual spawning recurs in a single bout as in gastropods like *Patella*, *Helcion*, *Gibbula* and *Littorina neritoides* (see Pandian, 2017) or in 8, 22 and 41 pulses in fishes, *Cynoscious striatus*, *Brevoortia aurea* and *Serranus atricauda*, respectively (Pandian, 2011a). Hence, the seasonal (SF)/annual fecundity is estimated by BF × the number of spawning pulses. In the tropics (~ 12°N), six species of the anemone fish *Amphiprion* spawn in 12.5 pulse/y, with BF of 573 eggs and SF of 7,188 egg/y. But in the sub-tropics (30°N), three species of anemone fishes spawn 4.0 time/y with BF of 3,900 eggs and SF of 15,601 egg/y. Hence, *the warmer tropic seems to increase the spawning pulse/frequency but decreases BF and SF, in comparison to those of sub-tropics.* Lifetime Fecundity (LF) is estimated by SF multiplied by number of years during the reproductive life span. For example, Life Span (LS) of the limpet *Patella vulgata* is 14 years; maturing at the age of 3+ years, the limpet spawns once a year for the remaining 11 years. The freshwater mussel *Quadrula patulosa* also matures at the age of 3+ years, but spawns once a year for the remaining 47 years of its life (see Haag and Staton, 2003). Maturing at the age of 27.5 years, the orange roughy *Hoplotethus atlanticus* also spawns once a year for the remaining > 100 years of its life (Pandian, 2011a). Relative Fecundity (RF) is the number of eggs shed for a unit weight of a female. Unlike BF and LF, the RF provides scope for comparative analyses of reproductive performance at intra- and inter-specific levels as well as temporal and spatial levels.

Potential Fecundity (PF) is the maximum number of oocytes commencing to differentiate and develop. Due to a number of factors like food supply, only a fraction of the PF is realized. Of them, the following are described: (i) *Food quantity*: (a) Being the prime factor in determining the levels of PF and RF, reduced ration decreases the RF from 91% of PF in well fed herring *Clupea harengus* to 60% in those receiving low ration (Ma et al., 1998). (b) Similarly, reduction in ration from *ad libitum* to 60% reduces PF from 3,581 at *ad libitum* to 2,033 eggs in the prawn *Macrobrachium nobilii* (Kumari and Pandian, 1991). (c) Reduced feeding frequency and consequent reduced ration not only reduce RF but also the egg size. For example, reduction in ration from *ad libitum* to 28% reduces spawning frequency from 30 times to 12, LF from 2,610 eggs to 727 eggs and egg size from 0.30 mg dry weight to 0.27 mg in the stickleback *Gasterosteus aculeatus* (see Table 4.1). Table 18.1 lists selected factors that alter realization of PF into RF in molluscs.

(ii) *Food quality*: Fed on cereal, spinach and tetramin, the polychaete *Dinophilus gyrociliatus* realizes the maximum RF of 4, 6 and 12 eggs, respectively (Prevedelli and Vandini, 1999). In the dipteran fly *Dacus dorsalis*, its ability to realize RF from PF decreases from 20.8×10^4 to 14.0×10^4 and 13.4×10^4 eggs, when fed on orange, papaya and guava juice, respectively (Vargas and Chang, 1991). For the parasitic hymenopteran ichneumonid, food quality decreases from the host egg to its larva and imago. Accordingly, the realized number of potential ovarioles decreased from 30, when the ichneumonid oviposited on a host egg to five with oviposition on a freshly emerged imago (Price, 1975). (iii) *Other factors*: In intra-ovarian (e.g. *Veneriserva pygoclava meriodionalis*, Fig. 18.1A) and extra-ovarian oogenesis (e.g. *Spirorbis spirorbis*, Fig. 18.1B) patterns, the number of pre-vitellogenic oocytes decreases to < 10% vitellogenic eggs in the former and 42% in the latter. The semelparous cephalopod *Ocythoe tuberculata* realizes only ~ 11.5% of PF, irrespective of changes in its size from 36 g to 126 g (Tutman et al., 2008). In the sea urchin *Strongylocentrotus droebachiensis*, the trends for maximum and minimum RF as a function of body weight run parallel throughout the tested range of body size. The maximal RF trend indicates that the urchin was under a favorable condition. Incidentally, a presumed trend for the PF is also shown in Fig. 18.1C. For more information on the RF of a dozen echinodermate species under favorable conditions, Table 1.14 of Pandian (2018) may be consulted. Reared at low (1,000/l) and high (9,000/l) densities, the egg shedding *Parvocalanus crassirostris* reduced RF from ~ 5 egg/♀/d to < 1 egg/♀/d. The reduction in RF may be more due to accumulation of nitrogenous excretory products than food supply (Alajomi and Zeng, 2014). With assured food supply, the parasites from different phyla are able to realize 10–1,000 times more RF, in comparison to the respective free-living counterparts (Fig. 13.1). In insects like *Scatophaga stercoraria*, RF rises from 30 to 100 eggs with increasing duration of copulation (Parker, 1978).

An interesting explorative study has revealed how the host density affects reproduction in the grasshopper *Melanoplus dawsoni*, a small hopper with

TABLE 18.1

Effect of selected factors affecting fecundity of molluscs (compiled from Runham, 1993 and updated by Pandian, 2017)

Species	Reported observations
Sexuality	
Lymnaea stagnalis	Rearing in groups facilitated cross breeding and advanced egg laying but reduced fecundity by 50%. Self fertilization delayed egg laying but doubled fecundity
Eubranchus doriae	Isolated aeolids died without laying eggs. When paired, fecundity depended on volume of water. The more crowded, the aeloid was less fecund
Bulinus contortus	Aphallics were more fecund (5609 eggs) than euphallics (4498 eggs)
Food Availability	
Thais lamellosa	Food abundance increased fecundity from 930 to 1428 eggs
T. emarginata	Food abundance increased clutch number and fecundity
Kathurina tunicata *Patella vulgata*	Despite 5 months starvations, gametes were produced reducing somatic growth
Mytilus edulis	Equal amount of gametes produced at the expense of somatic growth
Lymnaea stagnalis	Starvation delayed vitellogenesis
Biomphalaria glabrata	Prolonged starvation ceased oviposition and resorption of oocytes
L. peregra	Vitamin E profoundly stimulated egg production
Geukensia demissa	Spinach produced more eggs than algal food
Chromytilus meridionalis	Gametogenesis delayed in populations occupying high tides with less accessibility to water and food
Euprymna tasmanica	High mortality of eggs spawned by mothers fed on low ration
Temperature	
Lymnaea stagnalis	Increase in temperature stimulated oviposition
Macoma balthica	Increase in temperature stimulated spawning
Cardium edule	Temperature –14°C was required to stimulate gametogenesis
Photoperiod	
Melampus bidentatus	Photoperiod of 14 hours was required for gonad maturation
Lymnaea stagnalis	Long day (16 L : 8 D) snails were 9-times less fecund due to earlier death than short day (8 L : 16 D) snails due to their longevity
Sepia officinalis (Mangold, 1987)	Short photoperiod induced precocious sexual maturity. Light intensity inhibited maturation
Density	
Bulinus tropicus	High densities reduced fecundity due to accumulated waste
pH	
Amnicola limosa	Fecundity decreased by 66% in acidic waters (pH 5.8) containing < ~ 1 mg calcium/l, a critical limit for egg laying, in comparison to alkaline waters with pH 7.6
Shell Thickness	
Nucella emarginata	Thin shelled morphs were more fecund than thick shelled morphs

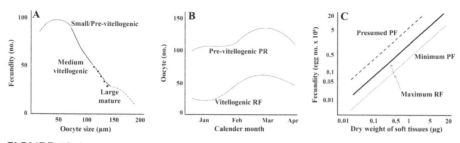

FIGURE 18.1

A. Progressive reduction in Realized Fecundity (RF) from pre-vitellogenic Potential Fecundity (PF) and concomitant increase in oocyte size of the intra-ovarian polychaete *Veneriserva pygoclava meridionalis* (modified from Micaletto et al., 2002). B. Reduction from pre-vitellogenic PF to vitellogenic RF during successive calendar months in the extra-ovarian polychaete *Spirorbis spirorbis* (modified from Daly, 1978). C. Minimum and maximum RF as a function of body size in urchin *Strongylocentrotus droebachiensis*. Presumed PF is shown in dotted line (modified from Thompson, 1979).

five nymphal instars lasting for 21–66 days. Its field density ranged from 6.7 to 8.9 hopper/m². Correspondingly, the mermithid prevalence also increased from 15 to 37.5% with an average of 27.5%. Experimental alterations of hopper density to 0.25, 0.5, 1.0, 1.25 and 1.5 times, considering field density of 8.9 hopper/m² as 1.0, resulted in reductions in the number of ovarian follicle (PF) and egg (RF)/♀ as functions of hopper density and parasitism. PF decreased from ~ 10.05 follicle/♀ at 0.25 hopper/m² to ~ 8.5 follicle/♀ at 1.0 hopper/m². Correspondingly, the realized fecundity also decreased from ~ 7.5 egg/♀ to 5.5 egg/♀, i.e. the reductions were 20% for potential fecundity but 27% for realized fecundity. Increase in hopper density reduced the resource availability to realize potential fecundity. At the mermithid-infected hoppers at 0.25 hopper/m² density level, the reduction was from 11 follicle/♀ to 8.25 follicle/♀, while it was from 5.75 egg/♀ to 1.5 egg/♀ at the hopper density of 1.0/m², i.e. the reduction was 20% for potential fecundity but as high as 27% for realized fecundity. A combination of hopper density and mermithid parasitism reduced potential fecundity from 10.25 follicle/♀ to 8.5 follicle/♀ and realized fecundity from 7.5 egg/♀ to 1.5 egg/♀. On the whole, the combination of increased density and parasitism reduced PF marginally by 20% but the RF by 82%. Hence, RF is more severely affected by parasitism than PF (Laws, 2009, see also Pandian, 2021).

18.2 Fecundity-Body Size

In major phyletics and eucoelomate minor phyletics, (1) the body size is the prime factor that governs a positive linear relation between it and fecundity. Irrespective of the body size being considered as age, length, width, segment number (annelids), carapace width (crabs), shell width (e.g. oyster) or weight, the relation holds good and it is equally good, when fecundity is expressed in number of eggs or cocoons (e.g. clitellate annelids) or capsule volumes in murician gastropod *Nucella crassilabrum*. The highest fecundity ranges from a few eggs to ten thousands in the acnidarian *Mnemiopsis macradyi* (Baker and Reeve, 1974) to a few thousands in penaeid shrimps (Fig. 18.2A) and to a few millions in pelagic spawning fishes. The larviparous oyster *Ostrea edulis* spawns twice a year, the first during summer and the second during September. The fecundity rises with increasing shell length in the 2nd, 3rd and 4th year old oyster (Fig. 18.2B). However, the positive linear relationship is altered by following factors: (2) The egg size is a prime factor that alters the levels in the linear positive relation between body size and fecundity in pelagic and benthic spawners. Typically, penaeids are pelagic spawners; in them, BF ranges from 36,500 eggs in the small (~ 10 cm body length) *Penaeus stylifera* to million eggs in the large (20 cm long) *Penaeus monodon* (Fig. 18.2A). In them, the body size and egg size determine the level of fecundity.

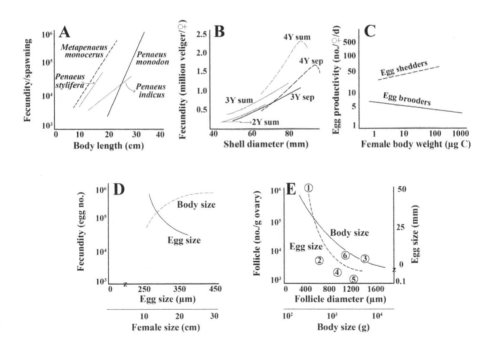

FIGURE 18.2

A. Batch fecundity as a function of body size in some broadcast spawning penaeids (CMFRI, 2013). B. Seasonal fecundity as a function of body size and age in summer- and autumn-spawning larviparous oyster *Ostrea edulis* (modified from Walne, 1964). C. Egg productivity as a function of body weight in egg-shedding and brooding crustaceans (modified from Kiorboe and Sabitini, 1995). Fecundity as functions of egg size and body size D. in some penaeids and E. teleost fishes (1 = *Serranus atricauda*, 2 = *Micromesistius australis*, 3 = *Gadus morhua*, 4 = *Cynoscion nebulosus*, 5 = *C. regalis*, 6 = *Hoplostethus atlanticus*).

For example, the egg size is small (260 μm) in larger *P. monodon* but large (380 μm) in smaller *P. stylifera*.

(3) *Shedders vs brooders*: A vast majority of crustaceans are egg brooders; however, calanoid copepods, penaeid shrimp and some euphausiids are egg shedders; on an average, the egg shedders spawn (296 eggs) nearly three times more number of eggs than that (96 eggs) of egg brooders (see Pandian, 1994). Compiling relevant information for the anecdysic 16 species of egg brooders and 29 species of egg shedders among copepods, Kiorboe and Sabatini (1995) found that productivity was 5.3 and 40.0 egg/♀/d in egg brooders and egg shedders, respectively. In other words, there are 7.5-fold differences in the number-specific fecundity and 2.5-fold difference in weight-specific fecundity. *Hence, there is a clear tendency for increase in egg productivity with increasing body size among egg shedders, whereas there is a decreasing one among egg brooders* (Fig. 18.2C). On the whole, egg brooders are trapped in reproductive senescence, while egg shedders have escaped from it (Pandian, 2016). In most crustaceans, the brooded eggs, following the completion of embryogenesis, are released either as non-feeding (e.g. nauplius) or feeding (e.g. protozoea) larvae. However, brooding lasts until the completion of the entire development and only the fully developed neonates are released in isopods and amphipods. In them, the relation *per se* increases with successive broods only upto the mid body size or middle age and subsequently decreases (Fig. 18.3A). Parthenogenesis turns the trend to an inverted U-shape (Fig. 18.6A, B), which is described in the ensuing pages. In the hermaphroditic viviparous polychaete *Neanthes limnicola*, the relation becomes almost horizontal (Fig. 18.3B). On enforced isolation of the hermaphroditic earthworm *Lumbricus terrestris*, the relation is linear but negative for want of adequate sperms (Fig. 18.3C). Hence, a number of factors alter the body-size-fecundity relation from linear and positive to linear and negative.

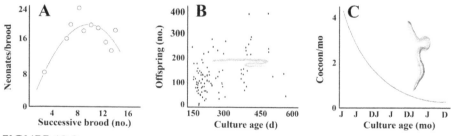

FIGURE 18.3

A. Neonate fecundity as a function of successive broods in a typical peracarid isopod *Quadrivisio bengalensis* (K.C.C. Nair and K.V. Jayalakshmy pers. comm). B. Offspring number as a function of age in self-fertilizing viviparous hermaphroditic *Neanthes limnicola* (modified from Fong and Pearse, 1992). C. Fecundity as a function of age in isolation enforced self-fertilizing hermaphroditic earthworm *Lumbricus terrestris* (modified from Lowe and Butt, 2005).

(4) *Hydration*: As described earlier (Fig. 17.2A), the timing of oocyte hydration significantly alters fecundity relation to the egg size and body size. Fecundity is decreased from 10^6–10^9 eggs in pelagic spawning fishes, in which the hydration occurs prior to spawning to less than a million in pelagic spawning penaeid shrimp, in which eggs are hydrated immediately after spawning. The hydration occurs during embryogenesis in benthic spawning fishes and toward the termination of embryogenesis in brooding crustaceans; these have only a few thousands eggs. In lobsters and *Ligia*, the level of hydration is decreased (Fig. 17.2A). This difference in timing of hydration in fishes and penaeids has a profound effect on fecundity–egg size–body size relation. As fully developed eggs are hydrated within the body of pelagic fishes, the trends for relative fecundity decreases with increasing follicular size and body size (Fig. 18.2E). Contrastingly, penaeid eggs are hydrated after spawning; in them, with increasing body size, fecundity increases, as expected but egg size decreases (Fig. 18.2D). Available values seem to suggest that more than the body size, the egg size determines fecundity.

(5) *Temperature, GT and RLS*: In females, oogenesis commences on attaining puberty. It may be ceased after a period and the female may pass through a period of menopause. Hence, the Life Span (LS) of an animal can be divided into (i) Generation Time (GT), when oogenesis has not yet commenced, (ii) Reproductive Life Span (RLS), when oogenesis recurs in iteropares and (iii) Post Reproductive Phase (PRP)/menopause, during which oogenesis is ceased. The fraction of GT as a percentage of LS may be reduced in animals encountering reduced food supply, increased predation and other environmental factors like latitude and related temperature. Due to page limitation, only two examples are described.

Being characteristic of arthropods, many crustaceans molt to grow and/ or reproduce at progressively extended intervals during their entire lifetime. Molting and spawning are two major events that involve cyclic mobilization of nutrient reserves from storage sites to the epidermis and gonad, respectively. Hence, these events determine the reproductive patterns in crustaceans. A remarkable publication by Beladjal et al. (2003b) reported that temperature elevation progressively decreases Generation Time (GT) in the Morrocan anostracan *Tanymastigites perrieri*; its GT decreases from 52% at 10°C to 21, 13, 14% and 9% at 20°C, 25°C, 30°C and 35°C, respectively. Available data for copepods indicate that GT ranges from 75% (e.g. *Limnocalanus johanseni*) to 90% (e.g. *Diaptomus sicilis*) for arctic copepods and 41% for *Mesocyclops leukarti* inhabiting Lake Kinnert (33°N). These findings hold good for amphipods also; in *Gammarus pulex*, the duration of GT decreases from 52 weeks at 5°C to 22, 17 and 3 weeks at 10, 15 and 20°C, respectively (Sutcliff, 2010). The GT of temperate peracarids average 46% for 16 isopod species and 54% for a dozen amphipod species (Pandian, 1994). For tropical amphipods, the GT values range from 15% in *Melita zeylandica* to 26% in *Quadrivisio bengalensis* (see Pandian, 2016). Arguably, tropical crustaceans grow faster and are more fecund than their temperate and arctic counterparts.

Life Span (LS) of crustaceans ranges from 5–6 days in cladoceran *Moina macrorura* to > 100 years in *Homarus americanus*. Limited information collected from several sources led Pandian (2016) to propose a model for GT as a fraction of LS of crustaceans. For the LS-GT model, the following were considered: (i) The LS averages a few days for branchiopods, cladocerans and copepods, a few months for isopods and amphipods, and a few years for most decapods. (ii) The model has also taken into consideration for parthenogenic, diecdysic and anecdysic crustaceans. (iii) Their respective averages were taken as points to draw the slanting lines. From Figure 18.4A, the following may be inferred: Within each of these groups, the LS-GT line is shorter and sharper for tropical crustaceans than those for the temperate and arctic counterparts. Accordingly, the horizontal trends are progressively shifted toward the right. Hence, the GT progressively increases from ~ 13–20% of LS for tropical crustaceans to ~ 50% for temperate and > 85% for arctic crustaceans.

The observed relation for GT has implication to RLS and consequently fecundity. For example, the RLS as a percentage of Life Span (LS) is 49, 69, 73 and 79% for the anostracan crustacean *Branchinectes orientalis*, *Tanymastigites perrieri*, *Branchipus schaefferi* and *Streptocephalus torvicornis*. The anostracods can produce a subitaneous egg and dormant cysts. In them, the number of clutches increases from 5.5 with 49% RLS in *B. orientalis* to 13.3 with 73% RLS in *B. schaefferi*. With increase in RLS, the number of cysts per clutch as well as the cumulative number of cysts increase (Fig. 18.4B), i.e. the longer the RLS, the greater is the fecundity. Another consequence of decrease in GT is the extension of RLS and increased fecundity. For example, the number of egg-bearing molt (indicated by inverted arrows) is decreased from 13 times in tropical daphnids to 4.4 and ~ 1 time(s) in their temperate and arctic counterparts. In fishes, the duration of gestation/external brooding also alters RLS. In *Poecilia reticulata*, gestation covers 15–20% of its lifetime but it parturiates juveniles that commence feeding immediately after their release (Fig. 18.4C). Brooding eggs in its bubble nest, the fighting fish *Betta splendens* requires 10 and 25% of its life span for incubation and GT. The long living *Salvelinus fontinalis* has a life span of 20 years, of which 50% is spent on the duration of incubation, non-feeding alevin, feeding fry and GT.

For different classes of Platyhelminthes, values are available for body size (Fig. 18.5A) and life span (Fig. 18.5B). Hence, these values provide an opportunity to test whether body size or GT determines fecundity. Thanks to the compilation of the then available data on LS and GT by Trouve et al. (1998) for 8 turbellerian, 10 monogenean, 16 digenean and 20 cestode species, the mean values for LS and GT were calculated. The GT as a fraction (%) of LS averages 47.7, 26.3, 17.6 and 12.3% for the turbellarians, monogeneans, cestodes and digeneans, respectively. The estimated mean fecundity values for these groups are 0.12, 9.6, 4,088 and 223,124 eggs. For platyhelminths, fecundity ultimately depends on body size, although GT may have a modifying role (Fig. 18.5C)

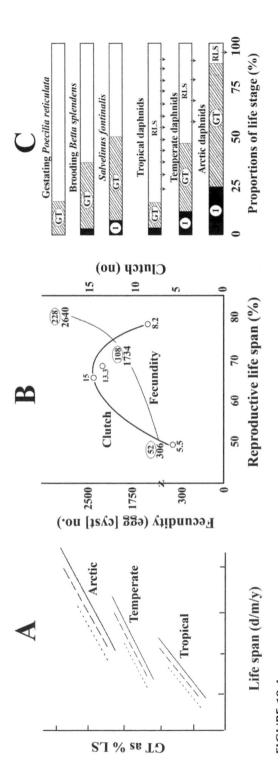

FIGURE 18.4

A. Generation time as percentage of life span in arctic, temperate and tropical crustaceans. Solid line = diecdysic brooders, broken line = anecdysic egg shedders and dotted line = parthenogenic brooders (modified from Pandian, 2016). B. Fecundity and cyst per clutch (within circles) and the number of clutches produced by four fairy shrimp species as a function of reproductive life span (drawn from data reported by Beladjal et al., 2002, 2003a, b, Atashbar et al., 2012). C. Effect of temperature on proportions of incubation (I), Generation Time (GT) and Reproductive Life Span (RLS) in arctic, temperate and tropical daphnids. Effect of Gestation (G), external brooding and free spawning on proportion of life stages in fishes (compiled from Pandian, 2016 and others).

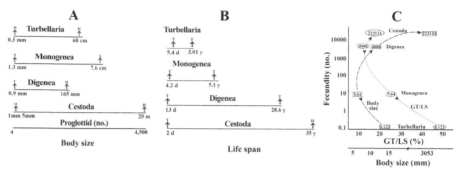

FIGURE 18.5

A. Body size and B. life span of platyhelminthic taxa. H = Hyman (1951b), T = Trouve et al. (1958), S = Sasal et al. (1999), M = Mackiewicz (1988). C. Fecundity as functions of body size and Generation Time (GT) as a percentage of Life Span (LS) in four classes of Platyhelminthes (drawn from data reported by Trouve et al., 1998).

(6) *Semelpares vs Iteropares*: In the semelparous polychaete *Nereis virens*, fecundity runs to million eggs but to a few thousands in the iteropare *Marphysa sanguinea*. With the life span of 6 months, the semelparous fish *Nothobranchus raschi* matures at the age of 2 months but spawn towards the end of the 5th month and dies at the age of 6 months (Pandian, 2014). (7) *Smaller vs Larger Size*: In small fishes and snails, in which spawning frequency occurs as frequent as hourly, daily or fortnightly. Among them, the trend can be almost horizontal, as in Owens pupfish *Cyprinodon radiosus*, which spawns 74 eggs after 200 courting/d (Mire and Mullett, 1994); it may be doom-shaped, as in the hermaphroditic self-fertilizing snail *Lymnaea palustris* (see Pandian, 2017). As indicated earlier, these are small snails, which allocate more for the reproductive output than somatic growth. (8) *Ploidy and Parthenogenesis*: For the body size-fecundity relation, parthenogens display a bow-like inverted U-shape trend, irrespective of the quantity (Fig. 18.6A) and quality (Fig. 18.6B) of food. Interestingly their fecundity rises with increasing algal size (Fig. 18.6C) but decreases with increasing density (Fig. 18.6C). Two reasons have been proposed to explain the inverted U-trend. Among them, the following, proposed for the first time, is more important. By producing diploid eggs, parthenogens are not able to double their egg size. For example, the tetraploid and hexaploid loach *Misgurnus anguillicaudatus* is unable to double and treble its egg size. Considering the size of 1 mm diameter in a diploid egg, the increase in egg size is merely 1.25 mm (37%) and 1.44 mm (44%) of the expected size (Table 18.2). Unlike in plants, increase in ploidy level is not followed with proportionate increase in cell number in animals, albeit cell size may increase. For example, 54% reduction in cumulative cell number occurs in 5-hours old tetraploid *Ctenopharyngodon idella* alevin, in comparison to that of the diploid (Cassani, 1990). Consequently, vitellogenesis in polyploids

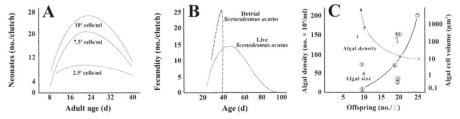

FIGURE 18.6

Parthenogenesis and fecundity: A. Inverted U-shaped trends for fecundity as a function of food quantity in *Ceriodaphnia cornuta*, B. food quality in parthenogenic copepod *Elaphoidella grandidieri* fed *Senedesmus acutus*. C. Effect of algal density and size on fecundity of the rotifer *Brachionus plicatilis* (modified and simplified from Richman, 1958, drawn using data reported by Nandini et al., 2011, Hirayama et al., 1979). 1 = *Nitzchia clostrium*, 2 = *Monochrysis lutheri*, 3 = *Dunaiella tertiolecta*, 4 = *Chlamydomonas* sp, 5 = *Cyclotella cryptica*, 6 = *Chlorella* sp, 7 = *Synechococcus elongatus*, 8 = *Eutreptiella* sp.

is not supported by the proportionate increase in number of vitellogenesis supporting somatic cells. As a result, the egg size is not proportionately increased in polyploid and diploid eggs produced by parthenogens. A second reason is that the parthenogens may progressively accumulate increasing number of deleterious mutations. For example, by undertaking up to 780 parthenogenic generations *Daphnia magna* may accumulate considerable deleterious genes. Not surprisingly, irrespective of changes in temperature and food supply (Fig. 18.6A, B), the inverted U-trend is maintained for the body size-fecundity relation. As a consequence, the older or larger females have a post-reproductive period of menopause. The parthenogenic marbled crayfish *Procambarus fallax* pass through menopause for 7–21% of its LS (Pandian, 2016).

TABLE 18.2

Effect of ploidy level on body weight, fecundity and hatching in *Misgurnus anguillicaudatus* (compiled from Arai et al., 1999)

Trait	Diploid	Tetraploid	Hexaploid
Body weight (g)	17.6	14	12
Lifetime fecundity (no.)	4,106	1,834	1,544
Relative fecundity (no./g)	238.0	140	127
Egg size (mm)	1	1.3	1.4
Hatching (%)	25.4	35.1	54.1
Survival at hatching (%)	4	66.4	83.2
Net progeny (no.)	669	374	698
Recruitment (%)	16.3	20.4	45.2

From the foregone description, the following may be summarized: (i) Environmental factors like adequate food supply and quality food, as well as favorable conditions enhance the scope to increase RF from PF. These factors may increase fecundity and genetic diversity among progenies, and thereby accelerate species diversity. In parasitic species from different phyla, an assured food supply dramatically increases fecundity including in eutelic nematodes. Iteropares characterized by greater LF and multiple spawning frequency distribute the eggs spatially and temporally and thereby enhances genetic diversity and species diversity. Most pelagic spawners are prolific and produce a large number of smaller eggs. In contrast, the demersal spawners shed fewer but larger eggs. The former increases the diversity and allows the scope for selection of the fittest to be recruited. It fosters species diversity. In a terrestrial habitat, motility is the prime factor in mate searching and selection. The greater the motility, the higher is the chance for siring more progenies. A shorter GT and longer RLS increase fecundity. Conversely, life history features like selfing hermaphroditism, diploid egg producing mitotic parthenogens and clonal species, in which sexual reproduction is either minimized or almost eliminated, fecundity is reduced leading to decrease in genetic diversity and diminishing the scope for species diversity. Not surprisingly, the cumulative species number of selfing simultaneous hermaphrodites, mitotic parthenogens and obligate clonals does not exceed ~ 4% of all animals.

B3
Embryogenesis and Development

Reproduction is central to all biological events. In many animals, the life cycle is indirect and involves one or more larval stages; in parasites, the cycle may incorporate one or more Intermediate Host(s) (IH). Hence, this account covers from embryogenesis to recruitment. For sessile and sedentary animals, the indirect cycle involving larval stage(s) is obligatory for dispersal. In them, production of a large number of smaller eggs ensures genetic diversity and survival of the fittest. The broadcasted eggs are a delicacy and are predated heavily. To avoid predation, an escaping route is to generate a small number of larger eggs with short non-feeding dispersive larvae. A second route is to produce less number of larger and smaller eggs that are brooded or gestated. With smaller number of larger lecithotrophoric or brooded eggs, development may be safer. But it reduces genetic diversity and hence species diversity.

19

Direct and Indirect Life Cycles

Introduction

In a direct life cycle, more or less fully developed neonates/juveniles are hatched or parturited. Instead, some animals produce larva(e), which differs from parents in appearance, feeding habit and in some cases develop in a different habitat. Fertilized eggs may be deposited or broadcasted followed by no parental care. They are a delicacy and highly nutritive. They pass through an indirect development. Many predators await spawning plume to feed on broadcasted eggs. A few nemertean species (e.g. *Carcinonemertes*) are narrowly specialized that they can feed only on berried eggs of crustaceans (Pandian, 2021). Unfortunately some eat their own eggs. For example, the cardinalfish *Apogon lineatus* cannibalistically devour 30% of their own eggs (Barlow, 1981, see also Vagelli, 1999). In this trade off, some animals produce a large number of smaller eggs followed by feeding and dispersive larval stage, while others produce a small number of larger eggs, and their non-feeding larvae pass through a relative shorter duration of a dispersive stage. Producing even less number of eggs, still others brood or gestate their eggs.

19.1 Taxonomic Larval Distribution

For the first time, the pictures for almost all the larvae, which are produced by animals that pass through an indirect life cycle, are assembled in Fig. 19.1. Notably, some taxa produce only one type larva, e.g. trochophore in polychaetes and molluscs, tadpole in amphibians. But the larva can be a nauplius, metanauplius or protozoea in crustaceans, and doliolaria, auricularia, bipinnaria, echinopluteus and ophiopluteus in crinoids, holothurians, asteroids, echinoids and ophiuroids, respectively. Interestingly, "the crustacean nauplia are the most abundant type (by number) of multicellular animals on earth" (Fryer, 1997). In Porifera, the larva can be a coeloblastula or amphiblastula. In parasitic flukes, a series of larvae is

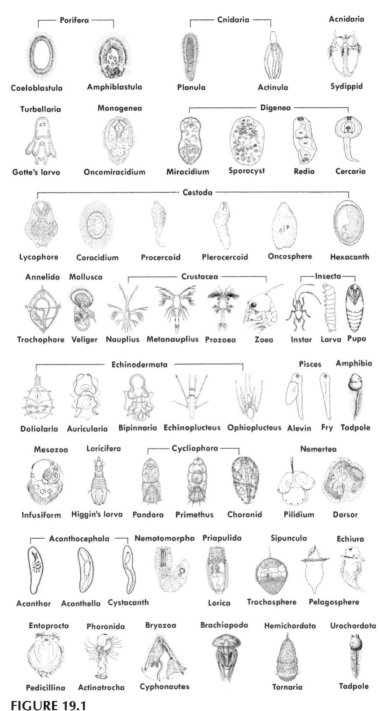

FIGURE 19.1

Larvae in different life cycles in animal phyla.

incorporated to prolong larval duration and appropriate infection at a greater risk. In monogeneans, the presence of a single oncomiracidium larval stage with high level of species specificity (stenoxenic) has not facilitated species diversity. But (i) flexible specificity (oioxenic) for both intermediate (IH) and Definitive Host (DH) and, (ii) incorporation of a minimum four larval stages namely (a) miracidium, (b) sporocyst, (c) redia and (d) cercaria and (iii) clonal multiplication of sporocyst and redia at the cost of IH have increased the digenean species number to 12,012 and distribution from aquatic to terrestrial hosts in Digenea. In the 4,671 speciose cestodes, self-fertilization seems to have limited species diversity. *Despite the choice limited to only one IH, the herbivorous food chain has exploded the diversity to 2,264 in Taenioidea alone.* But the choice of carnivorous food chain has reduced diversity to 1,266 species among a half dozen orders of aquatic cestodes (see Pandian, 2021).

19.2 Quantification

Estimation on the number of species passing through a direct or indirect life cycle was not difficult. Only two compromises had to be made. For Nemertea, Cantell (1989) indicated the occurrence of direct and indirect life cycle in 18 and 9 species, respectively. Hence, 867 species were assigned to direct life cycle (Table 19.1). In 2,000 speciose polycladid turbellarians, Rawlinson (2014) hinted at direct and indirect cycles in 4 and 8 species. Further, Norena et al. (2015) reported the exit of gametes in 2,050 turbellarian species belonging to Prolecithophora, Lecithoepitheliata, Typhoplanoida, Dalyelloida, Catenulida, Acoelomorpha, Proceriata and Macrostomatomorpha. The indirect life cycle is the rule in all parasitic Platyhelminthes. Considering these reports, 2,113 turbellarians species were assigned to direct life cycle and the remaining 25,587 species to indirect life cycle. For polychaetes, the values are drawn after due consideration to the survey reports of Wilson (1991) and Carson and Hentschel (2006). Among crustaceans, isopods (10,300 species minus 85 cryptonicoid parasite with indirect life cycle), amphipods (1,900 species) and cladocerans (620 species) pass through direct life cycle. All the other aquatic animals pass through indirect life cycle. Primitive insects are known to display an ametabolous direct life cycle. Metamorphosis is gradual and does not involve pupal stage in hemimetabolous insects comprised in 16 orders. On the other hand, holometabolous insects in 11 orders undergo visibly observable metamorphosis through the pupal stage. The species number for these insects was drawn from Blackmon et al. (2017). All values for Mollusca were drawn from Pandian (2017) and others. Remarkably, almost all sessile phyla in the Lophophorata and Urochordata undertake indirect life cycle (Table 19.1). Among minor phyletics, the parasitic Mesozoa, Myxozoa, Nematomorpha, Acanthocephala and Pentastomida pass through indirect

TABLE 19.1

Taxonomic distribution of direct and indirect life cycle in major and minor phyla (compiled from Pandian, 2011a, 2016, 2017, 2018, 2019, 2020, 2021, Blackmon et al., 2017)

Phylum	Indirect	Direct	Phylum	Indirect	Direct
Porifera	8,549	34	Placozoa	0	1
Cnidaria	10,852	4	Mesozoa	150	0
Acnidaria	166	0	Myxozoa	2,200	0
Platyhelminthes	25,587	2,113	Loricifera	34	0
Polychaeta	12,296	766	Cycliphora	2	0
Clitellata	0	3,849	Nemertea	433	867
Crustacea	41,649	12,735	Ganthostomulida	0	100
Myriapoda	0	11,880	Rotifera	0	2,031
Insecta	903,420	4,370	Gastrotricha	0	813
Chelicera	0	155,760	Kinoryncha	0	200
Pulmonata	0	24,000	Nematoda	0	27,000
Opisthobranchia	75	1,925	Nematomorpha	360	0
Others	80,000	0	Acanthocephala	1,100	0
Bivalvia	9,580	~ 6	Priapulida	19	0
Cephalopoda	0	800	Sipuncula	160	0
Scaphopoda	570	0	Echiura	230	0
Polyplacophora	930	0	Tardigrada	0	1,047
Others	0	293	Onychophora	0	200
Echinoderamta	6,941	59	Pentastomida	144	0
Pisces	31,933	1,857	Entoprocta	200	0
Amphibia	5,009	0	Phoronida	23	0
Caecilia	67	72	Bryozoa	4,856	844
Reptilia	0	9,545	Branchiopoda	391	0
Aves	0	10,038	Chaetognatha	0	150
Mammalia	0	5,513	Hemichordata	125	7
Subtotal	1,137,624	245,619	Urochordata	2,900	100
%	82.2	17.8	Subtotal	13,327	33,360
			%	28.5	71.5
Total	1,150,951	278,979	%	80.5	19.5

life cycle. The only exceptions are the parasitic nematodes, in which the life cycle is direct, as in their free-living counterparts.

The incidence of a direct and indirect life cycle occurs across almost all major and minor phyla, perhaps indicating their origin and disappearance in different taxa at various geological times. There are taxa like Clitellata among Annelida, ametabolous insects and Chelicerae among Arthropoda, Pulmonata and Cephalopoda among Mollusca, which pass through direct life cycle. So are the free living pseudocoelomates. Almost all members of Porifera, Cnidaria and Echinodermata are characterized by an indirect life cycle.

These estimates have revealed that 1,137,624 or 82.2% major phyletics pass through indirect life cycle (Table 19.1). Contrastingly, the corresponding value for minor phyletics is 28.5% only. The direct life cycle involving relatively less number of progenies is another reason for the reduced species diversity among minor phyletics. On the whole, of 1,429,930 species, for which information could be assembled, 80.5 and 19.5% animals are characterized by indirect and direct life cycle, respectively. *Taxa characterized by indirect life have greater scope to produce relatively more number of eggs. In them, the increasing genetic diversity and survival of the fittest accelerate species diversity.*

20

Brooding and Viviparity

Introduction

In animals, parental care commences from an egg guarding at one end of a spectrum to brooding, paternal gestation (e.g. pipefishes) and matrotrophic viviparity at the other end of it. Brooding allows protection in nest holders (e.g. some fishes) and ventilation in others (e.g. decapod crustaceans) and the delivery of nutritive fluids in ovoviviparous in yet others (e.g. Cladocera: *Moina, Polyphemus*). Studies on intra- and inter-specific translocation of brooded embryos from a brooding mother to another reveal that isopods like *Gammarus micronatus* can recognize conspecific embryos; *Corophium bonnelli* can distinguish the translocated equal-sized embryos of *C. volutator* and *Lembos webskeri* and replace them with conspecific embryos (Shillaker and Moore, 1987). Comparing yolk utilization efficiency of *Ligia oceanica* with that of some decapods, Pandian (1972) considered that some of these ovoviviparous crustaceans may receive up to 10% maternal nutritive fluids. In this account, behavioral brooding, as it occurs in reptiles, birds and others, is not considered. Only animals, in which eggs are brooded in a specific structure, are considered as brooders. A detailed analysis of Thorson (1950) revealed that fecundity of marine benthic invertebrates is in the range of 20 to 800 progenesis, 100 to 6,000 and 2,000 to a billion (including *Piaster ochraceus*, see Table 20.1, fish *Centropomus indecinalis*, Pandian, 2014) for vivipares, lecithotrophs/brooders and free-spawners, respectively (Fig. 20.1). Keller et al. (2007) reported that fecundity decreases from 4,355 eggs in gonochoric molluscs (e.g. gastropod *Pomacea canaliculata*) to 56 offspring in viviparids (e.g. *Viviparus georginus*). The broadcast spawning crustaceans produce 3–20 times more eggs than their respective brooding taxonomic counterparts (Pandian, 2016). This may also hold true for terrestrial animals, albeit the values for each of the three groups may not be as high as in those of aquatic animals. Representative examples on the effect of lecithotrophy (LEC), brooding and viviparity on egg size and fecundity of a few echinoderms, crustaceans and molluscs are listed in Table 20.1. For example, egg-shedding *Calanus finmarchicus* spawn 14–33 egg/pulse with a cumulative fecundity

TABLE 20.1

Effect of brooding and viviparity on fecundity and survival of some animals (compiled from Pandian, 1994, 2007, 2018, [†] estimated from Menge, 1975, see also Fig. 1.8B of Pandian, 2017, 2018)

Species	Fecundity (egg no.)	Survival (%)
Echinodermata		
Broadcaster: *Piaster ochraceus*	40×10^6	< 1
Vivipare: *Leptasteria hexactis*	45	8[†]
Crustacea: Euphausiids		
Broadcaster	296	< 1
Brooder	95	-
Mollusca	**Fecundity**	**Egg Size (µm)**
Lacuna vincta	54,432	88
L. pallidula	1,365	308
Onchidoris muricata	43,364	90
Adalaria proxima	4,224	180

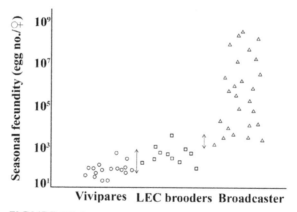

FIGURE 20.1

Effect of different modes of development, i.e. free spawning planktotrophics (PLK), lecithotrophics (LEC)/brooders and vivipares on seasonal fecundity of marine invertebrate animals. Double headed arrows indicate the overlapping (modified from Thorson, 1950).

of 71–1,779 eggs; however, the egg brooding *Eurytrema affinis* produce 1.0–5.5 egg/clutch and 5–133 eggs for its lifetime, suggesting the very high energy cost of brooding (see Pandian, 2016). The drastic decrease in the egg number from free spawners/layers to brooders and vivipares has a great deal of implication for genetic diversity and its impact on species diversity and selection of the fittest. Hence, this account focuses on a comparative picture of these three groups.

20.1 Taxonomic Survey

As it may be exhaustive, it is chosen first to estimate the approximate number of brooding and viviparous species and then subtract these values from the total number of animal species to arrive at figures for the free spawning/laying animals. From a survey and computer search, the approximate species number for the brooders and vivipares is listed in Table 20.2. Thanks to Ostrovsky et al. (2016), values for the matrotrophic vivipares are readily available

TABLE 20.2

Approximate number of brooding and matrophic viviparous species in major and minor phyla (compiled from Ostrovsky et al., 2016, Pandian, 1994, Wong et al., 2013, WoRMS*)

Phylum	Brooder	Vivipare	Phylum	Brooder	Vivipare
Porifera	0	34	Mesozoa	0	150
Cnidaria	2	5	Loricifera	-	1
Acnidaria	2	-	Cycliomorpha	0	2
Platyhelminthes	0	25	Nemertea	0	14
Polychaeta	4,617	19	Gnathostomulida	0	0
Clitellata	0	3	Rotifera	0	1
Crustacea	50,832	71	Gastrotricha	0	1
Insecta	10	6,475	Kinoryncha	5	0
Chelicera	3,386	1,753	Nematoda	0	36
Gastropoda	5	18	Acanthocephala	0	5
Bivalvia	0	42	Onychophora	0	86
Others	4	0	Bryozoa	?	844
Crinoidea	1	12	Urochordata	1,730	54
Holothuroidea	18	14	Subtotal: Minor	1,735	1,194
Asteroidea	4	6	%	3.7	2.6
Echinoidea	7	1	Subtotal: Major	58,899	16,174
Ophiuroidea	8	0	%	3.9	1.1
Pisces	0	1,857	Total species (no.)	60,634	17,368
Amphibia	3	72	%	3.9	1.1
Reptilia	0	254*			
Mammalia	0	5,513			
Subtotal: Major	58,899	16,174			
%	3.9%	1.1%			

except for reptiles. Yet, compromises had to be made for Platyhelminthes and Echinodermata. The reported values for matrotrophics are 450 and 18,000 for the polystomatid monogeneans and digeneans, respectively (Ostrovsky et al., 2016). But from a more reliable molecular phylogenic analysis, Littlewood et al. (2015) arrived at 12,012 species for Digenea; similarly, the value for the polystomatids is 150 only (see Pandian, 2020). Considering the total number of species for platyhelminths as 27,700 and description by Pandian (2020), the number of viviparous species is assessed as 8 for Turbellaria and 17 for Cestoda; all others are neither brooders nor vivipares. Ostrovsky et al. (2016) indicated that the number for the matrotrophic viviparity, for example, in Asteroidea as 10, but the values available for matrotrophics from Hyman (1955) and others are limited to zero. Hence, the number for matrotrophic species is taken as zero for all echinoderms. The value of 4,617 brooding polychaete species was arrived from the analyses of Wilson (1991) and Carson and Hentschel (2006). Many brooding polychaetes release the larvae at an early or later stage. So are the crustaceans. In Crustacea, 6.4% or 3,481 species are egg shedders (see Pandian, 1994); the number for matrotrophics is 71 species. And the remaining 50,832 species are assigned as brooders.

In their review, Wong et al. (2013) considered the brooding insects under three groups: (1) some 10 blattodean species providing protection, as defined in this account, (2) food provision for an individual larva, and (3) mass provision of food for a large number of larvae in an eusocial colony. Firstly, the last two types do not fall in line with this account's definition. Secondly, Wong et al. ignored many publications for group 2 (e.g. Muthukrishnan, 1994) and group 3 (e.g. eusocial insects, see Table 15.2). Incidentally, the food provision behavior of some wasps like *Sceliphron violaceum* is indeed interesting. In holes of trees, rocks or electrical sockets, *S. violaceum* larviposits. Its larva requires 200 mg paralyzed spiders as food; of this, a minimum of ~ 125 mg is required for successful completion of larval and pupal stages and to emerge as imago. The food provisioning behavior of the wasp dramatically differs prior to and after larviposition, when it has deposited ~ 65 mg food. On disturbance prior to larviposition, the wasp simply abandons the hole. However, for the same after larviposition, she makes enormous efforts to provide at least the minimal required 125 mg food (Pandian, 1985). Aside from it, the food provision either for an individual wasp larva or for a large number of larvae in eusocial colonies may not fall in line with the definition of brooding.

For minor phyletics, the number for viviparous species is 1,194 or 2.6% of 46,687 species. In major phyla, the number is 16,174 or 1.1% of 1,496,509 species (Table 20.2). On the whole, of 1,543,196 species, the vivipares with 17,368 species constitute only 1.1%. With the incidence limited to 1.1%, viviparity is unable to accelerate species diversity. Similarly, the number of brooders is also limited to 1,735 or 3.7% species among minor phyletics and 58,899 or 3.9% species among major phyletics. With incidence limited to < 4%, brooding has also not fostered species diversity. *On the whole, 95% of animals are free spawners/layers. As indicated, free spawning/laying accelerates species diversity, but brooding and viviparity limit species diversity.*

20.2 Vivipares: Size and Fecundity

Viviparity involves adaptations for internal fertilization, maternal gestation and parturition. Despite limited species diversity, the incidence of viviparity is reported from almost all phyla (Table 20.2) indicating the safest mode of completing embryogenesis. *From an analysis of the relationship between adult size and viviparity among marine invertebrates, Strathmann and Strathmann (1982) found that vivipares are usually smaller in size, in comparison to their respective taxonomic oviparous counterparts. This account brings to light that with increasing body size of vivipares, the number of progenies decreases in both aquatic and terrestrial animals.* For example, the number decreases from 300 progenies in the viviparous polychaete *Neanthes limnicola* (Pandian, 2019) to ~ 42–48 in *Poecilia sphenops* and *P. velifera* (George and Pandian, 1997) to 10 in the marine snakes (Lemen and Voris, 1981) and to one only in the pygmywhale *Kogia breviceps* (Fig. 20.2A). Interestingly, the largest bluewhale *Balenoptera musculus* gives birth to a calf once every 2–3 years and the calf measures 8 m in length and 2.7 ton in weight, and drinks 190 liters of fat-rich (35–50% fat) milk per day (Vivekanandan and Jeyabaskaran, 2012). Apparently, the large size of parturited offspring may be a reason for the reduction in the number of offspring in larger animals. In the terrestrial habitat too, the progeny number decreases from 10,000 in the parasitic nematode *Wuchereria bancrofti* (Pandian, 2021, Fig. 12.7B) to ~ 50 in a large scorpion, 4 in a larger

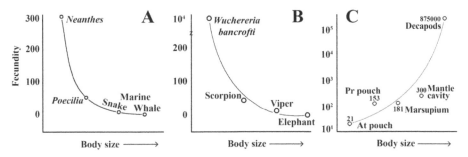

FIGURE 20.2

Fecundity as a function of body size in A. Aquatic and B. terrestrial vivipares and C. crustaceans brooding eggs in different sites of their body. Figures are drawn from data reported by Pandian, 1994, 2012, 2019, Madsen and Shine, 1992, Pandian, 2021, Lemen and Voris, 1981, Wikipedia. Note figures are not drawn to scale. Pr pouch = projected copepod pouch, At pouch = attached daphnid pouch.

Vipera berus (Madsen and Shine, 1992) and to one only in the largest elephant (Fig. 20.2B).

20.3 Brood: Size and Fecundity

As indicated, the brooders allow protection and ventilation alone. The brooding spaces are located inside the tube in polychates or a cloacal pouch, pallial groove, mantle cavity and marsupial demibranch in molluscs or in a pouch, chamber, intra-radial marsupium, tentacular crown in echinoderms. In contrast to vivipares, space provision increases with increasing body size, irrespective of the gestating sites. In brooding crustaceans, for example, fecundity increases from 21 in an attached daphnid pouch to 153 eggs in projected copepod pouch, 181 eggs in large marsupium-brooding peracarids and to 875,000 eggs in the largest pleopod-brooding decapods (Fig. 20.2C). This body size-fecundity relation holds true for other brooding species belonging to different phyla.

Brooding is a hall-mark strategy of crustaceans. During the checkered history of evolution, crustaceans have explored different sites for brooding. Based on the brood site, three brooding types are recognized; their advantages and limitations are summarized here under: (1) In a closed attached type, eggs are completely protected but limited by space and ventilation. It occurs in cladocerans (Fig. 20.3A) and copepods (Fig. 20.3B). For want of space and ventilation, holding too many eggs within the sac/pouch may not be possible. Besides, the closed dorsal pouch of *Polyphemus* may even hinder motility by altering the posture (Fig. 20.3C). The egg sacs are projected dorsally in anostracans (Fig. 20.3A), laterally in copepods (Fig. 20.3B) or ventrally in anostracans (Fig. 20.3D) and euphausiids (Fig. 20.3E). In them, motility ventilates but the closed sac limits the egg number, albeit allowing complete protection. The projection may retard the speed of locomotion but reduce the sinking rate. Partly open type occurs in mantle cavity of cirripedes or marsupium of peracarids. In them, the level of protection is reduced, space can be limiting but allows relatively better ventilation by irrigating water. (2) A distinct feature of peracarideans is the presence of a ventral brood pouch or marsupium. The marsupium is formed by large plate-like processes called oostegites appearing from certain thoracic coxae. Projecting the oostegites inwardly and horizontally to overlap with one another form the floor of the marsupium (Fig. 20.3F). The longer the female, the larger the marsupial space and the number of eggs brooded (Fig. 20.3G–H). The backward vibration of maxillipede projections produce antero-posterior water current and thereby provide ventilation to the brooded eggs. In comparison to the projected egg sac or open carriage appendages, the marsupium strikes an optimum between space and ventilation, on one hand and protection on the other.

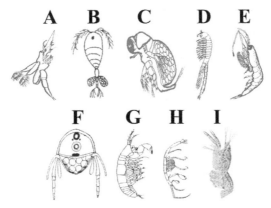

FIGURE 20.3

Brood site in crustaceans. A. Dorsal pouch in cladoceran, B. lateral pouch in copepod and C. in cladoceran *Polyphemus*, D. anostracan *Branchionecta*, E. euphausiid *Nematoscelis*, F. Transverse section of an isopod to showing the overlapping oostegites forming a marsupium, G. cumacean *Hemilamprops*, H. tanaidacean *Pseudotanais*, I. pleopod brooding *Procambarus*. All are free hand drawings from different sources.

(3) The open carriage of eggs on the (thoracic) pereopods or (abdominal) pleopods provides maximal ventilation and space but minimal protection. Fastening the eggs on thoracic appendages, in Stomatopoda, Leptostraca and some euphausiids (Fig. 20.3E) eliminates feeding during brooding. But a better option is to bear the eggs fastened onto the pleopods, whose fanning provides maximal ventilation. The frequency of pleopod fanning is so adjusted to increase ventilation, as the embryo approach the terminal stage of development. For example, with increasing number fanning from 3,000 time/h to 9,000 time/h, the water flow over the brooded eggs increases from 0.5 l/h to 2.5 l/h in *Macrobrachium nobilii* (Pandian and Balasundarum, 1980). However, the loss of brooded eggs in the pleopodal carriage type and required cost on ventilation and cleaning of eggs are some disadvantages. Due to fungal and other microbial infection as well as predation of eggs bynemerteans (e.g. *Carcinonemertes*), the pleopodal carriage incurs heavy loss of brooded eggs.

Table 20.3 lists fecundity and brood mortality in different crustaceans carrying eggs at different sites of their body. Albeit offering the maximal protection with minimal 8.5–19.0% brood mortality, the closed pouch type of egg carriage provides inadequate space and no ventilation. It occurs in < 15% of crustaceans. The open carriage on setosed pleopods provides the maximal space and ventilation, but it suffers the maximal (43%) brood mortality. Not

TABLE 20.3

Effect of brooding site on fecundity and brood mortality in crustaceans (compiled from Pandian, 1994)

Brooding site	Species (%)	Fecundity (no.)	Brood mortality (%)
Attached pouch	3.0	21	19.0
Projected pouch	12.0	198	8.5
Marsupium	34.0	181	23.8
Mantle cavity	11.0	300	-
Pereopods	1.5	405	-
Pleopods	31.5	875,000	43.2
Egg shedders	6.4	262,450	-

more than 31.5% of crustaceans have opted for egg carriage on pleopods. *The marsupial brooding, however, provides relatively more space and ventilation with < 24% brood mortality. More than 34% of crustaceans are marsupial type of egg brooders. Obviously, the egg carriage within the marsupium and on pleopods seems to have fostered species diversity.*

21

Feeding and Non-feeding Larvae

Introduction

Among taxa characterized by an indirect life cycle, egg size and lecithality decide whether the larva(e) are to become short living, non-feeding, dispersive type or long living, feeding and dispersive type. In them, egg size of Monogenea, for example, ranges from 63 µm in diameter in *Allomuraytrema robustum* to 600 µm in *Pseudodiplorchis americanus* (see Pandian, 2021) and larval duration of molluscs from 42 days in the *Pedicularia sicula* to 320 days in *Cymatium nicobaricum* (see Pandian, 2017). This chapter intends to assess the effects of egg size and lecithality on species diversity.

21.1 Egg Size

Egg size offers valuable information on the reproductive strategy of a species. As relevant information is available more for fishes, this account has chosen to explain the modulating role played by internal (e.g. body size) or external (e.g. food, temperature) factors on the egg size-fecundity relation in fishes. In general, fishes produce a large number of smaller eggs or small number larger eggs. For example, their egg size ranges from 0.25 mm in diameter to > 20 mm and Batch Fecundity (BF) from 33 eggs in the bagrid *Bagre marinus* to one million eggs in the snookfish *Centropomus indecinalis*. Most pelagic spawners are prolific and broadcasters of small but a larger number of eggs; the mean size of pelagic spawning 21 labrid species is 0.64 mm; they release a few thousand (e.g. *Centropyge ferrugatus*) to million (e.g. *Lethrinus nebulosus*) eggs. In contrast, demersal spawners lay only a few hundred (e.g. *Chandra ranga*) to thousand (e.g. *Nototropis leedsi*) eggs. The egg size of demersal spawners is large (e.g. 1.46 mm in *Tilapia zilli*). It is 1.5 mm for 48% fish species in British waters (Russel, 1976). Among pelagic spawners, Relative Fecundity (RF) increases from 136 egg/g body size in small *Brevoortia aurea* to 10,547 egg/g in large *Serranus atricauda*. Within the demersals too, it increases from 3 egg/g in *B. marinus* to 300 egg/g in *Clupea harengus*. Within pelagics, egg size increases from 0.52 µm to 0.64 µm, whereas it seems to decrease with increasing body size. Understandably, there is a trade off between egg size and fecundity in fishes and the trade off is limited by the volume of space available in the body cavity to accommodate ripe hydrated eggs, within which fecundity can be maximized by minimizing egg size or the reverse of it (Fig. 18.2E). Incidentally, smaller fishes opt for larger but small number of eggs. For a 3-fold increase in body length from 3.5 cm in *Apogonichthys waikiki* to 12.9 cm in *A. menesmus*, fecundity decreases by 20% in the Hawaiian apogonidan cardinalfishes. Indeed, small fishes make a heavy investment per egg (Barlow, 1981).

Within a species, changes in egg size are associated with latitude-dependent temperature and food supply. In the cobitid *Barbatula barbatulus*, egg volume is 2.5 times larger in the Finnish population than that in southern England. To reduction in food supply, the responses of arctic, temperate and tropical fishes seem to differ. In the arctic brook trout *Salvelinus fontinalis*, egg size varies by 35% in volume. With better food supply, the trout produces a large number of small eggs. But it spawns a small number of larger eggs on receiving low food supply. In response to reduced frequency of food supply, the temperate stickleback *Gasterosteus aculeatus* reduces fecundity and egg size (Table 4.1). The tropical damselfish *Acanthochromis polyacanthus* spawns a large number (3,000) of larger (5.76 mm³) eggs, when fed three-times more food for two months prior to spawning but a small number (2,200) of smaller (4.38 mm³) eggs at lower ration (see Pandian, 2011a). Variation in egg size can be as much 23% within a single brood and 13% among broods (e.g. sea urchin *Heliocidaris erythrogramma*, Pandian, 2018). In a linear relationship between body size and fecundity of polychaetes, about 26% of the variance in fecundity is influenced by body size (due to changes in body shape) and the remaining 74% by environmental factors (see Pandian, 2019). On the whole, egg size can vary within a single batch, between successive batches and body size as well as food supply and temperature within a species. Hence, egg size of the oviparous animals is determined by internal and external factors.

For terrestrial animals, insects are chosen to show how food quality and/or quantity affect the number and size of eggs. Table 21.1 lists representative examples from hemimetabolous orders Ephemeroptera, Orthoptera and Hemiptera as well as holometabolous orders Lepidoptera, Coleoptera, Hymenoptera and Diptera. Expectedly, reduction in levels of protein (e.g. algae, detritus), nitrogen (e.g. *Brassica oleracea*), lipid (e.g. *Cleome viscosa, Corotons parciflorus* leaves) and water (e.g. *Luffa acutangula* mature and senescent leaves) reduce fecundity in the selected insects. Among sanguivorous insects, reduction in blood meal reduces fecundity. Unfortunately, none of the authors, who have reported these reductions, have indicated that the reductions either increase or decrease the egg size. Fortunately, Bailey (1976) reported data for food intake, fecundity and size of larvae and pupa of the lepidopteran armyworm *Mamestra configurata*.

TABLE 21.1

Effect of food and its composition on fecundity of some insects (condensed and compiled from Muthukrishnan, 1994)

Food	Composition	Fecundity (egg no.)
Ephemeroptera: *Cloeon* sp		
Alga	43.5% protein	540
Baetis sp		
Detritus	17.8% protein	102
Orthoptera: *Aiolopus thallassimus*		
Cyperus rotundus	Water 10.7%	72
Cynadon dactylon	Water 7.0%	34
Hemiptera: *Acrosternum graminea*		
Cleome viscosa leaf	6.2% Nitrogen + 2.2% lipid	126
Coroton sparciflorus leaf	5.0% Nitrogen + 1.1% lipid	46
Lepidoptera: *Spodoptera litura*		
Brassica oleracea flower	6.4% Nitrogen	453
Leaf	3.2% Nitrogen	211
Coleoptera: *Raphidopalpa atripennis*		
Luffa acutangula mature leaf	2.9% Nitrogen + 74% water	205
senescent leaf	2.0% Nitrogen + 66% water	70
Hymenoptera: *Trichogramma japonicum*		
Corcyra cephalonica egg *C. cephalonica* reared on Green gram seed	3.1% Nitrogen	23
Sorghum seed	2.8% Nitrogen	12
Diptera: *Dacus dorsalis*		
Oviposited bottle filled with Orange juice	-	20.8×10^4
Papaya juice	-	14.0×10^4
Guava juice	-	13.4×10^4
Culex pipens		
Cannery blood	-	187
Human blood	-	58
Triatoma infestans		
Blood meal	1000 mg	50
	1500 mg	100

In lepidoterans, food consumption during the final instars determines fecundity, and size of larva and pupa. Increase in food consumption from ~ 400 mg (dry) to ~ 520 mg, not only increases fecundity from 300 to 800 eggs but also the size of larva (~ 100 mg to 145 mg) and pupa (~ 70 mg to 115 mg). Understandably, increase in food consumption increases egg size, from which larger larva and pupa appear (see also Fig. 24.1). Hence, environmental factors like food quality and quantity profoundly affect the number and size of insect eggs.

21.2 Lecithality

Aquatic animals, which spawn a large number of smaller, less lecithal eggs with a relatively shorter duration of embryogenesis, but a longer duration of feeding and dispersing larval stage(s), are known as planktotrophics (PLK). In contrast, those releasing a small number of larger, more lecithal eggs with a relatively a longer duration of embryogenesis and a shorter duration of dispersive larval stage(s), are called lecithotrophics (LEC). In Table 21.2, available values for the egg size for PLK and LEC of some taxa are assembled. Firstly, the egg size ranges widely within each taxon and within PLK as well

TABLE 21.2

Egg size and its ranges in planktotrophic and lecithotrophic aquatic animals (compiled from Fell, 1989, Pandian, 2011a, 2017, 2018, 2019, 2021)

Taxon	Egg size (µm)	
	Planktotrophics	Lecithotrophics
Porifera	40–100	-
Polychaeta	98–123	175–205
Opisthobranchia	80–88	165–205
Prosobranchia	179–225	321–390
Holothuroidea	121–210	70–750
Echinoidea	65–320	300–650
Asteroidea	110–190	320–1000
Ophiuroidea	70–165	70–950
Sipuncula	77–165	125–260
Phoronida	60	125
Bryozoa	70–110	100–400
Pisces	0.54–0.64*	1.5 – > 2.0*

* in mm

as LEC. Of course, the size ranges are clearly demarcated between some taxa like polychaetes. However, there is considerable overlapping between sizes of the PLK and LEC groups (e.g. echinoderms, see Sewell and Young, 1998 or Fig. 1.11 of Pandian, 2018). Therefore, it is difficult to fix a specific size or size range to distinguish PLK from LEC. One reason for this wide range can be traced to biochemical composition and energy content of eggs. Analysis of these components in echinoderm eggs has shown the following: Firstly, the PLK holds less energy (0.009 j/egg) in its smaller egg (16 nl) and completes embryogenesis in a relatively shorter duration. In contrast, LEC has more energy (4.4 j/egg) in its larger egg (4.4 nl) and passes through a relatively a longer embryogenic duration. Secondly, the PLK completes structural development required for motility, and food capturing and digestion within a relatively a shorter embryogenic duration and meets the metabolic cost using more protein (~ 62%). Contrastingly, the LEC construct structures required for motility alone during relatively a longer embryogenic duration using less (18%) protein for structure and the remaining for embryonic metabolism (see Pandian, 2018).

In echinoids, blastomeres can be totipotent up to a 4-cell stage; each of these 'totipotent blastomeres' is considered as an 'egg'. By virtue of their regulative development potency, experimental isolation of blastomeres is possible by cutting embryos into meridional twins and subsequent equitorial quadrupletes. Hence, these manipulations produce 'eggs' of different sizes within a single species. As of date, seven echinoid species including an irregular sand dollar *Dendraster excentricus* have been manipulated; of them, one species is an LEC; another one *Clypeaster rosaceus* is a facultative LEC and five others are PLKs. Table 21.3 shows that survival decreases among the twins from 95% in *Heliocidaris erythrogramma* to 4% in *Arbacia punctulata*. More importantly, the larval duration and size at metamorphosis of the isolated twin 'eggs' remain almost equal to those of intact eggs. In another sea urchin *Peronella japonica*, the development of isolated twins is also characterized as the LEC type, although the blastomere size is smaller than that of PLK species (Okazaki and Dan, 1954, see also Kitazawa and Amemiya, 2001). *P. japonica* and more specifically *H. erythrogramma* egg (400 µm), even when reduced to half the size (200 µm), which is smaller than the PLK egg size of 270 µm in *C. rosaceus*, still undertakes the LEC development mode. In the nemertean *Cerebratulus lacteus* also, the twin blastomeres develop completely normal but into miniaturized two PLK pilidium larvae (see Pandian, 2021). These reports clearly indicate that the lecithality or vitellogenic quantum has a genetic base. *Whereas the egg size is more regulated by environmental factors, the lecithality is genetically fixed.* This discovery is reported for the first time and may require more research inputs to confirm it.

Notably, there are a few poecilogonics, which generate dichotomic PLK and LEC egg-larval morphs either from the same egg mass of an individual or individuals of the same species. For more information on this rare poecilogonic reproductive mode, Pandian (2017, 2019) may be consulted.

TABLE 21.3

Effect of egg size manipulation on larval duration of some echinoids (modified from Pandian, 2018)

Egg size (µm)	Survival after manipulation (%)	Larval duration (d)	Size at metamorphosis (µm)
Echinoids: *Heliocidaris erythrogramma* LEC			
400	95	3.5	424
200	95	3.5	370
Echinoid: *Clypeaster rosaceus*			
270	33	20	4.75
135	25	18	4.67
77	13	19	4.55
Echinoid: *Strongylocentrotus droebachiensis* PLK			
161	58	36	102
76	< 46	36	72
63		39	-
Echinoid: *Echinarachnius parma* PLK			
142	50	38	101
71	41	42	91
Echinoid: *Dendraster excentricus* PLK			
110	28	11.62	14.8
55	20	11.66	14.6
Echinoid: *Arbacia punctulata* PLK			
80	23	60	-
40	4	68	-

Poecilogony is limited to six opisthobranch and one prosobranch species in Gastropoda as well as six spionid and one nereid species in Polychaeta. In the male heterogametic (XO) archiannelid *Dinophilus gyrociliatus*, the findings reported by Traut (1969a, b, 1970) are relevant. Its female simultaneously produce one small (40 µm) and three larger (80 µm) eggs; selective fertilization of larger eggs by sperms carrying X sex chromosome and small eggs by sperm bearing no sex chromosome nullifies the chromosomal mechanism of sex determination. But the question that remains to be answered is how eggs carrying X^1 and X^2 sex chromosomes introduce different levels of vitellogenesis resulting in eggs of different sizes?

In these echinoderms, the PLK larvae develop from oocytes that have accumulated more protein, while the LEC ones develop from oocytes that have accumulated more lipids. The transition from the PLK mode of development to LEC mode required four to seven million years in two clades of sea urchin inhabiting the two sides of the Isthmus of Panama (Zigler and Lessios, 2003). Hart et al. (1997) showed that it took two million years in asteroids. Some of these examples indicate that the transition from PLK to LEC occurred repeatedly in the history of different taxa. Hence, the switching from one to other mode of development is not rare in the geological scale of evolution. Among invertebrates, facultative PLK is known in two sea urchin species, four gastropods, a bivalve and a polychaete species. Another striking example for facultative PLK is provided by the snail *Alderia willowi*, which switches from PLK during winter and spring to LEC during summer (see Ostrovsky, 2013). In it, crossing between adults of PLK and LEC produced PLK progenies alone in F_1, indicating the dominant expression of PLK. However, both PLK and LEC were produced among F_2 progenies (West et al., 1984, see also Krug, 1998).

With regard to yolk utilization efficiency during embryogenesis, available values are limited to brooded lecithal eggs, except for the pelagic PLK eggs of *Solea solea*. Despite lecithality and brooding for a limited but varying stages among them, *Artemia salina* releases non-feeding lecithotrophic nauplius. Among others, the release of PLK larva is delayed from protozoea in *Crangon crangon* to zoea in *Eupagurus bernhardus*, a still later stage in *Homarus* spp and a juvenile manga in *Ligia oceanica*. The developing embryos of *Crepidula furnicata* and *L. oceanica* receive extra-embryonic and maternal nutrients, respectively. Hence, it may be interesting to know whether the delayed release of LEC larva/juvenile reduces the yolk utilization efficiency. Available information (Table 21.4) shows that (i) With increasing body size, egg size increases. (ii) The longer the delay, greater is the lecithality. Though, lecithality is genetically fixed within a species, the level of lecithality may differ among lecithal species; the duration of brooding is one factor that alters the lecithal level. (iii) The yolk utilization efficiency is relatively higher 76, 79 and 85% at 20°, 15° and 10°C for *S. solea*, which entirely release feeding PLK larvae; this is also true of *Clupea harengus* (iv) Receiving maternal nutrition, *L. oceanica* is able to increase the efficiency to 72% and (v) In others, the efficiency remains around 60–66%, irrespective of the stage, at which the PLK larvae are released from the brood.

In many parasitic taxa, the life cycle is indirect except in Nematoda (Table 19.1). Their riskier indirect life cycle may demand production of more eggs. The cycle incorporates one or more Intermediate Host(s) (IHs), except in Monogenea, Mesozoa and most Nematomorpha. The riskiest life cycle involving transfer of larva from a Definitive Host (DH) to IH and from IH to DH may demand production of even more eggs (Table 21.5). The parasites are able to meet the demand, as they are assured of a good food supply. Two life history features must be noted; they are (i) mode of transfer and (ii) clonal multiplication in one or more larval stages. In parasites, trophic transfer is the most common mode. However, it is penetrative in myxozoan Malacosporea, dermis-penetrating Monogenea, Pentastomida and a few nematodes. (i) A *penetrative mode seems to demand relatively less fecundity*

TABLE 21.4

Lecithality and yolk utilization efficiency in free spawned and brooded eggs

Species	Body size	Egg size (cal)	Larval type & release time	Yolk utilization efficiency (%)	Reference
Solea solea	~ kg	0.486	PLK larva	76	Pandian and Fluchter (1968)
Clupea harengus	~ kg	9.1 µg	Nourished by yolk	62 at 8°C 57 at 12°C	Blaxter and Hempel (1966)
Salmo fario	~ kg	-	Nourished by yolk	65	Gray (1928)
Crepidula furnicata	~ g	0.027	shellular nourishing + PLK Veliger	61	Pandian (1969)
Artemia salina	~ mg	0.018	LEC Nauplius	70 at 10°C 76 at 32°C	Hentig (1971)
Crangon crangon	~ g	0.0984	Brooded → PLK Protozoea	66	Pandian (1967)
Eupagurus bernhardus	~ g	0.16	Brooded → PLK Zoea	56	Pandian and Schumann (1967)
Homarus americanus	~ kg	6.4	Brooded → late Zoea	60	Pandian (1970b)
H. gammarus	~ kg	10.49	Brooded → late Zoea	60	Pandian (1970a)
Ligia oceanica	~ g	301	Maternal nourishing → Manga	72	Pandian (1972)

and results in less species diversity. For example, Malacosporea, involving a penetrative mode of infection, comprise only 2 genera, while Myxosporea, involving trophic mode of infection, include as many as 58 genera. The mean fecundity is 7,195 egg/♀/d for the penetrative nematode parasites, but is as much as 37,348 egg/♀/d for the trophic nematodes (Pandian, 2021). With reduction in fecundity, it is tempting to assign their eggs as lecithals. For example, the egg size range from 63 µm in *Allomuraytrema robustum* to 600 µm in *Pseudodiplorchis americanus* in Monogenea indicates the presence of low and high levels of lecithality. Firstly, the non-feeding oncomiracidium transmitted through the trophic route needs only to get attached with it sucker(s) on to the gills, while that penetrating the skin may require a longer time and energy to complete the process for successful infection. Secondly, the life span of oncomiracidium lasts for 33.3 hours in the dermal flukes but for 12.0 hours only in the branchial flukes. Apparently, the eggs of dermal flukes are more lecithal than that of the branchial flukes. The span is also limited to ~ 15 hours for miracidium (e.g. *Schistosoma mansoni*). But the digeneans can afford to produce relatively less number of smaller lecithal eggs, as they are clonally multiplied in sporocyst and redia. (ii) Surprisingly, *no parasitic larva*

TABLE 21.5

Life history characteristics of parasitic animal taxa

Taxon	Life cycle	Intermediate host	Mode of infection	Feeding mode of larvae	Fecundity	Remarks	Larva name
Branchial Monogenea	Indirect	No	Trophic	Yolk	High		
Dermal Monogenea	Indirect	No	Pentrative	Yolk	Medium		
Digenea	Indirect	††	Penetrative	Osmotrophic	Medium*		
Cestoda	Indirect	†† or †	Trophic	Osmotrophic	High		
Mesozoa	Indirect	No	Trophic	Osmotrophic			Infusiform
Myxozoa	Indirect	†	Trophic or penetrative**	Osmotrophic			Malaca- or Actinospores
Mermithid nematodes	Direct	No	Trophic	Sucking	High	Larval parasites	
Penetrative nematodes	Direct	† or ††	Penetrative	Sucking	Low		
Nematomorpha	Indirect	No, rarely, 1 IH	Trophic	Sucking	High	Larval parasites	
Acanthocephala	Indirect	††	Trophic	Osmotrophic	High		Acanthor, Acanthella, Cystacanth
Pentastomida	Indirect	No/yes	Penetrative?	?	Medium		

* clonal multiplication in sporocyst and redia; sporocysts are osmotrophs, redia and cercaria are parcel-feeders, † trophic in speciose Myxosporea; ** penetrative in less speciose Malacosporea, † one IH, †† two IH

is known to feed during its life span. Hence, they must all be lecithotrophic. But they have to produce a large number of eggs, as they encounter the risks involved in an indirect life cycle, especially in those cycles, which incorporate one or two IHs. In them, *the trophic mode of infection fosters species diversity.* Incorporation of non-feeding larvae may demand lecithality and consequent reduction in egg number. However, clonal multiplication in the first intermediate host and selection of the fittest clone in the second intermediate host has indeed fostered the species diversity in Digenea.

21.3 Incubation and Feeding Larvae

Lecithality imposes profound effect on proportions of lifetime and fecundity. As already indicated, the development of locomotory and food-capturing structures require a relatively shorter incubation duration in low lecithal eggs that are hatched into planktotrophic (PLK) and lecithal eggs hatching into lecithotrophic (LEC) larvae (e.g. echinoderms and phoronids, Fig. 21.1A). In monogeneans, LEC larvae are characterized by a shorter incubation and longer oncomiracidial larval duration in dermal parasites than those of branchial parasites. The former may require a longer larval duration to accomplish a dermal penetrative infection. Strangely, the feeding period is limited to a relatively longer larval stage (Fig. 21.1B) in ephemeropterans, plecopterans and lepidopterans. In the first two, the Life Span (LS) of their adults is transient; they die within 2–3 days after emergence, mating and oviposition; but in the last one, the span is little longer. Figure 21.1B shows approximate proportion of relatively a longer larval duration in the three

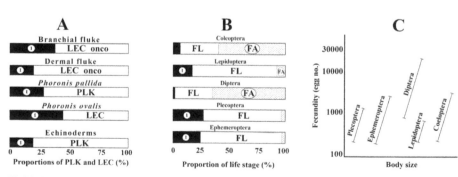

FIGURE 21.1

A. Effect of lecithality on duration of Incubation (I) in planktotrophic (PLK) or lecithotrophic (LEC) echinoderms and phoronids as well as dermal and branchial parasitic monogeneans (compiled from Pandian, 2018, 2020, 2021). Effects of feeding restricted to larval stage alone on B. proportion of lifetime and C. fecundity, in comparison to insects feeding throughout their lifetime (compiled from Brittain, 1990, Muthukrishnan, 1994 and others). FL = Feeding larva, FA = Feeding adult, onco = Oncomiracidium.

insect taxa, in comparison to that of dipterans and coleopterans, which feed during both the larval and adult stages. The differences in the feeding duration between the first three and last two, for example, have a significant implication to fecundity. Of course, fecundity increases with increasing body size in all the five taxa but the levels differ greatly (Fig. 20.1C). *The fecundity levels are far lower for ephemeropterans and plecopterans, in comparison to that of dipterans.* The same holds true for lepidopterans and coleopterans. Hence, internal factors, the body size and duration of feeding stages have a dramatic effect on fecundity. On the whole, both the egg size and number of eggs of insects are regulated by internal and external factors.

21.4 Taxonomic Survey

With incorporation of feeding/planktotrophic (PLK) larva in indirect life cycle, most animals are able to produce a large number of small feeding larvae but a small number of larger non-feeding larvae. Hence, a taxonomic survey was made to know the number of species producing feeding or non-feeding larvae. With no mouth and gut, all the larvae of Porifera, Cnidaria and Acnidaria were all readily assignable to a non-feeding mode of development (Table 21.6). All parasitic phyla characterized by an indirect life cycle including the parastic nematodes with direct life cycle but with incorporation of one or two Intermediate Host(s) (IHs) were also assignable to the non-feeding mode. For polychaetes, the numbers were drawn from Carson and Hentschel (2006), who classified 255 species into three groups, of which 135 (69.2%) and 60 (19.2%) species were designated as PLK and LEC, respectively. These percentage values were applied to the total number of polychaete species (of course, after subtracting matrophic number) (see Table 1.11 of Pandian, 2019). Crustaceans produce an array of different larvae, of which protozoea and zoea of decapods alone are feeding larval stages. Hence, 15,000 decapods were assigned to the feeding mode and the remaining (39,384) to a non-feeding mode. Both hemimetabolous and holometabolous insects have feeding larvae. For them, the value was drawn from Blackmon et al. (2017). In molluscs, archaeogastropods, bivalves (inclusive of larviparous, in which larvae are released subsequently, as in a few polychaetes), scaphopods and polyplacophora have a feeding trochophore with a subsequent feeding veliger. They were all assigned to a feeding mode. After a great deal of survey and computer search, a compromise had to be made for the speciose Meso- and Neo-gastropoda. From their analysis of 184 holothuroid species, 131 echinoid speices, 149 asteroid species and 132 ophiuroid species, Sewell and Young (1997) estimated the distribution of PLK and LEC in echinoderms. Their percentage values were applied to the respective species number of holothuroids (> 1,000 species), echinoids (> 800 species), asteroids (~ 1,800 species) and ophiuroids (2,604 species) to

TABLE 21.6

Taxonomic distribution of feeding and non-feeding larvae in major and minor phyla (compiled from Hyman, 1940, 1959, Fautin et al., 1909, Pandian, 2016, 2017, 2018, 2019, 2020, 2001, Blackmon et al., 2017, others)

Phylum/Taxa	Species number (no.)		Phylum/Taxa	Species (no.)	
	Feeding	Non-feeding		Feeding	Non-feeding
Porifera	0	8,549	Placozoa	0	0
Cnidaria	0	10,852	Mesozoa	0	150
Acnidaria	0	166	Myxozoa	0	2,200
Platyhelminthes	0	27,000	Loricifera	0	34
Polychaeta	6,480	2,883	Cycliophora	0	2
Crustacea	15,000	39,384	Nemertea	433	0
Insecta	903,420	0	Nematomorpha	360	0
Mollusca			Acanthocephala	0	1,100
Archaeogastropoda	4900	0	Priapulida	19	0
Other non-pulmonates	37750	37,750	Sipuncula	80	80
Bivalvia	9,574	-	Echiura	230	0
Scaphopoda	570	0	Pentastomida	0	144
Polyplacophora	930	0	Entoprocta	200	0
Echinodermata	5,435	1,412	Phoronida	11	11
Pisces	17,563	14,370	Bryozoa	2,428	2,428
Amphibia	5,076	-	Brachiopoda	261	0
Subtotal	1,006,698	142,366	Enteropneusta	115	0
% for 1,149,064	87.6	12.4	Urochordata	2,900	0
Grand total	**1,013,735**	**148,515**	Subtotal	7,037	6,149
% for 1,162,250	**87.2**	**12.8**	% for 13,186	53.3	46.7

assign values for PLK and LEC. The crinoids have non-feeding doliolaria larva. For fishes, of the then available values, 11 were pelagic with ~ 0.5 mm egg size and 8 were demersal spawners (Pandian, 2011a), i.e. 57.9% belong to PLK and 42.1% to LEC. Russel (1976) reported that ~ 48% fishes were demersal spawners with an average egg size of 1.5 mm in British waters. The values suggest that an average of 55 and 45% teleosts belong to PLK and LEC mode of development, respectively. For free-living minor phyletics, the values were either directly or indirectly drawn from those reported by Adiyodi and Adiyodi (1989, 1990) and others. For example, Ostrovsky (2013) listed only 11 bryozoan species to display the PLK mode of development. For want of relevant information, many compromises were made to quantify the number of feeding larva involved in the life cycle of minor phyletics. For all parasitic phyla, the larvae were assigned as non-feeding type (see p 228). In the remaining phyla, the number of ovipares (see Table 20.2) was subtracted from each phylum. Of the remaining number in Sipuncula, Phoronida and Bryozoa, 50% of the respective values were assigned to feeding larval types.

Firstly, this account does not claim that the assessed values are exhaustive or precise. Of 1,543,196 species, only 80.5% pass through an indirect life cycle (Table 19.1). However, information on feeding and non-feeding larval stages could be assembled for 1,013,735 species and for the remaining 137,216 species, information is not available. These approximate values reveal that (i) of 1,149,064 major phyletic species, 87.6 and 12.4% belong to the feeding and non-feeding mode of larval development, respectively (Table 21.6). (ii) Contrastingly, these values are 53.3 and 46.7% for the minor phyletic species. (iii) On the whole, of 1,162,250 species, 87.2 and 12.8% animals pass through a feeding and non-feeding larval stage, respectively. By producing a large number of smaller eggs, *the feeding mode of larval development may generate an enormous genomic diversity and let the only very few of the fittest to be recruited. Hence, this mode of development fosters and accelerates species diversity.* The other non-feeding mode by producing a small number of larger eggs generates less genomic diversity but allows more safety for the progenies. This has paved the way for limited species diversity, but not as much as for the feeding mode.

22

Parasites and Hosts

Introduction

According to Jennings (1997), parasites cannot survive in the absence of hosts. They are so ubiquitous that it led Windsor (1998) to suggest that ~ 50% of known animal species are parasites. This may imply that all living metazoans host at least one parasite species (Poulin and Morand, 2000). In humans, his livestock and agricultural crops, the most common parasites are helminthic flukes and nematode worms. Metazoan parasitism has evolved independently at least 60 times (Poulin and Morand, 2000). In contrast to nematodes, in which parasitism has evolved independently several times, the parasitic flukes collectively named neodermatants, that infect vertebrates, have arisen from a single evolutionary event (Collins, 2017). A remarkable feature is that in most cases, the adults are parasites and their larvae are free-living (e.g. monogeneans) but the larvae/juveniles are parasitic and adults are free-living in monstrilloid copepods, mermithids and nematomorphs; in digeneans, both larvae and adults are parasitic. The estimated infliction caused by schistostomes alone is in the range of 2 million and death of 0.2 million men/y. Globally, 1.7 million bovines are infected by liver fluke and the estimated loss on meat and milk is over US\$ 2.5 billion/y. Nematodes not only infect man and his livestock but also his agricultural crops. The estimated loss on agricultural crops by nematodes surpasses that of digenean parasites. Not surprisingly, voluminous literature is available on diagnosis, treatment and management of these flukes and worms. However, this account shall be devoted to whether parasitic or free-living life history trait has fostered species diversity.

22.1 Parasites: Location Sites

As adults, digeneans occur primarily in the digestive tract and associated organs (with access to the tract) such as the liver, lung (in birds), gall bladder and bile passage; the other sites are kidney, uterus, coelom, eye, head cavities and blood. However, considering different structural organizations in a wide spectrum of hosts, this account has brought the sites of parasitic location into five broad groups. Their advantages and limitations are briefly summarized hereunder: 1. Ectoparasites: The entry and exit pose no problem. Accommodated between scales, feathers or hairs, or buried into dermis (e.g. monogeneans), they cling and gain food. But they may easily be removed chemically (by mucus) or physical scratching (crustaceans) by aquatic hosts or scratching by terrestrial hosts. 2. Digestive system: The system also poses no problem for the exit and entry to parasites. It ensures continuous availability of semi-digested nutrients. In it, the host immune response against the parasites is greatly reduced. However, the disadvantages are (i) dislocation by propulsion of the digestive tract, (ii) exposure to digestive enzyme(s) and (iii) availability of low or no oxygen. Against the first, the gut parasites have adaptive structures (to retain them at the desired site) like suckers in the flukes, scolex bearing hooks, bothridia or phyllidia in cestodes and proboscis armed with hooks in acanthocephalans. Their syncytial chitinous integument provides safety against enzymes and the immune system. With ensured abundant food supply, they can allow the low efficient glycolysis to generate adequate ATPs. 3. Branchial chambers and trachaea provide an excellent site for parasites by providing soft tissue and/or blood as food, adequate oxygen supply and scope for exit and entry through the passage of water or air. But space and elimination by the force of gushing water pose problems. The dorso-ventrally flattened body of isopods seems to be a pre-adaptive feature. However, the monogeneans have suckers and isopods have hooks on their appendages. 4. Blood ensures adequate oxygen and food supply. While the blood fluke has solved the problem of entry by penetration and exit through urine or feces, nematodes obligately require transmission by the sanguivorous Intermediate Host (IH). 5. Miscellaneous internal sites include body hemocoelom for mermethids and rhizocephalans, body fluids for mesozoans and myxozoans, muscular tissues (e.g. cestodes) and other structures. The gonad is also the sites for a few myxostomid polychaetes and ophiuroids. For some ichneumonid hymenopteran parasites, the host can be nutrition rich eggs or larva/adult. Firstly within parasitic species, the shares for ectoparasites and endoparasites are 62 and 38%, respectively (Table 22.1). Secondly, within the endoparasites, the shares decrease in the following descending order: gut parasites 26.8% > branchial and tracheal site 5.5% > miscellaneous sites 5.0 and blood parasites > 0.5%. The easily accessible entry and exit is the key feature that has let the parasite to occupy a site either outside or inside the host. *On the whole, the easily accessible gateways for a parasite to enter and exit from the host are a decisively important feature that regulates species diversity.* Accordingly, the external surface of the body, gut and associated organs, branchial/tracheal chamber in the just described order foster and promote species diversity.

TABLE 22.1

Taxonomic distribution and approximate number of parasite species located in five sites in host animals (compiled from Pandian, 2016, 2021, Boxshall and Hayes, 2019, Kinne, 1983a, b)

Taxa	Ecto	Gut	Gill/Lung	Blood	Body
Cnidaria	0	0	0	0	0
Mesozoa	0	0	0	0	150
Myxozoa	0	0	0	0	22
Turbellaria	0	0	0	0	15
Monogenea	0	0	4500	0	0
Digenea	0	11,559	-	453	0
Cestoda	0	4,647	0	0	0
Mermithoids	0	0	0	-	3,827
Parasitic nematodes	0	8,360	0	0	0
Nematomorpha	0	0	0	0	360
Acanthocephala	0	1,100	0	0	0
Copepoda	4920	0	0	0	0
Branchiura	0	0	0	0	146
Rhizocephala	0	0	0	0	122
Isopoda	0	0	584	0	99
Ascothoracida	0	0	0	0	70
Acari	54,617	0	0	0	0
Gastropoda	85	0	0	0	0
Pentastomida	0	0	144	0	0
Polychaeta	20	0	0	0	10
Myxostomida	0	0	0	0	0
Total species (no.)	59,642	25,666	5,228	453	4,749
%	62.2	26.8	5.5	0.5	5.0

As they rapidly kill the hosts, the parasitic bacteria and viruses (e.g. COVID-19) can be epidemic and pandemic, respectively. Contrastingly, the parasitic fungi and animals gain energy continuously draining nutrients from the host as long as possible. In this process of energy draining, they may castrate and cause reproductive death of the host, while the host continues to serve as the source of nutrients. In zoological research, parasitic castration is a classical discovery. The ability of parasitic castration is more prevalent among crustaceans (Table 22.2). The castrating crustaceans includes ascothoracids (e.g. *Ascothorax ophioctensis*), rhizocephalans (e.g. *Sacculina granifera*), copepods (e.g. *Parachordeumium amphiurae*), isopods (e.g. *Bopyrus*,

TABLE 22.2

Parasitic castration in gonochoric and hermaphroditic crustacean, fish* and ophiuroid hosts*[†] (modified from Pandian, 2016; added from Jancoux, 1990)

Host	Parasite	Reported observation
Reproductive Death in Hermaphrodites		
Lysmata seticaudata	*Eophyrxus lysmatae*	Ovary partially destroyed. Restored at the loss of parasite
L. ambionensis	*Parabopyrella* sp	Ovary completed destroyed. Normal testes
Balanus glandula	*Hemioniscus balani*	Ovary destroyed. Testes function normally
Reproductive Death in Gonochores		
*Cyphochara gilbert**	*Riggia pranensis*	Oogenesis inhibited
*Cheilodipterus quinquelineatus**	*Anilocra apogynae*	Vitellogenesis inhibited. Physical presence inhibit males from mouth brooding
*Ophiocten sericerum**[†]	*Ascothorax ophioctensis*	Pleopod not setosed. Gonad destroyed
*Ophionotus victoriae**[†]	*A. gigas*	Gonad destroyed
*Amphipholas squamata**[†]	*Parachordeumium amphiurae*	Gonad destroyed. Reduced brooding capacity
*Ophiomitrella corynophora**[†]	Unidentified copepod	Gonad destruction
Palaemon squilla	*Bopyrus*	Castrated prawn with pleopods not setosed; hence, not berried
Gammarus sp	*Polymorphus minutus* larva	Castrated with no oostegete setose. Vitellogenesis inhibited
Portunus pelagicus	Sacculinid	Castrated crab resegmented but not broadened. Pleopods not setosed
P. pelagicus	*Sacculina granifera*	Partial castration of female and male. Reduced GSI. Fewer clutch. Not berriable
Caricinus maenas	*Sacculina carini*	Castrated, resegmented not broadened abdomen
Physical Presence Induces Sex Change		
Calianassa laticauda	*Ione thoracica*	The existing female chemical induces the second to become male
Varicorhinus bachatulus	*Ichthyoxenus fushanensis*	Newly arriving male induces existing male to become female
Physical Presence Inhibits Spawning		
Crepidula cachimilla	*Calyptraeotheres garthi*	Presence obstruct spawning and brooding. Parasite removal restored reproduction

Parabopyrella) and crabs (e.g. *Calyptraeotheres garthi*). The castration is inflicted on the hosts belonging to other free-living crustaceans (e.g. *Palaemon squilla*), gastropod (e.g. *Crepidula cachimilla*), ophiuroids (e.g. *Ophionotus victoriae*) and fishes (e.g. *Cheilodipterus quinquelineatus*). The infliction is extended to both female and male in gonochoric hosts but limited to females alone among hermaphroditic hosts. The reproductive death is brought by inhibition of vitellogenesis and/or eliminating the setose in the oostegites or ovigerous setae on pleopods and thereby berrying. In some like *Ione thoracica*, the parasite reverses the sex differentiation process to the opposite sex on the host. The very presence of the parasitic crab obstructs spawning in *C. cachimilla*. Incidentally, Hyman (1951b) hinted that inhabiting the gonads of the fruit fly, the allantonematid *Tylenchinema oscinellae* can also inhibit reproduction.

For want of complete information on parasitic infection in echinoderms, the values listed in Table 22.3 are not included in Table 22.1. As the information is interesting for the relatively not well-known echinoderms as hosts, the following is described. Commencing as external infection, some parasites penetrate deep into the body and coelom, and form galls or cysts inside the hosts. Within the body, parasites may be found in the coelomic or hemal cavities and a few are found in the gonad. From Table 22.3, the following may be inferred: 1. Internal body, especially the body cavities provide more favourable sites and foster more species diversity than those of ecto-parasites. This is in contrast to those reported for all other animals (Table 22.1). For holothuria, Jangoux (1990) listed the number of parasitic species as 8, 15 and 84 for nematodes, digeneans and gastropods, respectively. Of these, life cycle of the last two is indirect and infection must involve a larva. However, it is not clear how the sediment-feeding holothuroids (34.0%) and plant/sediment-feeding echinoids (24.4%) are more favoured hosts for parasites than the filter-feeding crinoids (12.2%) or carnivorous asteroids (19.5%).

TABLE 22.3

Parasitic locations among echinoderm hosts (modified from Jangoux, 1990)

Class	External	Internal					
		Gut	Inside body	Body cavities	Gonad	Total	
						(no.)	(no.)
Crinoidea	24	9	29	8	1	47	12.2
Holothuroidea	15	54	10	65	2	131	34.0
Echinoidea	27	42	24	22	6	94	24.4
Asteroidea	8	7	21	46	1	75	19.5
Ophiuroidea	7	3	25	4	6	38	9.8
Subtotal (no.)	81	115	109	145	16	385	-
%	-	29.4	28.3	37.7	4.1	-	-

Interestingly, the mouth of asteroids and ophiuroidsfaces the substratum but those of crinoids, holothurians and echinoids open either anteriorly or apically.

22.2 Taxonomic Survey

With regard to the Intermediate Host (IH), the following could be noted: (i) The molluscan database reveals that a single digenean species has been recorded from 41 first IH (FIH). (ii) *Lymnaea stagnalis, Planorbis planorbia, Radix peregra,* and *R. ovata* can serve as FIH for as many as 41, 39, 33 and 31 digenean species, respectively. Aside from the strictly host specific (stenoxenic 71%) monogeneans, the euryxenic *Benedenia hawaiensis* can infect as many as 24 host species namely *Priacanthus, Mulloidichthys, Parupeneus, Dascyllus, Amanses, Acanthurus, Synodus, Abudefduf, Chromis, Chaetodon, Alutera, Pervagor, Naso, Holocentrus, Scarus* and *Xanthichthys*: Due to page limitation, this survey is limited to parasitic species and not to intermediate host(s). Table 22.4 lists the approximate number of parasitic species from major and minor phyla, for which relevant reports are available. For the remaining phyla, parasitism is not thus far reported or the incidence is rarely reported (e.g. Rotifera), which may be sporadic and exceptional. Most values listed in Table 22.4 are adopted from Pandian (2016, 2020, 2021); the remaining values are drawn from Kabata (1984), Jangoux (1990), Boxshall et al. (2005) and Boxshall and Hayes (2019). Furuya and Tsunaki (2003) reported the incidence for 42 dicyemids in 19 Japanese cephalopods. But Kinne (1983b) provides a longer list of names of host species covering almost all 150 mesozoan species. For the phylogenetically enigmatic *Myxostomatida*, presently accommodated in Polychaeta, 144 species are described but the host species names are known only for < 30 species. Incidentally, the series on Disease of Marine Animals by Kinne (1983a, b, 1984, 1985) is informative and useful.

From the data summarized in Table 22.4, the following may be inferred: 1. Parasitism is manifested in acoelomorphic Platyhelminthes and hemocoelomatic Arthropoda among major phyletics and structurally simpler aorganomorphic Mesozoa and Myxozoa, and pseudocoelomate Nematomorpha, Acanthocephala and some nematodes in minor phyletics. 2. Parasitism is manifested in structurally simpler solitary Mesozoa and Myxozoa but not in the structurally simple colonial cnidarians and sponges, which are at the lowest order of coloniality. In fact, not a single species from the colonial Entoprocta, Bryozoa and Urochordata is reported as a parasite. 3a. Clearly, *coloniality neither allows the manifestation of parasitism nor let them to serve as hosts.* 3b. *Parasitism is also unable to manifest in structurally complete eucoelomates, which are, however, let to serve as hosts.* Besides the 96,276

TABLE 22.4

Life cycle involving larval stage(s) and intermediate host (IH) in parasites (compiled from Pandian, 2016, 2021, Boxshall and Hayes, 2019, Narendran, 2002 and others)

Phylum	Parasite species (no.)	Life cycle		Larval stage (no.)	Intermediate host (no.)
		Direct	Indirect		
Mesozoa	150	-	150	1:150	0
Myxozoa	2200	0	2200	1:2200	1:2200
Turbellaria	15	15	0	0	0
Monogenea	4500	0	4500	1:4500	0
Digenea	12012	0	12012	4:12012	2:12012
Cestoda with coracidium	1266	0	1266	3:1266	2:1264
Cestoda with oncosphere	3033	0	3033	1:3033	1:3033
Mermethid nematodes	3827	3827	0	0	0
Vertebrate nematodes	8360	8360	0	0	1–2:5432[†]
Nematomorpha	360	0	360	1:360	0
Acanthocephala	1100	0	1100	2:1100	2:1100
Crustacea: Direct cycle	585	585	0	0	0
Others: Indirect cycle	5386	0	5386	1–3:5386	0
Cryptoniscoids	99	0	99	3:99	1:99
Hymenoptera	344	0	344	2:344	0
Acari	54617	54617	0	0	0
Gastropoda	85	0	85	1:85	0
Pentastomida	144	0	144	1:144	1:144
Myxostomid polychaetes	10	0	10	1:10	0
Total species (no.)	98,093	67,404	30,689		
%		68.7	31.3		

[†] juvenile stage

parasitic species listed in Tables 24.1 and 24.3, there are 10,150 protozoan parasites (*parasite.org.au*) and unknown number of sarcophagan (Diptera) parasites. These are also ~ 50,000 hemipterans (*aavp.org*) and 4,100 nematodan (see Pandian, 2021) plant parasites, which are not included in this account. *On the whole, the number of animal parasites on animal hosts may not exceed 110,000 species or 7.0% of 1,543,196 cumulative animal species number. Hence, the suggestion of Windsor (1998) that ~ 50% of known animal species are parasites and that all living metazoan species host at least one parasite is not correct.* There are a number of species totally free from parasitic infection; even within the infected species, 100% prevalence is a rarity.

In their direct cycle, nematodes include four juvenile stages, of which Juvenile 2 (J_2) is infective on plants and J_3 on animals. Within animal parasites, 3,827 mermithids are parasitic as juveniles but their adults are free-living. Among vertebrate parasite species (8,360), a vast majority of them directly infect the Definitive Host (DH), while the others involve more frequently one IH and less frequently two IHs. Hyman (1951b) hinted that in the following taxa, one or two IHs are incorporated within the life cycle of: (i) Rhabditoidean Tylenchidae (282 species), and Allantonematidae (148 species), (ii) oxyuroidean Subuluridae (109 species), (iii) ascaroidean Anisakidae (244 species), (iv) strongyloidean Metastrongyloidae (289 species), (v) spiruroidean Spiruridae (183 species, Thelaziidae (266 species), Acuariidae (299 species), Physalopteridae (290 species), Camallanidae (347 species), Gnathostomidae (61 species); Cucullanidae (1,446 species, after detecting 219 species, for which the history is not known). (vi) Dracunculoidea (178 species), (vii) Filarioidea (740 species) and (viii) Trichiuroidea (550 species). The number of species for each order was drawn from Zhang (2013). On the whole, of 8,360 species, 5,432 or 65% vertebrate nematode parasite species may engage one or two IHs (Table 22.4). In Mesozoa, Myxozoa, Monogenea, oncospheric Cestoda, Nematomorpha and Pentastomida, inclusion of one larval stage is most common. Only Digenea, Acanthocephala and coracidium-involving Cestoda incorporate two IH(s). Of 96,276 parasite species (Table 22.4), 62% of them pass through direct life cycle; they comprise nematodes, turbellarians, mesozoans and isopod crustaceans. The remaining 38% parasites pass through indirect life cycle. Crustacean parasites can be ectoparasitic on skin (e.g. Lernaeiids), or gills (e.g. bobyrids) or endoparasitic (e.g. rhizocephalans). They comprise 6,075 species comprising 1. Cyamidae (32 species) and Bobyridae (652–99 cryptoniscoids, see Williams and Boyko [2012] = 553 species with direct life cycle, 2. Cyclopoidea (2,687 species) and Harpacticoidea (2,233 species) with indirect cycle involving 5 copepodid larval phases, 3A. Lernaeidea (129 species, 2 larval stage [LS]), Ascothoracida (70 species, 2 LS), Rhizocephala (122 species, 3 LS) and Branchiura (146 species, 3 LS, Olesen, 2018), and 3B. Cryptoniscoids incorporating one IH (Boxshall and Hayes, 2019).

In parasites, the inclusion of active motile LS may increase chances of successful infection but not more number of larval stage. For example, the frequency of LS number decreases from 8 host taxa with inclusion of one LS to one host taxa with inclusion of 4 parasitic LS. Not only the frequency of taxa but also the number of parasite species decreases from 10,487 with inclusion of one LS, except in digeneans, for which reasons are described below. With increasing risk involved in transfer from DH to FIH, FIH to SIH and SIH to DH, parasite species do not involve more than two IH. *The inclusion of two IH has decelerated species diversity in the 1,260 speciose cestodes with coracidium and 1,100 speciose Acanthocephala (see the box below).*

From Table 22.4, the underlisted values for the total number and incidence frequency of host taxa for the number of larval stages as well as IH.

Larval stage (no.)	Total species (no.)	Frequency of taxa (no.)		IH (no.)	Total species (no.)	Frequency of taxa (no.)
1	10,487	8		1	5,476	4
2	1,444	2		2	14,376	3
3	6,681	3				
4	12,012	1				

In fact, of 98,093 parasitic species (Table 22.4), 68.7% of them pass through direct life cycle.

With regard to digeneans, for the first time, Pandian (2020) brought to light the following: (1) In the First Intermediate Host (FIH), the digeneans undergo clonal multiplication in sporocyst and redia. However, the fittest clone(s) is selected in the Second Intermediate Host (SIH). Hence, SIH plays an important role in the life history of digeneans. (2) Of 12,012 species, 10,754 or 82.3% digeneans belonging to 22 families engage an SIH. (3) Being euryxenics, the digeneans are extremely flexible to DHs as well as IHs. An estimate has revealed that there are 33,014 potential host species that can be engaged as SIH by 10,754 digenean species. Hence, the cercaria of an average fluke species has the average choice to select one among 3.1 SIH species. The option is limited to one SIH species in five families but the choice is for two, three, four, five and six among the potential SIH in different digenean families (Fig. 22.1). In all, clonal multiplication in FIH, choice of the fittest clone in SIH and selection of 1 among 3.1 potential SIH have enormously increased genetic diversity in Digenea.

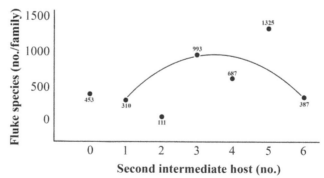

FIGURE 22.1

Number of fluke species/family distributed over the number of second intermediate host (from Pandian, 2020).

23

Sex Determination and Differentiation

Introduction

Sex determination, a delicate process, is the earliest and most basic 'decision' made by a developing embryo. The decision whether to become female or male is conveyed during embryonic development either immediately (e.g. echinoderm) or subsequently (e.g. fishes) by the sex determining primary gene(s). This signal is transmitted through cascade(s) of differentiation genes, which ultimately realize sexualization. Sex is genetically determined at fertilization. Pieces of *Sepia* gonad cultured in a non-hormonal medium developed the ovary and testis at the ratio of 1:1; this observation confirms the existence of genetic mechanism of sex determination and non-labile stable sex differentiation (Mangold, 1987). Sex differentiation is the process, through which the genetically determined sex is realized during the course of development. The process of determination is followed by differentiation and the realized sex ratios are the cumulative end products of these successive processes. But the determination and differentiation are highly diverse processes that have evolved independently a number of times (Hodgkin, 1990). Irrespective of the sex determining gene(s) harbored on one or more sex chromosomes, sex is determined only by gene(s) located on sex chromosome(s). However, the cascade of genes in the differentiation process can be altered by environmental factors. Clearly, the determination is entirely under the genetic control but the differentiation can be environmentally altered. Without distinguishing the difference between the two processes, some have enthustiastically gone to extend of naming sex determination as environmental in a few reptiles. In homeothermic birds and mammals, and possibly echinoderms, both determination and differentiation processes are entirely under the control of genes. In some poikilothermic vertebrates, the sex determination gene(s) may be located in more than one sex chromosomes or autosome(s). In them, sex differentiation is realized by a host of genes in different cascades. Unlike in mammals, the expression of these genes in these poikilothermic vertebrates can be switched on or off by environmental factors, resulting in a labile sex differentiation process. It is intended to know how species diversity is fostered by these processes.

23.1 Genes and Chromosomes

In mammals, sex is decisively and irrevocably determined by a single *Sry* gene located on a morphologically distinguishable Y chromosome, i.e. sex is determined by a single monogenic (XX/XY) chromosomal system so that genes located on autosomes have a small role in it. Among vertebrates, the only other known sex determining single gene is *Dmy/Dmrt1 by* in *Oryzias latipes* (Matsuda et al., 2002, Nanda et al., 2002) and *O. curvinotus* (Matsuda et al., 2003). Firstly, *Sry* and *Dmy* are not homologous genes. *Sry* encodes the transcription factor containing a High Mobility Group (HMG) box-domain but *Dmy* is similar to the *double-sex and male-3 related transcription factor (Dmrt1)* (Pandian, 2013). Secondly, the determination in this medaka is not decisive and irrevocable, as in mammals. In *O. latipes*, the sex differentiation process is so labile that it can hormonally be detrailed to produce 100% females (see Pandian, 2013). Thirdly, the *Dmy* is not the sex determining gene in other fishes including those belonging to the genus *Oryzias*. Not surprisingly, within a single genus, such as, *Poecilia reticulata* is male heterogametic (Kavumpurath and Pandian, 1993a, b) but *P. sphenops* is female heterogametic (George and Pandian, 1995). Hence, there is no common sex determining gene in fishes. Fourthly, unlike in homeothermic mammals, the sex determination gene(s) may be located in more than one sex chromosome in poikilothermic animals (e.g. fishes, crustaceans, insects, Table 12.2). Further, sex differentiation is realized by a host of genes in differentiation cascades. Unlike in mammals, the expression of these genes in fishes and polychaetes can be switched on or off by environmental factors, which result in labile sex differentiation.

Multiple chromosomes: Interestingly, the X and Y chromosomes of medaka are morphologically indistinguishable (Matsuda et al., 1998). In the stickleback *Gasterosteus aculeatus* also, they are morphologically homomorphic. In it, BAC (Bacterial Artificial Chromosome) containing *cyp19b* with 16.7 Mbp in the X chromosome assembly, hybridizes only the X but not the Y nor to any other location in the genome, suggesting that a part of the Y chromosome has been deleted. The large (6 Mbp) deletion on the Y is equivalent to 30% of the sequence content of the X chromosome. Similar findings have revealed that the sex determining region of the Y or W chromosome is a 'hot spot' for deletion, inversion, duplication, amplification, transposition, and other kinds of rearrangements, as well. For example, Y chromosome–autosome fusions have occurred in at least 25 fish species (see Pandian, 2012). These cytological changes and rearrangements have led to the manifestation of the existence of sex determining gene(s) in multiple numbers of sex chromosomes in fishes, and possibly crustaceans and insects. A look at Table 12.2 would reveal that fishes, crustaceans and insects have explored diverse number of multiple sex chromosome systems. However, the simple monogenic chromosomal system (XX/YY or ZZ/ZW) is in operation in 80–90% teleosts, and 71–77%

insects. Hence, *the operation of monogenic male heterogamety (XX/XY) or female heterogamety (ZZ/ZW) has fostered species diversity.*

Although polygenic sex determination system is shown to be evolutionary not stable, there are a few interesting exceptions. With a polygenic sex determination system, the platyfish *Xiphophorus maculatus* has three sex chromosomes in six different combinations, in which XX/XY and ZZ/ZW are mixed. From his long term experiments, Kallman (1984) resolved their sex determination system. In four (XX, XW, WW, XY) of the six possible mating combinations, the matings result in brood sex ratio of 0.5 ♀ : 0.5 ♂. However, the other mating combinations (WY, YY) result in biased sex ratios, i.e. the matings between WX females and XY males produce female biased broods (0.75 ♀ : 0.25 ♂) and those between XX females and YY males produce all male broods. Of 10 Malawian cichlids investigated by Ser et al. (2009), three are shown to carry a peculiar combination of male- and female-heterogamety, as listed below:

Metriaclima pyrosonotus			
Family *M1py*	XX/XY	*M. callonois*	ZW/XY
Family *Rmpy*	ZW/XY	*M. fainzilberi*	ZW/XY
Labeotropheus rewavasae ZW/XY			

Like the platyfish, these Malawian cichlids possess additional sex determining loci on different sex chromosomes. In *M. pyrosonotus*, the known two male- and female-heterogameties operate successfully in different families. Interestingly, when both sex determinant loci W and Y are present in a single individual, as in the cross between the platyfish with WX female and XY male, the W chromosome dominates over the Y chromosome.

Poecilidae comprises ~ 200 species belonging to 22–29 genera and with an exception of oviparous *Tomeurus gracilis*, all poecilids are vivipares. By virtue of a small size and short generation time, many species belonging to the genera *Poecilia* and *Xiphophorus* are preferred for experimental studies. The species sequence listed in Table 23.1 broadly indicates the possible course of evolution of multiple sex chromosomes and the combination of Y and W chromosomes. Two distinctly different trends are notable: (i) with exception of *P. sphenops*, genetic mechanism for deviation of sex ratio in *Poecilia* is limited to autosome(s) alone. (ii) Contrastingly, some *Xiphophorus* species have progressively reduced the importance of X chromosome and its replacement by W chromosome. (a) Thus, the Y chromosome is duplicated in *X. milleri*. (b) To weaken the X chromosome, AA/Aa autosomes are added in *X. nigrensis*. (c) The more dominant W chromosome is included, while retaining the X chromosome in *X. maculatus* and (d) Complete replacement of the X chromosome by the W chromosome in *X. hellerii*. On the other hand, *P. sphenops* seems to have at one stroke replaced the XX/XY male heterogamety by ZZ/ZW female heterogamety.

TABLE 23.1

Origin of three sex chromosomes and combinations of WY in some poecilids (modified from Pandian, 2015)

Species	Sex chromosomes	Reported observation
P. reticulata	X, Y	Autosomal genes override to generate XX ♂ and XY ♀. YY male and YY female are viable
P. cortezi	X, Y	Autosomal generation of XX male
P. velifera	X, Y	Autosomal generation of XX male
Xiphophorus xiphidium	X, Y	Autosomal generation is known
X. montezumae	X, Y	XX ♂, XY ♀, YY ♂ are rare but viable
X. milleri	X, Y, Y^1	2 Y chromosomes. XY^1 ♀, XY ♂, YY^1 ♂ viable
X. nigrensis	X, A, Y	XY ♂ but XX with 2 autosomal allelies A and a. XXAA ♀ and 95% and XXaa ♂ viable
X. maculatus	X, Y, W	XX, XW and YW ♀. XY and YY ♂. Autosomal alteration present
X. hellerii	Y, W	YW ♀, YY ♂
P. sphenops	Z, W	♀ heterogamety

Mitochondrial genes: In many respects, bivalves differ from other animal taxa. (1) In fishes and crustaceans, the maternally supplied mRNAs are located at the animal pole in the embryo and their PGCs migrate dorso-laterally toward posteriorly. Contrastingly, the mRNAs are located at the vegetal pole of the embryo and their PGCs migrate dorso-laterally toward anteriorly in the bivalves (see Pandian, 2017). (2) Many bivalves are not stable gonochorics (e.g. oyster species). (3) In some freshwater bivalves (*Elliptio*, *Anodonta*), sex is more a population specific trait rather than species specific trait. The mitochondrial lineage studies on the Manila clam, *Ruditapes phlippinarum* led Rawson and Hilbish (1995) to suggest that sex determination in bivalves is more a family trait than either population or species specific trait. Kenchington et al. (2002) reported that the matings of a female *Mytilus galloprovincialis* with different conspecific males produced the same sex ratio but those of the same male with different females produced different sex ratios. This finding has also been confirmed in other mussels *M. edulis* and *M. trossus*. Hence, sex determination in these mussels is independent of a male but a trait of the female parent. This important finding has led to the discovery of Doubly Uniparental Inheritance (DUI). Bivalve species with DUI carry two mitochondrial genomes, the first one called the M genome, is transmitted to sons only and the second one namely F genome is transmitted to both sons and daughters. In other words, sons and daughter inherit the F genome from the mother but M genome is inherited from the father to son only. Interestingly, hermaphrodites appearing from crossings of *M. galloprovincialis* have the M genome in the testicular tissues and F genome in the ovarian tissues of the ovotestis (Fig. 23.1A). The DUI has been detected from 36 species belonging to seven families of bivalves from freshwater (e.g. Unionidae) and marine (e.g. Mytilidae, Veneridae) habitats. While the role of chromosomal gene(s) and mitochondrial genome remains to be known, *the DUI may be restricted to a maximum of 9,856 bivalve species and is not known from any other animal taxa. Hence, the non-chromosomal DUI has not enriched species diversity.*

Autosomal genes: In many fishes (e.g. *Poecilia reticulata*, *Salvenilus fontinalis*), genetic differentiation is commenced prior to hatching (Fig. 23.1B). In others, the Gds (Genetic differentiation system) + A (Autosome) are postponed to alevin (e.g. *Oreochromis niloticus*), fry (*Betta splendens*, *Puntius conchonius*), juvenile (e.g. *Siganus guttata*, *Mugil cephalus*) or to puberty (e.g. *Anguilla anguilla*). Hence, it may be noted that the longer the delay of Gds, the greater is the scope for the involvement of Gds + A. In the Gds + A pathway, morphologically distinguishable sex chromosomes are identifiable. But, minor gene(s) located on one or other autosome may override and alter the differentiation to the opposite sex. Nevertheless, the autosomal gene(s) can affect the differentiation only in some individuals but not in all individuals of a labile species. Some of these minor autosomal genes are switched on by environmental factors like temperature, pH or hypoxia (see Pandian, 2015).

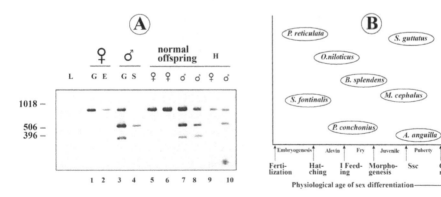

FIGURE 23.1

A. Doubly uniparental inheritance in gonochoric (Lanes 5–8) and hermaphroditic (Lanes 9–10) in pair matings of *Mytilus galloprovincialis*. Lanes 1 and 2 represent female (G) and egg (E) and Lanes 3 and 4 male (G) and sperm (S), respectively (modified and simplified from Saavedra et al., 1997). B. Temporal locations of labile period during physiological age of differentiation in selected fishes (modified from Pandian, 2013).

23.2 Sexualization and Endocrines

Sexualization: As described earlier, the Germ Cells Supporting Somatic Cells (GCSSCs) induce the bisexually potent Primordial Germ Cells (PGCs) to generate oogonia and spermatogonia in the presumptive female and male, respectively. As a result, the bisexual potency of the PGCs is progressively reduced to unisexual potency. Hence, sex is primarily determined by the gene(s) present in the sex chromosomes contained in the GCSSCs. Corresponding with the progressive reduction to unisexual potency of the PGCs, the tissues, organs and systems are also progressively sexualized during development from an embryo to puberty; for example, cheek growing hair in men and growing breasts in women. However, the brain, which controls and integrates the functions in an individual animal, is the most important organ to be sexualized. Thanks to Le Page et al. (2010), the available information reveals the following major differences between fishes and mammals in gonadal differentiation and brain sexualization: (i) Restricted to a small area, the brain is sexualized during the prenatal period in mammals but the sexualization continues in the entire brain throughout the development and adult life in fishes. (ii) The brain of fishes continues to grow during adulthood and regenerates after injury; consequently, the critical window for brain sexualization, which is limited to the prenatal period in mammals, is extended through the entire life span. (iii) Testosterone (T) is the male hormone in mammals, but it is 11-Ketotestosterone in fishes and (iv) there is only one aromatase gene with a high expression in the ovary and the associated function in mammals; however, there are two aromatase genes *cyp19a1a* in the gonad and *cyp19a1b* in the brain of fishes. Though brain sexualization limited to the prenatal stage may have facilitated viviparity with a more effective control and integration of functions, it has led to a kind of specialization and a point of no return (not to even to regenerate injury of the brain). It is not clear whether this led for conquest and colonization *terra firma* by higher vertebrates. *However, the limitation of brain sexualization to the prenatal stage may be one reason for the reduction in species diversity to 9,543 speciose reptiles and 5,513 speciose mammals from 32,510 speciose teleosts, in which sexualization is extended to almost the entire life span.*

Endocrines: In animals, hormones and neuroendocrines constitute the endocrine system. The latter arise from the nervous system and the former are synthesized in specific organs. They act as chemical messenger(s) of the genetic cascade that realize sexualization and maintain sex as well as reproductive cycles. These endocrines may be categorized under two broad groups namely (i) the vertebrate type of steroidogenesis present in vertebrates and echinoderms and (ii) all other invertebrate types found in polychaetes (see Pandian, 2019), and arthropods, as represented by crustaceans (see Pandian, 2016). In vertebrate reproduction, steroids play a major role.

Figure 23.2 shows the steroidgenic pathway, through which the male hormone Testosterone (T) and female hormone Estrogen (E$_2$) are synthesized. Unlike vertebrates, echinoderms do not possess a well defined glandular endocrine system. However, the role of E$_2$ and T in their sexualization and reproduction, as well as the syntheses of T and E$_2$ through the vertebrate type of steroidogenic pathway are described fairly well (see Pandian, 2018).

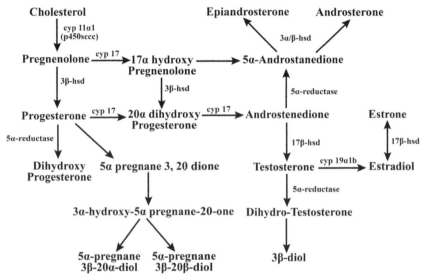

FIGURE 23.2

Steroidogenic pathways in echinoderms and vertebrates (from Diotel et al., 2011).

Among the invertebrate type, despite possessing a well developed circulatory system, the polychaetes do not have a glandular endocrine system. Like fishes, their sex differentiation process remains labile and protracted (see Pandian, 2019). Among arthropods, at least seven organs of crustaceans are identified to produce a dozen or so hormones and neuroendocrines, whose actions are targeted on as many as 10 organs. For example, the Gonad Stimulating Hormone (GSH) promotes the formation of Androgenic Gland (AG), which releases Androgenic Hormone (AH) and in its turn, it induces the development of testes and Secondary Sexual Characteristics (SSCs). In the absence of AG, the ovaries are spontaneously developed in a neuter and the individual becomes a female. Contrastingly, only in the absence of E$_2$, the males are spontaneously developed in fishes. In crustacean females, the presence of ovarian hormone, in its turn stimulates the development of SSCs.

Among protostomes, Mollusca represent the only phylum, in which the vertebrate type of steroids is considered to perform the reproductive functions. However, the following should be noted: (i) Cholosterol is the precursor for the synthesis of T and E$_2$. But it is present in inadequate quantities in many molluscs. (ii) The injected ^{14}C-acetate is not converted into the cholestrol in carnivorous cephalopods and gastropods. (iii) With no evidence for the presence of pregnenolane, the possibility may not exist for molluscs using steroids from other than cholestrol as a starter for biosynthesis of steroids (Fig. 23.2). (iv) The molluscan genome is shown not to contain the genes for any key enzyme that is involved in biosynthesis of the vertebrate type of steroids. Apparently, molluscs have to obtain these steroids from the surrounding water and/or by feeding on plants that contain these steroids (cf paper factor, Pandian, 2022, Section 2.3.1). Many laboratory experiments have convincingly demonstrated the remarkable ability of molluscs to absorb the vertebrate type sex steroids T and E_2. For example, the mud snail *Illyanassa obsoleta* requires only 8 hours to absorb 80% of ^{14}C-labled T from 3 ml solution. In *Mytilus edulis*, the E_2 uptake progressively increases from 0.4 ng/g body on the first day to 14.6 ng/g on the 13th day of exposure. In some molluscs, the presence of an egg-laying hormone and neuroamines is reported (e.g. dopamine, serotonin). Hence, molluscs may also be grouped under the invertebrate type. *On whole, the invertebrate types of endocrine system have fostered species diversity, as invertebrates surpass vertebrates not only by the number of species (as many as 95% of all known animals) but also by individual numbers.*

23.3 Surgical and Social Induction

Experiments involving surgical removal of the hormone secreting organ like the brain or gonad, or enforced isolation of individual(s) of the same sex have provided information on alteration of sex differentiation, and locations of Primordial Germ Cells and their derivatives Oogonial Stem Cells (OSCs) and Spermatogonial Stem Cells (SSCs). The former involves surgical removal or grafting of the selected organ or injection of a hormone. But social induction involves intact animals and provides clues for sex-dependent chemicals like pheromone arising from animal(s). The regenerative potency of nemerteans *Lineus ruber* and *Amphiporus lactifloreus* has provided an opportunity to remove the brain following bisection and trisection of the worm. Thanks to Dr. Jacques Bierne, the existence of Gonad Inhibiting Hormone (GIH) appearing from the brain is known from these primitive acoelomorphs (Table 23.2) and the GIH is conserved in almost all animals. Besides, the fact that the skin glands of *L. ruber* secrete sex-dependent secretion indicates that these glands are sexualized (Bierne and Rue, 1979).

In polychaetes, teleosts and crustaceans, the sex differentiation process is labile and diverse. But in molluscs and possibly also in echinoderms, it remains determinate and stable, and is not amenable to natural hormone(s) and environmental factors like temperature (Table 23.2). For example, isolated gonad pieces of *Sepia* cultured in hormone-free medium differentiated into

TABLE 23.2

Surgical and social induction on sex differentiation and changes in secondary sexual characteristics in nemerteans, molluscs and polychaetes (compiled from Pandian, 2017, 2019, 2021). J = juvenile, GIH = Gonand Inhibiting Hormone, JH = Juvenile Hormone, OH = Ootrophic Hormone, T = Testosterone, E_2 = Estradiol, TBT = Tributyltin

Species	Experiment	Inferences
Nemertea: *Lineus ruber* Bisectomic & trisectomic surgery to decerebrate & GIH	In ♀, ovary develops, faster oocyte growth, and formation oviduct & feminine skin glands In ♂, similar development occurs.	Retention of OSCs in ovary and SSCs in testis. Gonad is responsible for sexualization
Mollusca : Surgical		
Sepia	Gondal pieces cultured in non-hormonal medium develop into ovary and testis at 1 : 1 ratio	Sex differentiation is not amenable to external hormone(s)
Placopecten magellanicus	Injection of E_2 into J still produces 50% ♂. That of T produces 53% ♀	♂ and ♀ hormones do not change ♀ and ♂ sex ratio, respectively
Crassotrea gigas Protandric ♀	Injection of E_2 into J gonad induces no sex change	♀ hormone has no effect on sex differentiation
Mya arenaria	Exposure to synthetic TBT, present in anti-fouling paint	An androgen mimic imposes imposex by increasing male ratio
Polychaete : Surgical		
Platynereis dumerili	Decerebration and brain removal	Precocious sexual maturity
Nereis pelagica	Oocytes incubation with or without ganglia	Nereidin from ganglia inhibits oocyte growth
Ophyrotrocha puerilus puerilus	Decerebrated J attains ♂ phase but not ♀ phase. Grafting prostomium from ♀ into decerebrated J differentiates into ♀	Brain OH induces feminization; hence, brain is sexualized at 16-segment stage, when sex change occurs
Social		
Brania clavata	In the absence of a ♂ or its pheromone, ♀ changes sex to ♂	
Typosyllis prolifera	Isolated ♀ changes to ♂ No sex change, when ♀ : ♂ ratio is equal or unequal above 1 ♀ : 1 ♂	Labile gonochoric worms retain both OSCs and SSCs

female and male gonads at a 1 : 1 ratio. Unassailed by an injection of natural hormone, the testosterone (T) or estradiol (E_2), the process led to 50% females in the sea scallop *Placopecten magellanicus* and 50% of males in the protandric *Crassostrea gigas*, respectively. Only on exposure to the synthetic androgen-mimic tributyltin (TBT), *Mya arenaria* and others undergo imposex and increased male ratio.

In polychaetes, the differentiation process is labile but it is unidirectional. In nereids, the brain and ganglia secrete nereidin, the equivalent of Juvenile Hormone (JH) or GIH, which inhibits sexual maturity (Table 23.2). Decerebration of a juvenile *Ophryotrocha labronica* and implantation of the brain drawn from a sexually matured worm have shown that male sexualization is a stable process and sex can be changed from a female to male but not male to the female. On decerebration of J and implantation of prostomial brain from sexually mature female in the protandric *O. puerilus puerilus*, the Juvenile (J) directly differentiated into a female, bypassing the male phase. Ootrophic Hormone (OH) appearing from the brain was shown to induce feminization, suggesting that the brain is sexualized, as the worm grows to 16-segment stage, at which the male phase is changed into the female phase in this protandric worm.

Beside the lability, the differentiation process in teleosts is also diverse (Table 23.3). In castrated gonochores *Gambusia affinis*, *Oncorhynchus mykiss* and *Ctenopharyngodon idella*, following surgical removal of 96–100% testis including the surrounding mesentery, testis is regenerated including gonopodium in *G. affinis*. This regeneration indicates that their SSCs are stored elsewhere in the body. Castrated *O. mykiss* regenerates the testis, irrespective of E_2-treatment and even in XX neomales. The ovariectomized female regenerates the ovary only 67% of *Oreochromis niloticus* and 38% of *C. idella*. Clearly, the retention of OSCs elsewhere in the body varies from individual to individual female. Conversely, the removal of testicular tissues alone from the ovotestis of the protandric *Acanthopagrus schlegelii* eliminates the male phase but the ovary functions normally. Hence, SSCs are not stored elsewhere in the body of *A. schlegelii*. In the relatively more labile secondary gonochores, *Crenicara punctulata* and *Matriaclima livingstoni*, the enforced isolation of J or a few females of different sizes induce a sex change from female to male, but limited to only one α-male (Table 23.3). However, the sex change can be bidirectional in the anabantid *Macropodus operculatus*. In protogynics, on isolation of a pair of males, one of them changes sex to ♀, while the other remains as ♂. Reared in a group of small and large ♂, the small ones change sex to ♀. In isolated multiple ♀, the largest ♀ change sex to ♀, although an isolated ♀ remains as ♀ only.

In crustaceans, the process is also labile but more diverse (Table 23.4). Other than AG, MF and JH increase the male ratio in daphnids. Ablation of Mandibulan Organ (MO) enhances the reproductive functions in penaeids. Regarding social induction, the existing *Ione thoracica* female chemically induces the incoming conspecific undifferentiated juvenile to differentiate into the male. But the reverse happens in *Variocorhinus bachatulus*, in which the existing sex differentiated male changes sex to the female, when a new male arrives. Hence, the retention of PGCs with bisexual potency or both OSCs and SSCs is presumable. By almost completely depriving nutrients to the gonads, branchial parasitic isopods like *Bobyrus* destroy the gonads and AG, and thereby almost totally the OSCs in female and SSCs in their male hosts

TABLE 23.3

Surgical and social induction of sex change in labile teleost fishes (compiled from Pandian, 2013). SG = Secondary gonochore, P ♀ = Protogynic hermaphrodite

Species	Experiment	Inference
Surgical		
Gambusia affinis	Following 96% removal, testis and gonopodium regenerated	Adequate number of SSCs are retained in 4% testis
Oncorhynchus mykiss	Castrated male regenerates testis, irrespective of E_2-treatment. Gonodectomized neomale (XX) also regenerates testis	Irrespective sex reversal and E_2 treatment, only SSCs retained elsewhere (other than testis) are expressed
Acanthopagrus schlegelii P ♀	Testicular removal from ovotestis, ovary functions normally. Testis not regenerated	In ovotestis, OSCs are retained in ovary alone and SSCs in testis alone. PGCs are no more present in testis or ovary
Oreochromis niloticus	67% ovariectomized (including mesentery) ♀ regenerates ovary	OSCs are retained in ovary. Regeneration with adequate number of OSCs
Ctenopharyngodon idella	91% gonodectomized (including mesentery) ♂ regeneratea testis and 38% ♀ ovary	OSCs and SSCs are retained elsewhere; but SSCs in more numbers than OSCs
Belta spendens, SG	In ovariectomized ♀, regenerated ovary, even with E_2.	♀ retains OSCs in ovary only but SSCs elsewhere also. On ovariectomy, SSCs begin to express
Social		
Crenicara punctulata SG	Reared in combination of ♂ + ♀♀ ♀, of which α ♂ was removed	The large ♀ changed sex to α ♂
Matriclima livingstoni SG	Many ♀♀ reared in isolation	One large ♀ changed sex to α ♂. In them, SSCs are also retained
Macropodus operculatus SG	J reared alone ♂, ♂ reared individually isolation Reared in groups of 10 to 23 ♀♀	Matured into ♂ Changed sex to ♀♀ 2–3 ♀♀ changed sex to ♂♂
Centropyge spp P ♀	♂♂ reared together	One ♂ changed sex into ♀
Labroides dimidiatus P ♀	Reared large ♂ and small ♂	Small widowed ♂ differentiated into ♀
Thalassoma bifasciatum P ♀	J reared alone 3 J reared in a group	J differentiated into ♀ Largest J differentiates into ♂
Parapercis snyderi, P ♀	Single ♀ Multiple ♀♀♀	No sex change Largest ♀ changed sex into ♂
Trimma okinawa, P ♀	Polygynic mating system	Large one changes to ♂
Paragobiodon echinocephalus P ♀	Monogamous mating system	Sex changes brought equal sized ♀♀ and ♂♂

TABLE 23.4

Surgical, social, parasite or food induced changes on sex differentiation in crustaceans (compiled from Pandian, 2016). AG = Androgenic Gland, AH = Androgenic Hormone, MO = Mandibular Organs, MF = Methyl Farnesoate, JH = Juvenile Hormone, J = Undifferentiated juvenile

Species	Experiment	Inferences
Surgical		
Armadillium vulgare	Bilaterally AG ablated J differentiates into ♀ but AG implantation produces ♂	Through AH, AGs masculinizes J, Ovary is differentiated with no testis
Daphnia magna	Exposure to MF or JH increases male ratio	By reducing molting frequency
Litopenaeus vannamei	In MO ablated ♀, fecundity and hatching increases	Regulates testicular and ovarian maturation
Social		
Parasite	**Social (host)**	**Effects**
Ione thoracica	*Callianassa laticauda*	Existing female induces income into ♂
Variocorhinus bachatulus	*Ichthyoxenus fushanensis*	Arriving ♂ induces existing ♂ into ♀ *Retention of both OSCs and SSCs*
Parasite		
Palaemon squilla	*Bobyrus*	Castrated ♀ with non-setosed pleopods. Reproductive death of ♀ & ♂ gonad & loss of OSCs & SSCs
Cyphochara gibbert	*Riggia pranensis*	Oogenesis inhibited
Balanus glandula	*Hemioniscus balani*	Ovary destroyed. Testis functions normally. *In it, retention of SSCs in testis located elsewhere in the body*
Sacculinids Copepod: *Sphaeronellopsis monothrix*	Crabs Ostracod: *Pasterope pollex*	Destroys testis and possibly AGs and thereby eliminate SSCs
Food		
Copepods	Dinoflagellate *Blastodinium* spp	Castrates and feminizes copepodites. *The alga destroys (AGs) & thereby SSCs*
Calanus helagolandicus	Abundant food supply	Presumptive ♀ differentiates into ♂
Acrocalanus gracilis	Limited food supply	Presumptive ♂ differentiates into ♀. *In them, both OSCs and SSCs are retained till maturity*
Hippolyte inermis Protandric ♀	*Coconeis neothumensis*	*C. neothumensis* destroys AGs & thereby eliminates SSCs

e.g. *Palaemon squilla* (for more examples, see Pandian, 1994, 2016). However, the reproductive death of a gonad is limited to females alone in fish hosts like *Cyphochara gibbert* and in hermaphroditic cirripedes like *Balanus glandula*; in the latter, the testis and ovary are located at different locations in the host body. Regarding sacculinids and copepods, the story is a little different. Injection of aqueous extract of rootlets of the sacculinized *Loxothylacus panopei* into unparasitized *Rhithropanopeus harrisi* male resulted in pyknosis of the testis and destruction other organs including AGs. In the absence of AGs, depletion of AH and degenerated testis, the male host tends to exhibit feminine abdominal characteristics. Sucking the hemolymph of the host myodocopid ostracod *Pasterope pollex*, the copepod *Sphaeronellopsis monothrix* castrates its host and lays her eggs in a brood chamber of the ostracod host and the parasite's eggs are brooded by the host. Food quality and quantity have a profound effect on the sex differentiation process of crustaceans. For example, the dinoflagellate algae *Blastodinium* spp and *Coconeis neothumensis* castrate and feminize copepods and *Hippolyte inermis*, respectively. The latter is a protandric hermaphrodite and when fed on *C. neothumensis*, *H. inermis* directly differentiates into the female bypassing the male phase. Understandably, these algae destroy AGs and thereby eliminate the formation of testis and SSCs, as well. In other copepods, levels of food supply alter differentiation of the presumptive female into a male and male into a female (Table 23.4). Abundant food supply seems to enhance the activity of AGs and AH to skew sex ratio in favor of the male. The reverse is true for limited food supply.

24

Metamorphosis and Recruitment

Introduction

The life cycle of more than 82% taxa is indirect (Table 20.1) and involves one or more larvae (Fig. 20.1), which differ from the respective adults in appearance, feeding habits and habitats. The sessile aquatic animals settle at a selected demersal site either during or after metamorphosis. Hence, recruitment, which perpetuates a population/species, is preceded by metamorphosis and settlement. This chapter describes the role played by environmental factors on metamorphosis, settlement and recruitment.

24.1 Metamorphosis

It is a crucial event in the life history of animals passing through an indirect life cycle. It involves structural, functional and behavioral changes inclusive of (i) regression of structures that are relevant only to larva, (ii) dedifferentiation of larval structures into those suitable for an adult life and (iii) development of *de nova* structures and functions essential to the adult. In most animals, metamorphosis is gradual, for example, hemimetabolous insects, teleost fishes. In free-living crustaceans like Caridae, Anomura and Brachyura and parasitic Monstrilloida, Ascothoracica and Rhizocephala, it is drastic and strong. However, available relevant information is limited to a few like the holometabolous insects. Hence, the description of these aspects has to be limited to two examples only; these are holometabolous insects and amphibians.

A number of environmental factors like food quality, ration, rearing density, temperature and others greatly affect the larval duration. The duration for the frog *Rana tigrina* ranges from 16 days at 37°C to 96 days, when fed plant food. The same holds true for *Chironomus circumdatus*; in it, the larval duration ranges from 16 to 28 days (Table 24.1). But the metamorphic duration remains constant at 6 days for the tadpole reared on different feed,

TABLE 24.1

Effects of environmental factors on larval and metamorphic duration in *Chironomus circumdatus* and *Rana tigrina* (compiled from Sakaraperumal and Pandian, 1991, Pandian and Marian, 1985e)

Factor	Tested range	Duration	
		Larval	**Metamorphic**
Chironomus circumdatus			
Ration	*Chlorella*: 13–25 × 10⁶ cells.ml at 27°C	16–28 days	16 hours
Temperature	22–37°C at 25 × 10⁶ cell/ml	8–20 days	6–24 hours
Rana tigrina			
Food quality	Animal or plant	24–96 days	6 days
Ration	10–100% of *ad libitum*	24–78 days	6 days
Rearing density	116 tadpole/aquarium	24–37 days	6 days
Water depth	2.5–5.0 cm	25–30 days	6 days
Temperature	22–37°C *ad libitum*	16–50 days	3–8 days

ration and densities. In *C. circumdatus* also, it remains constant at 16 hours at 27°C, irrespective of changes in ration from 13 × 10⁶ cells/ml to 25 × 10⁶ cell/ml. The metamorphic duration is increased from 3 days at 37°C to 8 d at 22°C for the frog and from 6 hours at 37°C to 24 hours at 22°C for chironomus. Hence, the larval duration is regulated more by external factors but that of metamorphosis by internal factors, except for temperature.

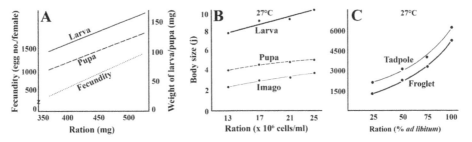

FIGURE 24.1

A. Effect of ration on fecundity and weight of terminal larva and pupa of the lepidopteran armyworm *Mamestra configurata* (redrawn from data reported by Bailey, 1976). B. Body size of terminal larva, pupa and imago of *Chironomus circumdatus* as a function of ration levels. C. The same for the tadpole and froglet of *Rana tigrina* (drawn from data reported by Sankaraperumal and Pandian, 1991, Pandian and Marian, 1985e).

In the lepidopteran armyworm *Mamestra configurata*, increasing ration from ~ 300 mg (dry weight) to ~ 530 mg food, the Terminal Larval (TL) size increased from 100 mg (live weight) to 160 mg (Fig. 24.1A). Correspondingly, the pupal weight also increased from 70 to 120 mg. In the dipteran *Chironomus*

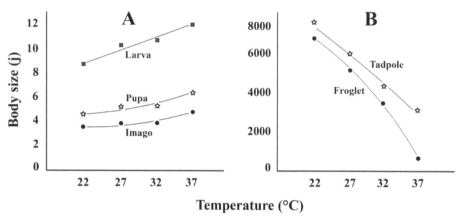

FIGURE 24.2

Effect of temperature on body size of (A) terminal larva, pupa and imago of *Chironomus circumdatus* and (B) the same for the tadpole and froglet of *Rana tigrina* (drawn from data reported by Sankaraperumal and Pandian, 1991, Pandian and Marian, 1985e).

circumdatus too, the TL size at 27°C increased from 7.8 j at the lowest ration (13 × 10⁶ *Chlorella* cell/ml) to 10.2 j at the highest ration (25 × 10⁶ cells/ml). Correspondingly, the increases were from 4.0 to 5.1 j for the pupa and from 2.4 to 3.7 j for the imago (Fig. 24.1B). In the frog *Rana tigrina* also, the terminal (climax) tadpole size increased from 2,026 j at the lowest ration (25% of *ad libitum*) to 6,103 j at the highest ration (*ad libitum*). Correspondingly, the froglet size also increased from 1,335 j to 5,199 j (Fig. 24.1C). Therefore, increasing ration increases the body size of TL/tadpole as well as imago/froglet.

At the ration size of 25 × 10⁶ cells/ml, the TL size for the *C. circumdatus* increased from 8.6 j at 22°C to 12.0 j at 37°C. Correspondingly, its pupal (4.5 to 6.2 j) and imago (3.5 to 4.9 j) size also increased (Fig. 24.2A). Contrastingly, the tadpole size decreased from 8,901 j at 22°C to 3,679 j at 37°C (Fig. 24.2B). As a consequence, the froglet size also decreased from 6,882 to 3,106 j. However, Metamorphic efficiency (Me) (size of imago/froglet ÷ size of TL/tadpole × 100) ranges between 82 and 85% for the frog reared at 22°C and 37°C. On the other hand, the efficiency progressively increased from 66% at the lowest ration to 85% at the highest ration. In the chironomus too, it increased from 30.6% at the lowest ration to 36.3% at the ration size of 21 × 10⁶ cells/ml and rapidly increased to 72% at the highest ration. From these, the following (around 40% for the chironomus and 80% for the frog) may be inferred: 1. The wide differences between the efficiency values reported for the dipteran insect and amphibian frog indicates that the holometabolous insects undertake strong drastic metamorphosis, in comparison to the gradual metamorphosis by the frog. Whereas the size of TL, pupa and imago rises with increasing ration as well as temperature (up to 37°C), the size of the tadpole (at climax) and froglet decreases with increasing temperature, although the size increases with rising ration. Hence, to change in temperature, the response of chironomus—as indicated by the body size—differs from that of the frog.

An attempt was made to assemble available data on metamorphic efficiency of animals characterized by an indirect life cycle. The just hatched herring (*Clupea herangus*) larva is sustained by the left over yolk; the non-feeding larva undergoes gradual metamorphosis. Blaxter and Hempel (1966) indicated the size of the just hatched larva and that in which yolk is completely utilized as 9.06 and 5.20 mg; these values suggest 57.4% metamorphic efficiency. As in ephemeropterans, *Chironomus circumdatus* also has a transient adult life span of 2–3 days. The lepidopteran adults have a little longer fluid-feeding adult. The efficiency values available for *Spodoptera litura* is 22.1 and 24.8% at lower and higher rations (Seth and Sharma, 2002). As indicated earlier, the hymenopteran wasp *Sceliphron violaceum* larviposits only after depositing 65 mg (maximum 200 mg) paralyzed spiders as feed for its developing larva. Marian et al. (1982) provided a range of rations from 50 to 100% of *ad libitum* and found that the efficiency was 31% at 50 ration and 38% at *ad libitum* ration and the efficiency averaged 31.9%. For the hemimetabolous orthopteran grasshopper *Poekilocerus pictus*, the value reported is around 90% (Delvi and Pandian, 1979). *Hence, reduced ration decreases metamorphic efficiency in almost all taxa. On the whole, the available data suggest that with a change from gradual to strong, drastic metamorphosis, the efficiency decreases from ~ 90% in hemimetabolous insects to 68–84% in the frog, ~ 57% in herring, 35–40% in a dipteran, 31–38% in a hymenopteran and to 22–25% in lepidopterans. Notably, the efficiency is the lowest in holometabolous insects.* For want of adequate information, it may be premature to draw any generalization, especially in the context of species diversity.

24.2 Settlement

In animals, sessility occurs only in aquatic habitats. It occurs across almost all major phyletics and in many minor phyletic species. The approximate number of sessile species is about 44,580. Of 353,573 aquatic animal species (Table 24.2), sessility constitutes 12.9%, which clearly indicates that *sessility in animals deters species diversity*. Settlement of pelagic larvae is of both academic and economic importance. For, the settling or fouling larvae creates problems in habors, ships and on coolant screens in power stations. Hence, these larvae have attracted much attention, especially those like polychaetes and barnacles, which foul gregariously from colonies. A preamble is required to distinguish metamorphosis from settlement. They are temporally separate processes. The former is defined as the process, by which a larva undergoes a series of changes to terminate the larval phase. But the settlement is

TABLE 24.2

Approximate number of sessile animals. 1 = Pandian (2021),
2 = Gibbons et al. (2010), 3 = WoRMS, 4 = Pandian (2018)

Phylum	Species (no.)
Porifera[1]	8,553
Chidaria	10,388
Polychaete (see Table 3.2)	2,000
Ostracoda + Cirripedia	11,204
Gastropoda – Crepidulidae[3]	155
Bivalvia (Mytilidae)[3]	459
Crinoidea[5]	700
Rotifera[1]	1950[†]
Entoprocta[1]	200
Bryozoa[1]	5,700
Brachiopoda[1]	391
Hemichordata[4]	25
Urochordata[4]	2,855

[†] 4% motile rotifers; 245 speciose[2]
medusozoan hydrozoa + 223 scyphozoan species[3]

the process, by which a planktonic larva explores and selects a suitable substratum, toward which it moves to finally settle. Larval settlement on a large spatial scale is primarily determined by hydrodynamics; however, the successful settlement by competent larvae (starved and aged larvae may not be competent) on smaller spatial scale is mediated by abiotic and biotic cues. These cues may originate from host plants/animals, bacterial microfilms or habitats (Table 24.3). The ease with which the microscopic settling larvae can visually be recognized by the formation of a milky white calcareous tube in serpulid polychaetes has facilitated many publications. Bryan et al. (1998) recognized that alcohol extract of dried aqueous *Bugula neritina* leechate carries the compound responsible for attraction of *Hydroides elegans* larvae to settle. Subsequent experimental investigations have shown that one or other polysaccharides, fatty acids and amino acids can also attract settlement. Recent studies have shown that the microbial film (Beckmann et al., 1999), and chemical signals like the polar aliphatic aminoacids emanating from the biofilm (Hadfield et al., 2014) are not individually responsible for the final settlement but the sorbent-like substratum acting as a co-factor is also responsible to induce final settlement of *H. elegans* larvae (Pandian, 2019).

More interesting information is available for the distance travelled and duration required for settlement. The estimated pelagic larval duration for some gastropods ranges from 42 days in *Pedicularia sicule* to

TABLE 24.3

Factors that induce or inhibit settlement of polychaete larvae (from Pandian, 2019)

Species	Factors/Reference
Polydora ligni	Starvation reduces settling ability
Hydroides elegans	Aged larva loses settling ability
Spiorbis borealis	Dark substratum attracts settlement
Capitella sp	Organic rich sediments (but not hydrogen sulfide (Cuomo, 1985) attracts settlement
Capitella sp	Juvenile hormone arising from sediment attracts settlement
S. borealis	*Fucus serratus* attracts settlement
S. rupestris	*Lithothammoni* attracts settlement
Spirobranchus giganteus	*Diploria strigosa* attracts settlement
H. elegans	*Bugula neritina* attracts settlement
H. elegans	Levels of glycojuvenate secreted by rod bacteria attract settlement
Janua brasiliensis (experimental)	Extracellular polysaccharides and glycoprotein attract settlement
Phragmatopoma californica (experimental)	Fatty acids: cis eicosapentaenoic acid, palmitic acid, palmitoletic acid attract settlement
H. elegans (experimental)	Amino acids: glycine, glutamine, aspartic acid arising from leechate of *B. neritina* attract settlement

207 days in *Cymatium nicobaricum*. However, many gastropod larvae do not metamorphose in the absence of their preferred food or substratum. For example, the prosobranchs *Rissoa splendida* and *Bittium reticulatum* settle preferentially on the alga *Cystoseira*. The duration required for transport and settlement of a gastropod larva across the Atlantic is ~ 170 days (Sheltema, 1966). Incidentally, the trans-Atlantic voyage of Christopher Columbus to discover America took 86 days. The probability for a larval transportation across the Atlantic and settlement is 1 in 80 billion (Sheltema, 1966).

24.3 Recruitment

In animals, the level of recruitment determines the population size. From the follicular stage onward to natality, an important milestone in recruitment, a taxon is subjected to continuous reductions at (i) potential to realized fecundity, (ii) fertilization and (iii) hatching. For each of these categories, only piecemeal information is available. The following may be kept in mind: (i) As indicated (e.g. Table 4.1), reduction in food supply can considerably reduce the realized fecundity. In the field also, competition for resources among females and parasitic infection can also reduce it (e.g. *Melanoplus dawsoni*, p 190). (ii) In almost all eutelic pseudocoelomates, for want of adequate number of sperms, 30–50% of subsequent batches of eggs of nematodes remain unfertilized in eutelics as well as in some major phyla. In many sessile animals, eggs are retained within parent body; the emitted sperms have to travel a long way prior to being captured by tentacles (e.g. Polychaetes: *Nicolea zostericola*, *Neoplea septochaeta*), lophophore (e.g. Bryozoa) or by incoming water current (e.g. mesogastropod *Fissurella nubecula*). In the bryozoan *Lophopus crystallianus*, only 115 sperms are able to reach the ovary to fertilize 18 ocytes (see Ostrovsky, 2013). (iii) Hatching success is significantly affected for those animals, whose digestion-resistant eggs are dispersed by birds and other animals. Among crustaceans, the fraction of subitaneous and diapausing digestion-resistant eggs, that are able to hatch, is low (3–18%), in comparison to those of copepods (50–84%) (Pandian, 2016).

The second component in recruitment is mortality of incubated eggs and brooded embryos and free-spawned eggs, larvae and juveniles. Considering crustaceans, brood mortality ranges from 12% for pouch brooders to 24% for marsupium brooders and 43% for pleopod brooders (Pandian, 1994). In asteroid echinoderms, the mortality ranges from 66.7% in *Leptosynapta clarki* to 92.5% in *L. hexactis* (see Fig. 1.8 of Pandian, 2018). From his survey in the natural field, Seigel (1993) reported mortality of treefrog *Rana sylvatica* as ~ 5% for incubated eggs and ~ 35% for the tadpole in the relatively safer treeholes. However, no information on survival for the froglet was reported. Being small, the vivipares and their smaller progenies are subjected heavy predation (Reznick et al., 2006).

A number of mathematical models have been developed to assess or predict recruitment (e.g. a robust supervised classification method). Figure 24.3 indicates some life history stages, at which recruitment is reduced in free-living and parasitic animals. However, a computer search for information on actual recruitment revealed the availability of limited data from the egg to recruitment stage in natural fields. Using a direct counting method to estimate the number of oviposited eggs and emerged young ones for a period of two years, Mathavan and Pandian (1977) reported that of 7.27 million eggs oviposited by the dragonfly *Brachthemis contaminata* from September to November, only 23,990 adults emerged (from November to May), i.e. the recruitment from an egg to emergence was only 0.03%. Using a genetic marker method, Kristiansen et al. (1997) found that of 18 million eggs spawned by the cod *Gadus morhua* in the western Norwegian coast, only 120 young ones survived after a period of one year, i.e. the recruitment was only 0.00067% (see also Marcogliese, 1995). Hence, recruitment can range from 0.03 to 0.00067% in free spawning free-living aquatic animals. In humans, birth rate (~ 'recruitment') is 1.60, 1.77 and 2.24/female in China, the USA and India, respectively (Gupta, *The Hindu* dated September 11, 2020). In the context of cereal production (Table 27.2), these values clearly urge the need for

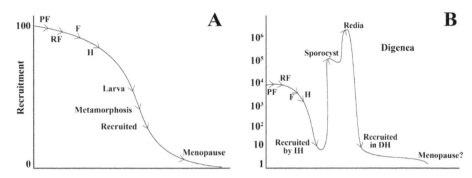

FIGURE 24.3

Models for the proportion of recruitment in A. free-living animals involving indirect life cycle (based on dragonfly, Mathavan and Pandian, 1977) and B. Parasitic animals involving indirect life cycle and incorporation of one or more intermediate host(s), in which clonal multiplication occurs (based on Digenea, see Pandian, 2020). PF = Potential fecundity, RF = Realized fecundity, F = Fecundity, H = Hatching, IH = Intermediate host, DH = Definitive host.

the earliest stricter implementation of family planning in India. There are also some reports for parasites. Dronen (1978) reported that of 58,320 egg/m²/y released, only 3.25% miracidia successfully infected the First Intermediate Host (FIH) the snail *Physa virgata*; the snail emitted 2,100 cercaria/m²/y; of them, only 1.55% gained access into the dragonfly nymph, the Second Intermediate Host (SIH). Finally, 0.03% of the fluke's eggs were recruited into the Definitive Host (DH). The cestode *Bothriocephalus rarus* oviposits at the rate of 26.5 proglottid/d (each containing 8.57 egg) for a period of 30 days. The transmission efficiency is 2.2% for the coracidium larvae to infect the copepod *Macrocyclops ater* (FIH) and 3.8% for the plerocercoid larva to infect the red spotted newt *Notophthalmus viridescens* (DH) (Jaroll, 1980). However, the transmission efficiency values from planktonic SIH to DH ranges from 0.0005% for the digenean fluke *Paronatrema* sp to 0.16% for *Brachyphallus cretanatus* (Marcogliese, 1995). Firstly, these recruitment/transfer efficiency values may indicate the magnitude of 'struggle for existence, natural selection and survival of the fittest'. Secondly, the transfer efficiency values are higher in relatively smaller aquatic systems like the Pond Idumban or Colorado River but in the larger oceanic system, it is low.

Incidentally, it may be noted that there are publications reporting recruitment ranging from 16% in diploid to 45% in hexaploid Japanese loach *Misgurnus anguilicaudatus* (see Table 18.2). In the grasshopper *Poekilocerus pictus* inhabiting the natural fields of Bangalore, India, the recruitment is ~ 30%. *P. pictus* accumulates the poisonous glycosides by feeding on *Calotropis gigantea* and it is predated by *Mantis religiosa* alone. Hence, *C. gigantea* → *P. pictus* → *M. religiosa* remains as an isolated food chain within the highly complicated food web. These incidences may be more an exception than rule.

It has been a time-tested practice to use *C. gigantea* as a natural fertilizer for paddy plants by the South Indian farmers. Being a natural pesticide, *C. gigantea* killed all the root parasites. By the time the paddy is fruitening, the glycosides are all decomposed. Consequently, neither the straw (feed for cattle) nor paddy seeds contain even traces of glycosides.

Part C
Past, Present and Future

Since the origin of life, earth has witnessed independent emergence of new taxa multiple numbers of time. It has also witnessed extinction of a large number of microorganisms fossilized as fuels, and macroorganisms with hard structures as imprints, ambers and fossils. The statement from the past tells us of the emergence and extinction of taxa are ongoing processes and unavoidable by-products of evolution. However, efforts are being made by man to conserve endangered animals and plants. With the arrival of molecular biology, it has become possible to sequence the genome and store it in a computer, from which only the same can be downloaded. On the other hand, the genome can be stored in eggs and dormant eggs by animals, seeds by plants and spores by microbes—all of them are downloaded as their respective offsprings and that too even after hundreds of years from resting eggs by animals, thousands of years from seeds by plants and millions of years from spores by microbes. To meet his avarious, ever increasing demand, man continues to extract more and more energy from fossil fuels at an increasingly faster rate. This has concentrated carbon dioxide in the atmosphere (as oxygen did it during the geological past), which has led to global warming and ocean acidification, collectively known as climate change. The change continues to reduce the life span and reproductive output, alter sex ratio and geographical distribution of organisms. Yet, green shoots and new hopes are beginning to appear in the form of temperature-resistant individuals and temperature-insensitive strains and species. Hence, the earth shall continue to witness burgeoning life forms, hopefully inclusive of man.

25

Message from Fossils

Introduction

Evolution is an ongoing process, and speciation and extinction are its unavoidable by-products. Extinct organisms have left their remnants in the form of fossil fuel by microbes, coal by plants, imprinted by soft-bodied plants and animals, amber by plants gum secretion over insects and fossil by skeleton/shell of animals. They all have a statement to tell us that extinction of organisms is not new or rare, and shall continue to occur so long evolution proceeds.

25.1 Geological Time Table

Using the carbon dating method, geologists have developed a procedure to estimate the age of earth during the geological past. Accordingly, the earth's age is considered as 4 billion years. More than 2.0 billion years ago, organisms discovered sex, which was successfully manifested in them (Butlin, 2002). Almost all the oceanic invertebrate phyla have been flourishing since 360 million years ago. The period between those times to the present is divided into three namely Paleozoic, Mesozoic and Cenozoic eras (Table 25.1). The Paleozoic era is further divided into four epochs, and each of the Mesozoic and Cenozoic eras consists of six epochs. A look at Table 25.1 reveals that (1) The earth began to warm up during Ordovician, continued it during Mississippian, and Oligocene (at least once during Palaezoic, Mesozoic and Cenozoic eras). Hence, climate change and global warming, that are encountered these days, are not new to the earth. (2) The extinction of organisms is also not new. In fact, extinction of trilobites was already abundant during the Cambrian. It was widespread among plants and animals during the Permian. Cretaceous witnessed the extinction of dinosaurs. (3) Speciation is also not new. Considering only vertebrates, their appearance began during the Ordovician; amphibians appeared during the Devonian, reptiles during

TABLE 25.1

Geological time table with remarks on origin, evolution and extinction of organisms as well as climate change (based on Wallace, 1991 and others)

Epoch	Million Y ago	Remarks
Era : Paleozoic (from 600 to 360 million years)		
Cambrian	600	Appearance and evolution of organismic life – most invertebrate phyla – abundant trilobite fossils – **Mild climate**
Ordovician	500	Appearance of vertebrates – Dominance of invertebrates and algae, mosses. **Warming climate**
Silurian	425	Appearance of land plants and animals *Continents increasingly becoming drier*
Devonian	405	Ascendance of teleosts – Appearance of amphibians and vascular plants. **Frequently glaciation**
Era : Mesozoic (from 359 to 160 million years)		
Mississippian	355	Many sharks and amphibians, large scale trees and gymnosperms **Climate warm and humid**
Pennsylvanian	310	Appearance of reptiles and gymnosperms. Dominance of amphibians and insects
Permian	280	Widespread extinction of animals and plants. Cooler and drier climate **Widespread glaciations. Atmospheric CO_2 and O_2 reduced**
Triassic	220	Appearance of dinosaurs, Gymnosperms dominant. Extinction of seed ferns. Deserts appear
Jurassic	181	Appearance of birds, mammals and flowering plants. Rapid evolution of dinosaurs
Cretaceous	135	Appearance of monocots, oak and maple forests, and modern grasses and cereals. Massive extinction of dinosaurs
Era : Cenozoic (from 159 to 1 million years)		
Paleocene	65	Appearance of placental mammals
Eocene	54	Appearance of hoofed mammals & carnivores – **Erosion of mountains**
Oligocene	36	Appearance of modern mammals and monocotyledons. **Warmer climate**
Miocene	25	Appearance of anthropoids, apes – Rapid evolution of mammals
Pliocene	11	Appearance of man – Declining forests – spreading grasslands. **A lot of volcanic activity**
Pleistocene	1	Age of man – Large scale extinction of plant and animal species – **Repeated Glaciations – End of Ice age – Warmed climate**

the Pennsylvanian, dinosaurs during the Triassic, birds and mammals during the Jurassic and Man during the Pliocene. During the checkered history evolution, speciation and extinction have repeatedly occurred, in response to continuously changing the earth's climate and habitats.

25.2 Extinction: Examples

There are taxa, which appeared as early as during the Cambrian, that are continuing to evolve up to now (e.g. Cyclostomatid bryozoans, see Pandian, 2021). On the other hand, calcarean sponges that originated during the Devonian became extinct during the Eocene, except for the remnants of a few species belonging to *Petrosoma, Murrayana* and *Minchinella* (Hyman, 1940). Strikingly, dinosaurs, which originated during the Triassic, were completely extinct during the Cretaceous epoch. Aside from them, the best example for extinction appeared from the 20,000 speciose Echinodermata. Of them, only 7,000 species (35%) are living today. In many respects, the echinoderms are a little different. Among the higher metazoans, they are characterized by radial symmetry but their larvae are all bilaterally symmetrical. Except for the holothuroids, echinoids and crinoids, the others face substratum. Hence, they must obligately remain benthics. The sub phylum Pelmatozoa within Echinodermata included Crinoidea and a few others. Of them, the blastoideans are known to have existed from the Ordovician to the Permian (Fig. 25.1). Among its close relatives of Crinoidea, Cladida, Flexibilia

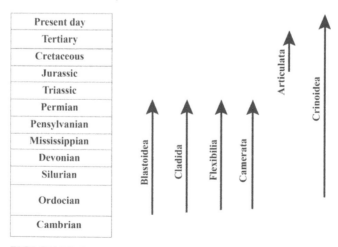

FIGURE 25.1

Origin and extinction of some pelmatozoan echinoderms during the geological past (compiled from Rozhnov, 2002, fossilcrinoids.com, Echinodermata, Uppsala University).

(10 families, 47 genera) and Camerata (29 families, 129 genera), which began to appear during the Ordovician, also became extinct during the Permian. The short lived 630 speciose Articulata, which began to appear during the Jurassic, did not last beyond the Tertiary. Only 35% of crinoids (700 species) have survived from the Cambrian to the present. Among echinoderms, the life cycle of crinoids alone involves non-feeding (lecithotrophic) doliolarialarva. Hence, their fecundity must have been low. *With low fecundity and clonality, crinoids seem to have led to a relatively greater extinction.*

25.3 Extinction and Its Causes

The story of dinosaurs is known to school children. Hence, it may be a good idea to trace the causes for their extinction. The Jurassic Park has taught one that dinosaurs and gigantism are synonymous. However, the newly described dinosaur *Kongonaphon kely* from Madagascar was as small as a bug and measured 10 cm tall and 30 cm long. *K. kely* was alive around 237 million years ago, when Madagascar was directly attached to India as part of the Gondwana (Kammerer, 2020). Hence, gigantism may not be a reason for the extinction of dinosaurs. The size of *K. kely* suggests that some dinosaurs and pterosaurs had extremely small ancestors. A series of hypotheses relates the extinction to hypothermic killing of all dinosaurs including their incubated eggs. Unlike viviparous sea snakes, no aquatic dinosaur (*New Scientist*, September 11,2014) is known to have been a vivipare (except *Dinocephalosaurus orientalis, blog.everythingdinosaur.co.uk*). Like marine turtles, they all had to migrate to the sea shore, where their deposited eggs, underwent incubation and on hatching, they returned to the sea home. These hypotheses suggest that the temperature of the earth dropped drastically to cool it toward the end of the Cretaceous epoch. According to them, the earth was suddenly struck by a great asteroid, which raised a cloud of dust that blocked the sun for many months and effectively impeded photosynthesis. Others consider that it was not an asteroid but three or four comets that struck the earth. Whether it is an asteroid or comets that led to block the sun and impede photosyntheses, and thereby disrupt the food chain resulting in massive nemesis, these suggestions are circumstantial, but solid. Yet, if the disruption of food chain is traced as the cause for extinction of dinosaurs, it should have also equally affected other terrestrial reptiles.

A second line of suggestion is related to temperature-dependent sex differentiation in poikilothermic vertebrates. In the evolutionary process, genes meet the environment by closing one door and opening the other (as Hindus believe it). For example, lecithality of eggs is genetically fixed but the egg size is determined by environmental factors. Similarly, sex is determined by gene(s) but its differentiation process is amenable to the environment that can alter the genetically determined sex to the opposite sex. The responses of the sex differentiation process by poikilothermic fishes were considered fewer than three patterns by Ospina-Alvarez and Piferrer (2008). In the second pattern, the differentiation is skewed in favor of the female ratio, with decreasing temperature (see Pandian, 2015). This is also true for the poikilothermic reptiles (Crews and Bull, 2008). The Cretaceous epoch seems to have witnessed abrupt cooling of the land mass beyond a critical level due to the strucking by an asteroid or comets. Hence, irrespective of their aquatic or terrestrial habitat, their incubated eggs in land were also subjected to hypothermia during the Cretaceous epoch. Due to hypothermia, the reversely sex differentiated females were hatched from more and more eggs. These females were doomed to roam in search of less and less males, which ultimately led to their demise. This hypothesis may not be acceptable, as it is unable to explain the continued existence and evolution of other reptiles, whose incubated eggs must have also been subjected to hypothermia.

26

Conservation

Introduction

The negative anthropogenic activities have driven many animals to a point of near extinction. For example, the activities are solely responsible for the extinction of one species every four years from 1,600 to 1,900, one species per year after 1,900 and currently > one species per day (see Kratochwil and Schwabe, 2001). Presently, efforts are being made to conserve them. *In situ* conservation demands legislation and its implementation by governments but it may be of local and temporal importance. The engraved stone edifices of the Indian Emperor Asoka (270–232 BC) reveal that he was the first to ban inland fishing operations during the spawning season. Employing scientific indices, the IUCN, an independent organization under the umbrella of UNESCO has recognized some as the red listed endangered animals demanding urgent and strict implementation of laws, requiring different measures of conservation. On the other hand, *ex-situ* conservation of animals requires scientific techniques, which may be of global importance. Apart from these, organisms have also developed their own strategies to conserve themselves for a short or longer duration.

26.1 Diapause: Survival Strategy

To tide over an unfavorable environmental situation, animals have developed a number of survival strategies. Of them, diapause facilitates survival over relatively a shorter duration but may not be relevant to conservation. However, an account is included to provide a comparative picture. During direct (e.g. rotifers) and indirect (e.g. insects) development, diapause may intervene to extend the life span. It is an adaptive strategy to survive over an unfavorable season and/or habitat, and synchronize hatching/emergence during a period favorable to development, growth and reproduction. Since its first discovery in insects, voluminous literature is available. According to the life stage, at which diapause occurs, it can be categorized into (a) embryonic (at cleavage stage: e.g. *Artemia parthenogenica*; half-developed, e.g. grasshopper *Melanoplus diflerentialis*, fully developed and ready to hatch, e.g. wheat blossom midge *Sitodiplosis mossellana*), (b) larval (crustacea: e.g. cyclopid copepodids, gypsy moth *Lymantria dispar*), (c) pupal (e.g. flesh fly *Sarcophaga crassipalpis*) and (d) adult (reproductive resting, e.g. Isopoda: *Jaera ischiosetosa*; Decapoda: *Crangon crangon*) stages. Gill et al. (2017) designated aestivation as summer diapause and hibernation as winter diapause in adult animals. Diapause can be obligatory, when it involves every individual and every generation and results in univoltonism. But facultative diapause may result in multivoltonism; in it, a few individuals may enter it. However, it alternates with next generation, in which all individuals enter it.

Depending on the site, the incidence of aestivation can be tidal and seasonal. The transient tidal aestivation occurs almost twice a day in intertidal anemones (e.g. *Halclava, Metridium*), cirripedes (e.g. *Balanus*) and molluscs (e.g. Gastropoda: *Littorina*; Bivalvia: *Mytilus*). But the transient tidal aestivation is not usually considered along with the long-lasting diapause. Table 26.1 summarizes the incidence of diapause in major and minor phyla. Diapause commences its first appearance from pseudocoelomate Rotifera and continues to occur in parasitic nematodes and hemocoelomate tardigrades. However, its incidence is restricted to limnic and terrestrial habitats. The remaining minor phyla are mostly marine inhabitants and in them, diapause is not known to occur. This is also true of the exclusively marine Echinodermata among major phyla. Interestingly, diapause occurs only in limnic and terrestrial inhabitants in all phyla, except in marine crustaceans.

The tight budgeted diapause-destined animals mitigate the energy cost of diapause by accumulation of reserves and/or metabolic suppression. To accumulate more reserves, they (a) eat more, (b) enhance food utilization efficiency and (c) divert nutrients away from somatic growth to storage, or use combinations of all three. Because of their high energy content, low hydration state and relatively high yield of metabolic water, triacylglyceride lipids are chosen as the most preferred reserve. The second option is the polysaccharide glycogen. The liver of vertebrates, hepatopancreas of arthropods and molluscs, and flight muscles of insects serve to store substantial quantities of glycogen. Prior to the diapause, the corn stalk borer *Sesamia nonagrioides* larva, for example, eats considerably more than the non-diapause-destined larva. During the diapause, the metabolic level is suppressed but the suppression is selective; it is reduced for processes like development, growth and/or reproduction but is increased to others like stress resistant mechanisms. Still, the net result is the suppression of an overall metabolic level. The suppression level ranges from 15% in the flying monarch butterfly to 90% in the flesh fly pupa. It facilitates conservation of reserves at low temperatures during winter. But in tropics, summer diapausers have to be better prepared with larger reserves. Though the metabolic suppression conserves nutrients, reserve utilization is a dynamic process and not uniform among diapausing animals. For example,

TABLE 26.1

Incidence of diapause in major and minor phyletic species (compiled from Hyman, 1951b, Pandian, 1975, 1994, 2016, 2019, 2020, Hahn and Denlinger, 2010)

Phylum/Class	Reported observations
Rotifera: Mostly limnic	
Monogononta	Embryonic diapause in dormant eggs
Bdelloidea	Dormancy in adults
Nematoda: Mostly in limnic and terrestrial parasites	
Ascaroidea, Trichuroidea	Obligate embryonic diapause. Hatched on infection of host(s). Juveniles diapause in *Bursaphelenchus xylophilus*
Gastrotricha	
Hermaphrodites may produce dormant eggs	
Gnathostomulida	
Embryonic diapause: subitaneous eggs first and subsequently dormant cysts	
Tardigrada: In limnic – limno-terrestrial	
Macrobiotus richtersi	Embryonic diapause in resting eggs Encysted resting adults
Turbellaria: Limnic species	
Mesostoma ehrenbergii	Subitaneous eggs first and dormant cysts last
Digenea: Terrestrial	
Fasciolidae	Cercaria encysted on plants
Cestoda: Terrestrial	
Taenia solium	Encysted hexacanth in pork
Annelida: Limnic and terrestrial	
Aeostomatidae	Embryonic diapause in cocoon; it lasts from autumn to winter in *Stylaria lacustris*
Tubificidae	Aestivation: e.g. *Rhyacodrilus hiemalis*
Lumbricidae	Hibernation: e.g. *Lumbricus terrestris*
Crustacea	
Anostraca (limnic)	Subitaneous eggs and diapausing embryonic cysts
Ostracoda (limnic)	Subitaneous eggs and diapausing embryonic cysts
Cladocera (limnic)	Diapausing embryonic ephippia
(marine)	Diapausing embryonic cysts
Copepoda (limnic)	Diapausing resting eggs
(marine)	Diapausing embryonic cysts
Isopoda, Amphipoda, Decapoda (marine)	Diapausing resting adults Diapausing resting adults

TABLE 26.1 Contd. ...

...TABLE 26.1 Contd.

Phylum/Class	Reported observations
Insecta : Terrestrial and limnic	
Gryllulus	Embryonic diapause
Sesamia nonagrioides	Larval diapause
Culex pipens	Pupal diapause
Drosophila melanogaster	Adult diapause
Mollusca	
Seasonal aestivation limited to limnics and terrestrials alone (e.g. *Pila*)	
Vertebrata	
Pisces	Limnic air-breathing teleosts (e.g. *Channa*) & coelacanths (e.g. 'living fossil' *Protopterus aethiopicus*)
Amphibia	Terrestrial toads (*Xenopus, Scaphiopus*)
Mammalia	Hibernating (overwintering) polar bear

the adult diapausing female *Culex pipens* first depletes the glycogen store and then switches to lipids, as the leftover lipids are subsequently used for egg production. Contrastingly, lipids are used first and are followed by other substrates in pupa of the flesh fly. When all the reserves are depleted, the animal may die either during the diapause or during the post-diapause stage. Others like the blow fly *Calliphora vicina* are capable of averting diapause, if the larva imminent to enter the diapause is too small. Within direct life cycle of the identified minor phyla, it occurs either during the embryonic or adult (e.g. bdelloid rotifer, tardigrade *Bursaphelenchus xylophilus*). The incidence of diapause is limited to a few species in Platyhelminthes, Annelida and Vertebrata. This is also true of molluscs, in which aestivation occurs. But the incidence is ubiquitous among crustaceans and insects. Interestingly, eggs arise from alternate ovaries in many phyla, in which ovaries are present in one or more pairs (e.g. Nemertea, Gastrotricha). In the decapod *Orconectes*, copulatory morphs appear following summer and winter molts but not following spring and autumn molts. Oostegite setae (e.g. Amphipoda), oostegites (e.g. Isopoda) and pleopod setae (Decapoda) are not developed in resting female crustaceans (Pandian, 1994, 2016). In the dormant state, adults of some minor phyletics can endure supercooling or withstand drying for relatively longer durations. For example, the bdelloid *Callidina quadricornia* can be revived after drying for 59 years in a dry herbarium. Plant parasitic nematodes can survive as galls over a period of 4–27 years. In the dormant/tun state, the tardigarde *Macrobiotus* can survive over 4–12 years (see Pandian, 2021).

The duration of the diapause ranges from 3–4 months in sawflies to 9–10 months or longer in temperate species. An extreme example is the wheat blossom midges, in which it lasts from 3 years in *Contarinia titici* to 12 years in *Sitodiplosis mossellana*. In the wax-coated encysted diapause, it may extend the duration up to 10 years (e.g. *Fasciola hepatica*, ground pearl hemimetabolous nymph of *Margarodes vitiums*, Gill et al., 2017). Muthukrishnan (1994) lists the life span of some diapausing insects from 1 year in the siphonopteran *Xenopsylla cheopis* to 6 years in the gryllobattoidean *Gryllabatta armata*. The encysted dormant state can also extend the adult life span up to 3.0 years in the amphipod *Echioniscus testudo* or 2.5 years in the bdelloid rotifers *Macrotrachela quadricornifera* or to 19 years in *Callidina quadricornia* and even up to 120 years, as anhydrobiont tardigrades (Pandian, 2021). In cladocerans and copepods, the diapause duration lasts up to 125–332 years. This is elaborated next. In all, diapause ensures survival during unfavorable conditions and extends the life span, irrespective of its occurrence during the embryonic, larval or adult stage and named as diapause, dormancy, resting stage, aestivation and hibernation in the forms of eggs, resting eggs, mictic eggs or cysts.

26.2 Conservation Strategy

With the discovery of dormant eggs and the like, many animals have developed a type of conservation strategy to survive over longer durations in the process of overcoming unfavorable environmental situations. These dormant eggs are named differently as gemmules in sponges, statoblasts and hibernacula in entoprocts and bryozoans, resting eggs, cysts and ephippia in crustaceans, and cysts in turbellarians (Table 26.2). The cysts contain much less water (e.g. 0.7 µg water/g cyst of *Artemia*) than that of resting eggs. An ephippium holds many dormant resting eggs. Gemmules and statoblasts are clonally produced structures. But the resting eggs, cysts and ephippia are either parthenogenically or sexually produced. Understandably, most of these dormant forms are produced by limnic taxa. Surprisingly, some marine sponges (Hexactinellida), entoprocts (Pedicellinidae), bryozoans (ctenostomatids) and crustaceans (e.g. cladocerans) also produce them.

The crustaceans display an array of strategies to produce subitaneous eggs and dormant eggs sexually and/or parthenogenically. Within anostracans, *Artemia* can produce cysts both sexually and parthenogenically; however, the other anostracans like the fairy shrimp *Streptocephalus dichotomus*, notostracans (e.g. *Triops cancriformis*) and spinicaudates (e.g. *Eulimnadia texana*) can produce cysts sexually alone. Like *Artemia*, some cladocerans can also produce ephippia either parthenogenically or sexually (e.g. *Daphnia pulex*). Interestingly, *D. ephemeralis* produce ephippia sexually alone, while *D. pulicaria* produce them parthenogenically alone. More interestingly, some anastrocans (e.g. *Cheirocephalus stagnalis*) produce subitaneous eggs as well as cysts. In copepods too, *Limnocalanus macrura* produce dormant resting eggs alone in some habitats but both subitaneous and dormant in others.

TABLE 26.2

Taxonomic distribution of dormant forms at different life stages (compiled from Fell, 1993, Pandian, 1994, 2016, 2020, 2021[†]).
[†] Marine taxa

Taxa	Product	Limnic taxa
Clonals		
Porifera	Gemmules	Keratosa, Hexactinellida[†]
Entoprocta	Statoblasts Hibernacula	Urnatellidae Pedicellinidae[†]
Bryozoa	Statoblasts Hibernacula	Plumatellida Ctenosomata[†]
Parthenogens		
Anostraca	Cysts	e.g. *Artemia*
Cladocera	Ephippia	e.g. *Daphnia pulicaria, D. pulex*
Ostracoda	Cysts	Darwinoids: *Heterocypris incongruens*
Sexuals		
Rotifera	Amictic eggs	Monogononta
Turbellaria	Cysts	*Mesostoma ehrenbergi*
Anostraca	Cysts	*Streptocephalus dichotomus*
Notostraca	Cysts	*Triops cancriformis*
Spinicauda	Cysts	*Eulimnadia texana*
Cladocera	Ephippia	*Daphnia ephemeralis, D. pulex*
Ostracoda	Cysts	*Eucypris virens, H. incongruens*
Calanoida	Resting eggs	*Diaptomus minutus*
Tardigrada	Encysted eggs	*Bertalanius volubilis*

In others like turbellarians, subitaneous eggs are produced in the first few clutches but dormant eggs in the last one prior to death (e.g. *Mesostoma ehrenbergi*). Unfertilized subitaneous amictic eggs and fertilized mictic eggs are produced at 4:1 ratio in monogonont rotifers.

Although the incidence of dormant cysts is reported from Turbellaria, Rotifera, and Branchiopoda and Ostracoda among crustaceans, more information is available for the 620 speciose cladocerans and ~ 12,000 speciose copepods. Understandably, parthenogenesis has enriched the cladocerans with more density than that of sexually reproducing calanoids. For example, in an equal volume (ml) of water, the cladoceran *Alona pulchella* can be accommodated in 55 numbers (Nandini et al., 1998), but only 9 copepod *Parvocalanus crossirostris* (9,000/l, Alajomi and Zeng, 2014). Arguably, *copepods have gone for species diversity, while cladocerans for number diversity.* This is reflected in their abundant production of dormant ephippia. Table 26.3 lists the density of sedimented resting eggs/ephippia found in freshwater and

TABLE 26.3

Abundance of parthenogenic cysts of anostracans, ephippia of cladocerans and sexual dormant eggs of spinicaudates and copepods (modified from Pandian, 2016)

Taxon	Density (no./m²)	Taxon	Density (no./m²)
Freshwater anostracan cysts		Freshwater Spinicaudata cysts	
Hoplopedium gibbereum	170,000	*Lynceus biformis*	241,184
Branchinella kugenumaensis	29,574	*Eulimnadia brauerina*	31,969
Streptocephalus virens	16,000	Freshwater copepode dormant eggs	
Marine anostracan cyst		*Diaptomus pallidus*	650,000
Artemia monica	7,300,000	*D. sanguineus*	45,000
Freshwater cladoceran ephippia		Marine copepod resting eggs	
Ceriodaphnia pulchella	520,000	*Acartia clause*	3,200,000
Daphnia pulex	40,000	*A. erythreae*	780,000
D. dentifera	30,000	*A. tonsa*	370,000
Bosmina longirostris	20,000	*Centropages abdominalis*	148,000
D. ephemeralis	11,000	*Tortanus forcipatus*	62,000
D. pulicaria	10,000	*Labroides aestiva*	43,400
Marine cladoceran ephippia		*A. steueri*	42,000
Penilla avirostris	122,000	*Ce. hematus*	36,500
Evadne tergestina	20,000	*Calanopia thomsoni*	24,000
Podon polyphemoides	4,000	*L. wollastoni*	20,500
Total: 777,000 or 86,333/species		Total: 5,421,000 or 451,750/species	

cysts/ephippia in the coastal zone. (1) After ignoring a couple of unreliably low values, it has been possible to assemble the density values for two spinicaudates, four anostracans, nine cladocerans and a dozen copepods. (2) A single value of seven million cyst production by *Artemia monica* indicates that *A. monica* can produce nearby 10-times more number of cysts by parthenogenesis than the mean number (71,858) of cysts produced sexually by three anostracan species. (3) Considering the total number of 777,000 ephippia produced by nine cladocerans, i.e. 86,333 ephippia/species, the freshwater cladocerans produce 2.3 times (105,167 ephippium/species/m²) more number of ephippia than that (48,667 ephippium/species/m²) of marine cladocerans. (4) Surprisingly, marine copepods produce ~ 1.4 times (472,600 resting eggs/species/m²) than that (~ 347,500 resting egg/species/m²) produced by their freshwater counterparts. Apparently, the production cost of sexually produced resting eggs (45,750/species/m²) by copepods is not as costly as that of (86,333 ephippia/species/m²) the mostly parthenogenically produced ephippia. But, it must also be noted that each ephippium holds a few resting eggs. It is also not known whether the number of eggs held in an ephippium of marine cladocerans is less than that of limnic cladocerans. Hence, some of these generalizations may have to wait until relevant information is reported in future for all the species.

26.3 Sleeping Strategy

During the sleeping duration in cysts, resting eggs and ephippia, these crustaceans not only survive but may also be dispersed. The eggs may be dispersed horizontally by running waters, or animals, as the eggs are resistant to digestion. Their unsynchronized hatching phenology is responsible for temporal distribution. The sleeping period within the dormant eggs ranges from 5 years in the marine *Podon polyphemoides* to > 125 years in freshwater *Daphnia pulicaria* (Table 26.4) and averages 59 years for the known cladocerans. The duration ranges from 2 years in the marine copepod *Tortanus discaudatus* to 332 years in freshwater *Diaptomus sanguineus* and averages 98.5 years.

Following drying or dehydration, life begins to sleep within the dormant eggs of animals and seeds of plants. The sleeping duration lasts until they are woken up by rehydration. For example, as rehydration proceeds, a host of biochemical activities is initiated with ~ 0.3 g water/g cysts of *Artemia* (see Pandian, 1994). Surprisingly, an inverse relation became apparent, when available values for the dormant egg size were plotted against sleeping duration (Fig. 26.1A). Accordingly, the duration is prolonged from < 2 months for the clonal gemmule of the sponge *Ephydratia mulleri* of 2 mm size to 4 years for the parthenogenic *Artemia franciscana* cysts of 300 μm, 35 and 125 years for the ephippia of *Bosmina* sp (225 μm) and *Daphnia*

TABLE 26.4

Viability and sleeping duration of crustacean cysts and dormant eggs (from Pandian, 2016)

Species	Viable period (y)	Species	Viable period (y)
Cladoceran ephippia		Dormant eggs of copepods	
Podon polyphemoides	5	*Tortanus discaudatus*	2
Ceriodaphnia pulchella	14	*Diaptomus minutus*	30
Bosmina longirostris	35	*D. oregonensis*	30
Daphnia longirostris	50	*D. sanguineus*	332
D. galeata mendotae	> 125	Mean	98.5
D. pulicaria	> 125	Anostracan cysts	
Mean	59	*Artemia salina*	4

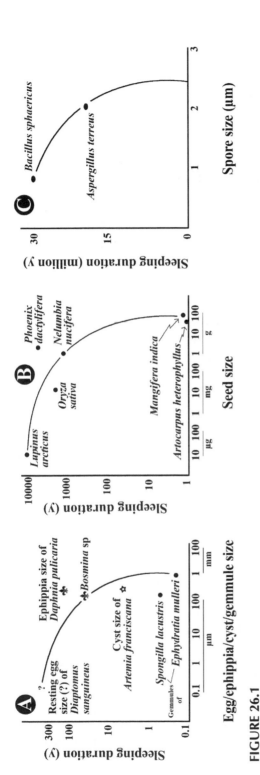

FIGURE 26.1

Sleeping duration as function of A. egg/ephippia/cyst/gemmule size in animals, B. seed size in plants and C. spore size in microbes (drawn from Fell, 1993, Pandian, 2016, Guisande and Gliwicz, 1992, Fageria et al., 2013, Wikipedia, Al-Farsi and Lee, 2001, *Lupinus arcticus*, Symbios Res & Restoration, 2003, BBC Earth News.

pulicaria (271 µm), respectively and to the smallest (a computer search did not yield the required data) sexually produced resting eggs of *D. sanguineus*. Three issues may have to be resolved by future research: (i) whether the duration depends on clonal, parthenogenic or sexual products? (ii) more importantly, whether the dormant products switch from protein to lipid reserve to meet metabolic energy demand; for example, glycerol and trehalose serve as reserve for *Artemia* cyst (see Pandian, 1994) but may be protein in *Daphnia pulex* (see Fink et al., 2011) and (iii) whether it depends on the level, to which metabolism is suppressed; for example, the suppression level is one of five thousand times of the normal level in *Artemia* cyst (see Pandian, 1994). Interestingly, the duration does not last longer than 4 years in the cyst of *Artemia salina*, whose relative *A. monica* produce 7 million cysts. On the other hand, with the duration of 332 years, *D. sanguineus* produce 45,000 eggs only. Similarly, *Ceriodaphnia pulchella* with the largest number of 520,000 ephippia has a shorter sleeping duration of 14 years. Hence, there is a type of compensation for the shorter sleeping duration with production of a larger number of dormant eggs.

A few lines may have to be added on collection and revival procedures of life from seeds and spores. The paddy seeds that were offered to the mummies of the Egyptian Pharaohs some 2,000 years ago, were earlier mixed with the thorny phytoplankton *Cocolithopore* (filtered through a muslin cloth) to prevent ants eating the seeds. Using polyclonal antibodies, a calcoflour and fluorescent optical brightner, Dr. Chandra Raghukumar and her colleagues at the Natural Institute of Oceanography, Goa, India isolated gonial spores of *Aspergillus terreus*, collected from 5,700 m depth at Chagos Trench off Sri Lanka in the Indian Ocean and revived life from the spores. The sleeping duration in the spore was assessed from the sinking rate of the spore to the depth and growth of the sediment over the sunken spores (Raghukumar et al., 2004, Damare et al., 2006). At California Institute of Technology, Cano and Borucki (1995) cut opened an amber-containing the insect *Propelebeia dominicana*; from its gut, they collected spores of *Bacillus sphaericus* and revived life from the spores.

More surprisingly, an inverse relation between the sleeping duration and seed size was also became apparent for the sessile flowering plants (Fig. 26.1B). Accordingly, the duration is prolonged from seeds of mango (~ 300 g) and jack fruit (< 30 g) from less than 1–2 years to 1,220 years in lotus seeds (1.2 g), 2,000 years in paddy seeds (< 19 mg) and 10,000 years in seeds of the Norwegian *Lupinus arcticus* (< 100 µg). Incidentally, it is not known whether the duration is also temperature dependant, as many tropical plant seeds (e.g. field bean, groundnut) begin to germinate, even while they are within the pod (see Pandian, 2022). It is needless to state that a seed is an amazing discovery of plants. So are the spores of microbes. Most surprisingly, an inverse relation between the duration and spore size also became apparent for the microbes (Fig. 26.1C). The duration is prolonged from 18 million years in the relatively larger fungal gonial spores (2.0 µm) of *A. terreus* to

30 million years in the smaller bacterial spores (0.8 μm) of *B. sphaericus*. Briefly, the sleeping duration lasts for a few hundred years in gammules/cysts/ephippia/resting eggs of animals, but for a few thousand years in seeds of plants and for a few million years of spores of microbes. *During the checkered history of evolution, the in-built conservation mechanism of the sleeping duration has been progressively and drastically reduced. Hence, conservation measures are more urgently required for animals than for plants and microbes.*

26.4 Scientific Development

In animals, the offspring can be recovered from their preserved gametes and/or embryos. Cryopreservation of gametes and embryos involves a series of complex and dynamic physio-chemical processes of temperature and water transportations between the 'cryopreserved' and the surrounding medium. Lecithal eggs and embryos are not amenable to cryopreservation, although fish blastomeres can be preserved by vitrification with a viability ranging from 20% in *Sillagoja ponica* to 96% in *Cyprinus carpio*. However, sperms of most animals are amenable to cryopreservation. Protocols are available for successful cryopreservation for ~ 200 species of commercially important fishes (Pandian, 2015). In the mussel *Crassostrea gigas*, 98% of the cryopreserved sperms remain fertilizable (see Pandian, 2017). A couple of alternative methods are available for sperm preservation. Techniques have been developed to preserve the sperms in a powder form, which facilitates long-term preservation and easier transportation. Wakayama and Yanakimachi (1998) reported successful preservation of the mouse sperms in a powder form. On restoration, the sperms were, however, found tail-less. When their heads were microinjected into eggs, litters were born. The easier and more readily practicable method seems to obtain sperm from post mortem preserved specimen. From accidentally kept post mortem specimens of the catfish *Heteropneustes fossilis* at −20°C for 240 days, 3% live fertilizable sperms could be obtained (Koteeswaran and Pandian, 2002). Confirming this finding, live and fertilizable sperms could also be drawn from specimen post mortem preserved at −20°C for period of 30 days in the rosy barb *Puntius conchonius* (Kirankumar and Pandian, 2004) and 40 days in the Buenos Aires Tetra (BT) *Hemigrammus caudovittatus*. In tetra, the sperm count and fertilizability could be improved from 8.6×10^3/ml and 19% fertilization to 6.8×10^4/ml and 24% fertilization, when specimens were glycerol-packed prior to preservation at −20°C (David and Pandian, 2006). These cadaveric sperms were capable of fertilizing not only homospecific but also genome-inactivated heterospecific eggs. There is considerable scope to further improve this simple and widely practicable method of sperm preservation. Some landmark achievements in *ex-situ* conservation techniques for fishes are listd in Table 26.5.

TABLE 26.5

Landmark achievements in *ex-situ* conservation of fishes (modified from Pandian, 2015)

Scientists	Achievement
Bercsenyi et al. (1998)	Restored androgenic F_1 ♀ and ♂ goldfish from the surrogate carp egg, using cryopreserved sperms of goldfish
Kirankumar and Pandian (2004)	Restored androgenic F_1 ♀ and ♂ rosy barb from the surrogate tiger barb egg, using cadaveric sperms of rosy barb
Lee et al. (2002)	Restored clones of F_1 ♀ and ♂ zebrafish transferring nucleus from embryonic fibroblast G_1 cell cultured over a long period into an enucleated egg of zebrafish
Kobayashi et al. (2003, 2007)	Restored allelogenic chimeric F_1 ♀ and ♂ rainbow trout, using cryopreserved *Gfp*-labeled PGCs of trout
Okutsu et al. (2007)	Recovered diploid chimeric allogenic F_1 ♀ and ♂ rainbow trout using cryopreserved trout sperm and triploid alevin masou (Fig. 25.3A)
Higaki et al. (2010)	Restored dark spotted, dark-strifed F_2 ♂ and pure allogenic ♀ zebrafish following a PGC transplantation from gold colored zebrafish. The PGC was drawn from a nucleus-removed embryo that was cryopreserved by vitrification

Nevertheless, sperm preservation alone cannot restore a species, as its eggs are not amenable to preservation. At present, there are two alternative techniques to restore a species; they are (i) interspecific androgenesis and (ii) allogenesis/xenogenesis (see Section 9.4) involving the transplantation of Primordial Germ Cells (PGCs) or one of their derivatives namely (a) Oogonial Stem Cells (OSCs) and (b) Spermatogonial Stem Cells (SSCs). Of these, SSCs are more readily available in greater numbers and almost throughout the mature life span of males. The protocol for induction of interspecific androgenesis involves the following three steps: (i) exposure of the surrogate recipient egg to UV-irradiation for a specific period of time to completely inactivate its genome; complete disappearance of the body color of the surrogate species is used as a marker to determine the optimal exposure duration of UV-irradiation, (ii) activation of the genome-inactivated eggs by introduction of live or cadaveric sperms of the donor species, which is to be restored and (iii) diploidization of the embryo at its specific age to thermal shock at the pre-determined temperature level and duration; for example, it is 41°C for 2 minutes in rosy barb (Fig. 26.2A). However, the third step can be avoided, when dispermic activation is achieved, as in tetra (Fig. 25.2B). To generate double sperms, the milt has to be first incubated for an optimal duration in polyethyl glycol (PEG). Being an extremely hydrophilic molecule, PEG has the property of fusing similarly charged membranes. However, the fusion may or may not be perfect and the level of fusion varies widely. Consequently, fertilization success is reduced to ~ 1.7% (Kirankumar and Pandian, 2004,

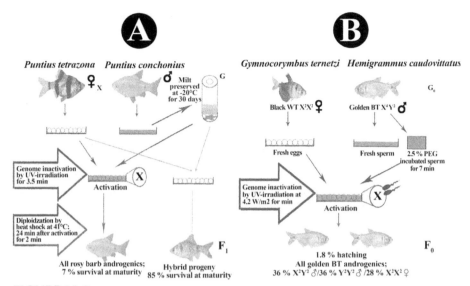

FIGURE 26.2

A. Induction of allo-androgenics of *Puntius conchonius* using cadaveric sperms for monospermic activation of genome-inactivated (surrogate) eggs of *P. tetrozona* (from Pandian and Kirankumar, 2003). B. The same for *Hemigrammus caudovitatus* but using dispermic activation of genome-inactivated eggs of *Gymnocorymbus ternetzi* (from David, 2004).

David and Pandian, 2006). The interspecific or intergenic androgenesis has been successfully achieved in barb, tetra, loach, zebrafish and trout (Table 26.5). The newly extinct bizzare gastric brooding frog *Rheobatrachus silus* has been revived by implanting a nucleus from a 'dead cell' into a genome inactivated egg of another frog species (*https://newsroom.unsw.edu.au/*). In fish, the method has at least one important limitation, i.e. the need for compatibility between the size of sperm head and micropyle diameter.

The PGCs are the prime source for gamete production; transplantation of the preserved PGCs, OSCs or SSCs from a donor species (to be restored) into the blastoderm or alevin of the recipient surrogate species has been shown to generate the donor's gamete from surrogate gonads. For example, the Atlantic trout *Oncorhynchus mykiss* offspring are generated from the sterile triploid Japanese masu *O. masou*, in whose alevin the donor trout's PGCs had earlier been transplanted (Fig. 26.3A). Presently, Japanese scientists are actively endeavoring to generate the long living (> 20 years) large (450 kg) *Thunnus orientalis* (~ 10 kg, sexually maturing at age of 4+) tuna using the surrogate recipient smaller *Scomber japonicus* (2.9 kg at the age of 3+) (Fig. 26.3B).

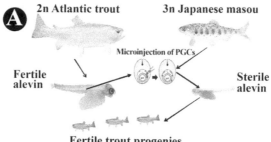

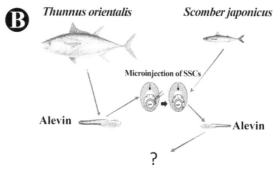

FIGURE 26.3

A. Production of xenogenics by implantation of spermatogonial stem cells (SSCs) of 2n fertile Atlantic trout *Oncorhynchus mykiss* into 3n sterile alevin of Japanese masu *O. masou* (drawn from Okutsu et al., 2007). B. The Japanese design to induce xenogenic by implanting SSCs into the alevin of a smaller (*Scomber japonicus*) into the alevin of large (640 kg) tuna *Thunnus orientalis*.

27

Climate Change

Introduction

Due to anthropogenic activity, the earth and its organisms are being more frequently threatened by bacterial epidemics and viral (e.g. COVID-19) pandemics, the serotonin-induced swarming of desert locust and environmental factors like violent earthquakes, tsunamis, cyclones, storms and pollutants. But these are all transient localized episodes. There is no historical evidence to show that any of these factors either singly or in combination have wiped out an animal species, albeit they may drastically reduce the population size of the affected taxa. Contrastingly, climate change is a long lasting phenomenon that covers the entire earth. Not surprisingly, research on the unprecedented increase in atmospheric carbon dioxide (CO_2) concentration and consequent global warming and ocean acidification during recent years have become the hottest research area. This chapter intends to present an emerging scenario in the context of species diversity.

27.1 Air-Water Interaction

The earth is surrounded by atmosphere consisting of ~ 78% nitrogen, ~ 21% oxygen, 0.3% carbon dioxide (CO_2) and others. With the beginning of the industrial era in 1750 and the accompanied ever increasing energy extraction from fossil fuels has concentrated the atmospheric CO_2. In its turn, the concentration has led to global warming and ocean acidification, which are collectively known as climate change. Incidentally, global warming has also begun to melt the polar caps, which may lead to immersion of coastal areas into the sea. Since 1750, the level of atmospheric CO_2 has risen from 280 ppm to 385 ppm in 2010 and is predicted to go up to 550 ppm by 2050 (Table 27.1). During the last 250 years, the levels of other greenhouse

TABLE 27.1

Changes in climate features during the last 30–50 years and predicted changes by 2050 (from Pandian, 2015)

Climate features	Last 30–50 years	By 2050
Atmospheric CO_2 (ppm)	385	550
pH of oceanic waters (unit)	– 0.1	– 0.1 to – 3.5
Sea surface temperature (°C)	+ 0.4	+ 1.5
Coral bleaching (times)	+ 2	+ 1.5
Sea level rise (mm/y)	1	8*
Hypoxic aquatic system (no.)	400	+ 280
Wind speed %/1°C increase	3.5	+ 280

gases have also increased from 715 ppb to 1,774 ppb for methane and from 270 ppb to 320 ppb for nitrous oxide (IPCC, 2013). As a consequence, the global mean temperature increased at the rate 0.2°C/decade over the last 30 years. Most of the added energy is absorbed by surface waters of the oceans (up to 700 m depth), where temperature increased by ~ 0.6°C over the last 100 years and is continuing to increase (see Pandian, 2015).

Besides absorbing atmospheric temperature, oceans also absorb CO_2, as it combines with water chemically. Covering 70% of the earth's surface and holding 97% of its water, they serve as a buffer to CO_2 concentration. Consequently and thankfully, the daily uptake of atmospheric CO_2 by the oceans is 22 million metric ton (mmt). Since the advent of the industrial era, the oceans have absorbed 127 billion metric ton (bmt) carbon as CO_2 from the atmosphere. The CO_2 absorbed by the oceans ranges between 25 and 40%, i.e. a third of atmospheric carbon emission. Without this 'ocean sink', the atmospheric CO_2 concentration would have by now increased to 450 ppm and a consequent increase in temperature on land.

Hydrolysis of CO_2 increases the Hydrogen ion (H^+) concentration with concomitant reductions in pH and carbonate (CO_3^{-2}) concentration. This process of reducing sea water pH and concentration of carbonate ion is called 'ocean acidification'. Consequent to the acidification process, the mean pH levels to the world's oceans have declined by 0.1 unit and 0.3–0.4 unit reduction is expected by 2050. The decrease in sea water pH and carbonate ion concentration is one of the most persuasive environmental changes in the oceans and poses one the most threatening challenges to marine organisms. The progressive reduction in the availability of carbonate ion (CO_3^{-2}) renders the acquisition of biogenic calcium carbonate ($CaCO_3$) by calcifying organisms energetically costlier, but may not totally inhibit the acquisition. In fact, the reduction in pH is more critical for the calcifying poikilothermic organisms than the increase in sea water temperature (see Pandian, 2015).

27.2 Aquatic Animals

The undesired progressive elevation in temperature may inflict unimaginable negative effects on aquatic systems and animals. It may rapidly dry and lead to the disappearance of millions of smaller ponds and puddles. Inflicting more pronounced and long lasting stratification in larger aquatic systems, it may squeeze oxygen and reduce its availability to animals. As relevant information is available for poikilothermic fishes, they are considered as representatives of aquatic animals. In response to temperature elevation, fishes may reduce, expand or shift spatial distribution to a cooler environment. In the North Sea, for example, cod is being replaced by sea bass. The planktivorous horse mackerel *Trachus trachura* has expanded its spatial distribution, accompanying that of plankton. Long term monitoring of the life span and reproductive status of *Perca fluviatilis* in the Baltic Sea has revealed the reduction in life span from 6.0 years in 1988 to 4.2 years in 2000 and Gonado-Somatic Index (GSI) from 6.75 to 4.50 (see Pandian, 2015). The reduction in GSI can be traced to delay and reduction in the hormone level and other events during the maturation process. For example, (i) elevated temperatures inhibit the luteinizing hormone and ovulation at 10°C in *Salvelinus fontinalis* in subarctics, (ii) impair maturation, reduce vitellogenesis and proportion of ovulating females from 98% at 5–7°C to 57% at 13–14°C in *Salmo salar* in cooler temperate zones and (iii) reduce fecundity sharply from the peak at 25°C to a very low level at 27°C in *Stegestes beebei* in the warmer temperate zone and decrease aromatase activity, estradiol level and fecundity at 30°C in *Acanthochromis polyacanthus* in the tropics (see Pandian, 2015).

In response to increasing temperature, motiles like fishes may shift to cooler habitats but sessiles like cnidarians may succumb to death. Within the North Sea, the benthic plaice *Pleuronectes platessa* has shifted northward to deeper waters. However, the best example is the pelagic oil sardine *Sardinella longiceps*. Thanks to the up-welling around the Kerala coast in Southwest India, sardines constitute massive fisheries of India. The length of the Indian west coast is 3,340 km, of which the Southwest coast (8°N to 14°N, 75°E to 77°E, ~ 1,140 km²) has been the 'home' of the sardine; Sea Surface Temperature (SST) ranges between 27°C and 29°C in the Southwest coast. With increasing SST climbing at 0.04°C/decade, the warmer tongue (27°C–28.5°C) of the surface waters begin to expand beyond 14°N enforcing the sardine to extend its range to the northern latitude (22°N, 90°E) both in the West and East coasts of India. During last 30 years, the sardine's spatial distribution has expanded by ~ 2,200 km on the coastal length at the rate of 88 km/y. Considering the peak distribution of the sardine up to 15 km from the shore, the fish has extended its range of distribution to more than five times, i.e. from 17,000 km², until the 1980s to 90,000 km² along the entire Indian coast in 2010s (Vivekanandan, 2011). Dr. E. Vivekanandan noted that despite the geographic expansion, the cumulative landings of sardines are progressively decreasing. It is not clear whether the migration-driven spatial expansion has led to stress in sardines. If so, cortisol, the stress hormone, may accumulate in them, ultimately leading to the production of more number of phenotypic males and corresponding reduction in females (Fig. 27.1). Incidentally, the temperature elevation of 0.32°C in SST in the Bay of Bengal has reduced the number of spawning females from 4,917 during 1980s to 2,355 during 2000s in the threadfin bream *Nemipterus japonicus* (Vivekanandan, per. comm). It is for the Central Marine Fisheries Research Institute, Kochi, India to assess the sex ratio of the entire landed sardine and to trace the cause for the progressive decline in landings of the sardine.

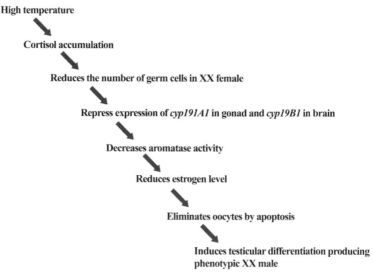

High temperature

Cortisol accumulation

Reduces the number of germ cells in XX female

Repress expression of *cyp191A1* in gonad and *cyp19B1* in brain

Decreases aromatase activity

Reduces estrogen level

Eliminates oocytes by apoptosis

Induces testicular differentiation producing phenotypic XX male

FIGURE 27.1

Cortisol-germ cell-aromatase-apoptosis in temperature sensitive fishes (from Pandian, 2015).

Corals, especially their symbiotic zooxanthellids are known for their extreme sensitivity to changes in temperature (Vivekanandan et al., 2009). Exposure to higher temperatures just a few degrees above the long term of an average at any location can cause corals to become stressed, bleach due to the loss of zooxanthellids and die. In the Gulf of Mannar (India) known for the abundance and diversity of corals, coral bleaching has been reported at least once during April and May in every year (*coralreefwatch.noaa.gov*). Prevalence of coral disease and bleaching increased from 6.9 times during 2007 and 10.6 times during 2011 (Edward et al., 2012). Incidentally, 4,000 teleost species are ultimately associated with corals. Declines in live coral cover and changes in the species composition of the remaining corals will also decrease species diversity and abundance of coral fishes. However, no investigation is yet available to show the loss of a specific coral species reducing the abundance of one or more fish species.

As most molluscs possess shell(s), they are the preferred animals for investigation on the effects of ocean acidification. The shell(s) not only protect them from predation, but also facilitate spawning and growth, as well. Not surprisingly, acidification not only affects the shell size, thickness and strength but also fertilization and growth. Consequently, the molluscan response to the acidification ranges from decrease in fertilization success and embryo development at pH 8.0 in *Placopecten magellanicus* to shell dissolution, reduced growth and increased mortality at pH 7.4 in *Pinctada fucata martensi* and to 5–50% growth reduction at pH 7.93 in *Haliotis rubra* (see Pandian, 2017).

27.3 Terrestrial Crops and Animals

The carbon dioxide (CO_2) concentration of the atmosphere is predicted to increase from the present level of 385 ppm to 550 ppm by 2050 (IPCC, 2013). Consequently, the temperature may rise by 3–4°C. Parmesan and Yohe (2003) indicated the global advancement of spring events 2.3 d/decade and a species range shift of 6.1 km/decade toward the poles. More importantly, agricultural crops also produce Green House Gases (GHGs) inclusive of CO_2. During crop production, carbon emission is in the range of ~ 956–4,700 kg C/ha for paddy and maize. But for dry crops like millet and sorghum, the value (~ 900 kg C/ha) is ~ 4 times less than that for the wet crops like paddy. Of a total cereal production of 1,668 million metric ton (mmt), dry crop cereal production is < 1% (Table 27.2). India and China are the only countries that produce millet. Many more countries may have to encourage and support millet production to reduce carbon emission level during crop production. Further, a 70% increase cereal food production is required to feed the predicted world population of 9.8 billion people by 2050. China is the largest producer of cereal crops in the world. Occupying the third rank, India produces 21, 13 and 36% of the world's production of paddy, wheat and millets, respectively. These two countries hold 37% of the world population. China produces cereal grains at significantly higher rates compared to India. Hence, India must contain its burgeoning population more strictly at the earliest. Being the second largest producer of cereal grains, the USA with a smaller population is in a commanding position to supply wheat and maize to other countries.

Though cereals constitute the staple Indian diet, protein-rich pulses are important ingredients to the mostly vegetarian Indians. Legume crops differ from cereals on two counts: (i) their unique ability to fix atmospheric

TABLE 27.2

Largest producing countries of agricultural crops in million
metric tons in 2016 (modified from Wang et al., 2018)

Country	Maize	Paddy	Wheat	Millet	Total
USA	985	-	63	-	448
China	232	211	132	2	577
India	-	159	94	10	263
Russia	-	-	73	-	73
Brazil	64	-	-	-	64
France	-	-	30	-	30
Indonesia	-	77	-	-	77
Bangladesh	-	53	-	-	53
Argentina	40	-	-	-	40
Vietnam	-	43	-	-	43
Total	721	543	392	12	1668
(%)	43.0	32.6	23.5	< 1.0	

nitrogen and enhance fertility of different croppings and farming systems and (ii) unlike cereal crops pollinated by wind, they obligately require insect pollinators for seed production. There are reports suggesting that legumes will be less affected by climate changes (Bahl, 2015). With advancement of spring during the last 45 years, flowering plants bloomed progressively earlier, which ultimately may result in more and more legumes not being pollinated. Especially, the behavioral response of pollinating insects like bees avoiding extreme temperatures may significantly reduce pollination services. Most pollinators are insects. As they are small and poikilothermic, global warming and increasing temperature is critical for the development of the life cycle and activity pattern (Pudasaini, 2015). Among insect taxa, the importance of the pollination role decreases in the following descending order: Apidae > Diptera > Lepidoptera in all, common and rare plants (Table 27.3). Due to mostly monogamy (see Section 15.4), and the inability to withstand pesticides, the thermally sensitive honey bees are unable to forage beyond 12°C, 24°C and 36°C in arctic, temperate and tropical zone, respectively. Consequently, the role of pollination played by honey bees may be decreased with progressively warming global temperatures, especially in the tropics, where atmospheric temperature is likely to shoot over 36°C. It is likely that their role may be replaced by temperature resistant dipterans and lepidopterans. It is tempting to recall the Hindus Geeta, which states: "What is yours today shall be of others tomorrow". Hence, the important role played by honey bees today may be replaced by other insects tomorrow.

TABLE 27.3

Insects taxa as pollinators to all, common and rare plant species

Insects	All plants (%)	Common plants (%)	Rare plants (%)
Apidae	79	74	97
Diptera	65	72	40
Lepidoptera	22	25	13

27.4 Green Shoots and New Hopes

With regard to green shoots providing new hope to overcome the ongoing climate change, two examples are described: one on calcification of shells in sessile molluscs in acidic waters and another on sex differentiation in motile fishes. Production and maintenance of shell(s) by molluscs require energy. Where and when possible, they acquire calcium carbonate at additional cost by increasing feeding (e.g. *Mytilus edulis*). Alternatively, net calcification can be maintained at the expense of growth, reproduction and/or defense, for example, increasing shell thickness or strength against predation. On exposure to CO_2 levels at present (390 ppm) and predicted (750 ppm), *Mercenaria mercenaria* and *Argopecten irradians* reduced shell thickness to different levels. However, even on exposure to 1,994 ppm CO_2, i.e. acidic water, no shell dissolution occurred in *M. edulis*, which ate more at a faster rate and met the additional cost of acquiring calcium carbonate. *Saccostrea glometra*, on exposure to acidic water (856 ppm CO_2), produced more robust larvae. Hence, *M. mercenaria* and *A. irradians* may head toward endangerment. But *M. edulis* and *S. glometra* represent the green shoots proceeding toward adaption to elevated temperature and acidic waters. In these days of climate change, their lineages may flourish and diversify (see Pandian, 2017).

With regard to temperature effects on sex differentiation in motile fishes, the available information is briefly summarized: (i) In the Nile tilapia *Oreochromis niloticus*, sensitivity to sex differentiation is not a species specific trait but an individual specific trait. (ii) In medaka *Oryzias latipes*, sex is determined by *Dmy* gene but its differentiation is derailed by temperature. Its wild *HN1* strain is more sensitive to temperature elevation than *HdrR* strain. The differentiation process remains unaffected in XY medaka male embryos but the exposure of XX embryos to higher temperature induces sex reversal to phenotypic males. (iii) There are fish species, which remain insensitive to changes in temperature. The exposure of the Atlantic habitat *Hippoglossus hippoglossus* to elevated temperatures did not alter sex ratio from unity (Hughes et al., 2008). Similarly, the sex ratio of *Thymallus thymallus* in all 12 different thermal combinations did not significantly vary from unity. Also in *Salmo trutta*, no evidence was found for temperature-induced sex reversal in any of the 14 ecologically relevant temperature experiments that the fishes were successfully exposed to over five consecutive years. It was found that no paternal or maternal effects on family sex ratio within samples of 26 genitors (6 female + 20 male) (Pompini et al., 2013). On the whole, thermal sensitivity to sex differentiation is an individual specific trait in tilapia, strain and sex specific trait in medaka and species specific trait in many fish species. Those individuals, strains within a species and those species insensitive to temperature elevation shall flourish and diversify during the days of climate change, while those sensitive to change in temperature may head toward endangerment. This is the law of evolution. Even after catastrophic massive extinction of dinosaurs during the Cretaceous, the reptilean species diversity (9,545 species) remains almost as high as that (10,038 species) of birds, which did not suffer such massive extinction. Hence, with the appearance of more green shoots, which are resistant to temperature elevation, the mother earth shall continue to witness burgeoning life forms and its species diversity, hopefully inclusive of man.

28

References

Abadia-Chanona, Q.Y., Avila-Poveda, O.H., Arellano-Martinez, M. et al. 2018. Reproductive traits and relative gonad expenditure of the sexes of the free spawning *Chiton articulatus* (Mollusca: Polyplacophora). Invert Reprod Dev, DOI: 10.1080/07924259.2018.1514670.

Abebe, E., Decraemer, W. and De Ley, P. 2008. Global diversity of nematodes (Nematoda) in freshwater. Hydrobiologia, 595: 67–68.

Acosta, R. and Prat, N. 2010. Chironomid assemblages in high altitude streams of the Andean region of Peru. Fund App Limnol, 177: 57–79.

Adiyodi, K.G. and Adiyodi, R.G. 1989. *Reproductive Biology of Invertebrates: Fertilization, Development and Parental Care*. Oxford & IBH Publishers, New Delhi, 4B, p 527.

Adiyodi, K.G. and Adiyodi, R.G. 1990. *Reproductive Biology of Invertebrates, Fertilization and Development and Parental Care: Clitella–Urochordata*. Oxford & IBH Publishing, New Delhi, 5, p 536.

Adoutte, A., Balavoine, G., Lartillort, N. et al. 2000. The new animal phylogeny: Reliability and implications. Proc Natl Acad Sci, USA, 97: 4453–4456.

Adrianov, A.V. and Maiorova, A.S. 2010. Reproduction and development of common species of the peanut worms (Sipuncula) from the Sea of Japan. Russ J Mar Biol, 36: 1–15.

Adrianov, A.V., Maiorova, A.S. and Malakhov, V.V. 2011. Embryonic and larval development of the peanut worm *Phascolosoma agassizii* (Keferstein 1867) from the Sea of Japan (Sipuncula: Phascolosomatidea). Invert Reprod Dev, 55: 22–29.

Agarwal, D.C. 2013. Average annual rainfall over the globe. Physics Teacher, 51: 540–541.

Akesson, B. 1976. Morphology and life cycle of *Ophryotrocha diadema*, a new polychaete species from California. Ophelia, 15: 23–35.

Alajomi, F. and Zeng, C. 2014. The effects of stocking density on key biological parameters influencing culture productivity of the calanoid copepod, *Parvocalanus crassirostris*. Aquaculture, 434: 201–207.

Alam, S.M.I., Sarre, S.D., Gleeson, D. et al. 2018. Did lizards follow unique pathways in sex chromosome evolution? Genes, 9: 239, doi: 10.3390/genes9050239.

Alexander, G. 1951. The occurrence of Orthoptera at high altitudes with species reference to Colarado Arcididae. Ecol Soc Am, 32: 104–112.

Al-Farsi, M.A. and Lee, C.Y. 2011. Usage of date (*Phoenix dactylifera* L.) seeds in human health and animal feed. In: *Nuts and Seeds in Health and Disease Prevention*. (eds) Preedy, V.R., Watson, R.R. and Patel, V.B., Academic Press, pp 447–452.

Almeida, F.F.L., Kristofferson, C., Taranger, G.L. and Schulz, R.W. 2008. Spermatogenesis in Atlantic cod (*Gadus morhua*): A novel model of cystic germ cell development. Biol Reprod, 78: 27–34.

Alonzo, S.H. and Mangel, M. 2004. The effects of size-selective fisheries on the stock dynamics and sperm limitation in sex changing fish. Fish Bull, 102: 1–12.

Amin, O.M., Sharifdini, M., Heckmann, R. and Ha, N.V. 2019. On three species of *Neoechinorhynchus* (Acanthocephala: Neoechinorhynchidae) from the Pacific Ocean off Vietnam with the molecular description of *Neoechinorhynchus* (N.) *Dimorphospinus* Amin and Sey, 1996. J Parasitol, 105: 606–618.

Anderson, R.M., Mercer, J.G., Wilson, R.A. and Carter, N.P. 1982. Transmission of *Schistosoma mansoni* from man to snail: experimental studies on miracidial survival and infectivity relation to larval age, water temperature, host size and host age. Parasitology, 85: 339–360.

Angeoloni, L., Bradburry, J.W. and Burton, R.S. 2003. Multiple mating, paternity and body size in a simultaneously hermaphrodite *Apysia californica*. Behav Ecol, 14: 554–560.

Anonymous, 2013. SoyStats 2013. U.S. Department of Agriculture.

Appeltans, W., Bouchet, P., Boxshell, G.A. et al. 2011. World Register of Marine Species (2011). *marinespecies.org*.

Arai, K., Taniura, K. and Zhang, Q. 1999. Production of second generation progeny of hexaploid loach. Fish Sci, 65: 186–192.

Argue, B.J. and Dunham, R.A. 1999. Hybrid fertility, introgression and backcrossing in fish. Rev Fish Sci, 9: 137–195.

Atashbar, B., Agh, N., Beladjal, L. et al. 2012. Effects of temperature on survival, growth, reproductive and life span characteristics of *Brachinecta orientalis* Gosars, 1901 (Branchiopoda, Anostraca) from Iran. Crstaceana, 85: 1099–1114.

Bachtrog, D., Mank, J.E., Peichel, K. et al. 2014. Sex determination. Why many ways of doing it? PLoS ONE, 12: e1001899.

Bahl, P.N. 2015. Climate change and pulses: Approaches to combat its impact. Agri Res, DOI: 10.1007/s40003-015-0163-9.

Bailey, G.G. 1976. A quantitative study of consumption and utilization of various diets in the bertha armyworm *Mamestra configurata* (Lepidoptera: Noctuidae). Can Entomol, 108: 1319–1320.

Baker, L.D. and Reeve, M.R. 1974. Laboratory culture of *Mnemiopsis mccradyi*. Mar Biol, 26: 57–62.

Balachandran, C., Dinakaran, S., Chandran, M.D.S. and Ramachandra, T.V. 2012. Diversity of aquatic insects in Aghanashini River of Central Western Ghats, India. Natl Conf Mgmt Wetland Ecosys, Mahatma Gandhi University, Kerala, pp 1–10.

Banta, A.M. 1939. Studies on the physiology, genetics and evolution of some Cladocera. Carnegie Institution, Carnegie, Washington D.C., 182: 11–200.

Barlow, G.W. 1981. Patterns of parental investment, dispersal and size among coral reef fishes. Environ Biol Fish, 6: 65–85.

Beaumont, H.M. and Mandl, A.M. 1962. A quantitative and cytological study of oogonia and oocytes in the foetal and neonatal rat. Proc R Soc, Biol Sci, 155B: 557–579.

Beckmann, M., Harder, T. and Qian, P.Y. 1999. Induction larval attachment and metamorphosis in a free amino acids: mode of action in laboratory bioassays. Mar Ecol Prog Ser, 190: 167–178.

Beladjal, L., Vandekerckhove, T.T.M., Muyssen, B. et al. 2002. B-chromosomes and male biased sex ratio with paternal inheritance in the fairy shrimp *Branchipus schaefferi* (Crustacea, Anostraca). Heredity, 88: 356–360.

Beladjal, L., Peiren, T., Vandekerckhove, T.T.M. and Mertens, J. 2003a. Different life histories of the co-occurring fairy shrimp *Branchipus schaefferi* and *Streptocephalus torvicornis* (Anostraca). J Curr Biol, 23: 300–307.

Beladjal, L., Khattabi, E.M. and Mertens, J. 2003b. Life history of *Tanymastigites perrieri* (Anostraca) in relation to temperature. Crustaceana, 76: 135–147.

Bell, G. 1982. *The Masterpiece of Nature: The Evolution and Genetics of Sexuality*. University of California Press, Berkeley, p 635.

Benazzi, M. and Benazzi-Lentati, G. 1993. Platyhelminthes-Turbellaria. In: *Reproductive Biology of Invertebrates: Asexual Propagation and Reproductive Strategies*. (eds) Adiyodi, K.G. and Adiyodi, R.G., Oxford & IBH Publishing, New Delhi, 6A: 107–142.

Benbow, M.E. 2009. Annelida, Oligochaeta and Polychaeta. In: *Encyclopedia of Inland Waters*. (eds) Tockner, K. and Likens, G.E., Academics Press, pp 124–127.

Bennet, L. 2018. Sea Turtles: Chelonidae and Dermatochelyidae. Smithsonian Institution. *https://ocean.si.edu/ocean-life/reptiles/sea-turtles*.

Bercsenyi, M., Magyary, I., Urbani, B. et al. 1998. Hatching out goldfish from common carp eggs. Interspecific androgenesis between two cyprinid species. Genome, 41: 573–579.

Bertolani, R. 2001. Evolution of the reproductive mechanisms in tardigrades—A review. Zool Anz, 240: 247–252.

Bierne, J. and Rue, G. 1979. Endocrine control reproduction in two rhynchocoelan worms. Int J Invert Reprod, 1: 109–120.

Birstein, V.J. 1991. On the karyotype of the *Norhabdocoela* species and karyological evolution of Turbellaria. Genetika, 83: 107–120.

Bishop, D.D. and Sommerfeldt, A.D. 1990. Not like *Botryllus*: indiscrimate post-metamorphic fusion in a compound ascidian. Proc R Soc, 226B: 241–248.

Black, D.A. and Howell, W.M. 1979. The North American mosquitofish: a unique case in sex evolution. Copeia, 1979: 509–513.

Black, J.M. 1996a. Introduction: pair bonds and partnership. In: *Partnerships in Birds: The Study of Monogamy*. (ed) Black, J.M., Oxford University Press, UK, pp 1–20.

Black, J.M. 1996b. *Partnerships in Birds: The Study of Monogamy*. Oxford University Press, UK, p 438.

Blackmon, H., Ross, L. and Bachtrog, D. 2017. Sex determination, sex chromosomes and karyotype evolution in insects. J Heredity, 108: 78–93.

Blackstone, N.W. and Jasker, B.D. 2003. Phylogenetic considerations of clonality, coloniality and mode of germline development in animals. J Exp Zool, 297B: 35–47.

Blake, J.A. 1996. Family Spionidae. In: *Taxonomic Atlas of the Santa Maria Basin and Western Santa Barbara Channel*. (eds) Blake, J.A., Hilbig, B. and Scott, P.H., Mus Nat Hist, Santa Barbara, 6 Part, 3: 81–223.

Blaxter, J.H.S. and Hempel, G. 1966. Utilization of yolk by herring larvae. J Mar Biol Ass UK, 46: 219–234.

Blaxter, M. and Koutsovoulos, G. 2015. The evolution of parasitism in Nematoda. Parasitology, 142S: 26–39.

Boi, S., Fscio, U. and Ferraguta, M. 2001. Nuclear fragmentation characterizes paraspermiogenesis in *Tubifex tubifex* (Annelida, Oligochaeta). Mol Reprod Dev, 59: 442–450.

Bosch, I., Rivkin, R.B. and Alexander, S.P. 1989. Asexual reproduction by oceanic planktotrophic echinoderm larva. Nature, 337: 169–170.

Boxshall, G.A., Lester, R.J., Grygier, G. et al. 2005. Crustacean parasites. In: *Marine Parasitology*. (ed) Rhode, K., University of New England, CSIRO Publishers, Victoria, pp 123–169.

Boxshall, G. and Hayes, P. 2019. Biodiversity and taxonomy of the parasitic Crustacea. In: *Parasitic Crustacea*. (eds) Smit, N.J. et al., Springer Nature, Switzerland, pp 73–134.

Brahmachary, R.L. 1989. Mollusca. In: *Reproductive Biology of Invertebrate*. (eds) Adiyodi, K.G. and Adiyodi, R.G., Oxford & IBH Publishing, New Delhi, 4A: 280–348.

Brehm, K. 2010. *Echinococcus multicularis* as an experimental model in stem cell research and molecular host-parasite interaction. Parasitology, 137: 537–555.

Brittain, J.E. 1990. Life history strategies in Ephemeroptera and Plecoptera. Ent Ser, 44: 1–12.

Brunn, A.F. 1957. Deep sea and abyssal depths. Geol Sci, America, 1: 641–675.

Bryan, J.P., Kreider, J.L. and Qian, P.Y. 1998. Settlement of the polychaete *Hydroides elegans* on surfaces of the cheilostome bryozoans *Buguia neritina*. Evidence for a chemically mediated relationship. J Exp Biol Ecol, 320: 171–190.

Buckland-Nicks, J., Williams, D., Chia, F.-S. and Fontaine, A. 1982. Studies on polymorphic spermatozoa of a marine snail. 1. Genetics of apyrene sperm. Biol Cell, 44: 365–314.

Bull, C.M. 2000. Monogamy in lizards. Behav Processes, 51: 7–20.

Butler, R.A. 2019. 2019: The year rainforests burned. *https://news.mongabay.com/2019/12/2019-the-year-rainforests-burned/*.

Butlin, C.D. 2002. Evolution of sex: the costs and benefits of new insights from old sexual lineage. Nat Rev Genet, 3: 311–317.

Calow, P. 1987. Platyhelminthes, Rhynchoceoela with sporal reference to the triclad turbellarian. In: *Animal Energetics*. (eds) Pandian, T.J. and Vernberg, F.J., Academic Press, 1: 159–183.

Cammen, L.M. 1987. Polychaeta. In: *Animal Energetics: Protozoa through Insecta*. (eds) Pandian, T.J. and Vernberg, F.J., Academic Press, 1: 217–260.

Cano, R.J. and Borucki, M.K. 1995. Revival and identification of bacterial spores in 25- to 40-million-year-old Dominican amber. Science, 268: 1060–1064.

Cantell, C.-E. 1989. Nemertina. In: *Reproductive Biology of Invertebrates*. (eds) Adiyodi, K.G. and Adiyodi, R.G., Oxford & IBH Publishing, New Delhi, Vol 4A: 147–177.

Caplins, S.A. and Turbeville, J.M. 2015. High rates of self-fertilization in a marine ribbon worm (Nemertea). Biol Bull, 229: 255–264.

Carefoot, T.H. 1987. Gastropoda: In: *Animal Energetics: Bivalvia through Reptilia*. (eds) Pandian, T.J. and Vernberg, F.J., Academic Press, 2: 90–172.

Carlisle, D.B. 1961. Locomotary powers of adult ascidians. Proc Zool Soc Lond, 136: 141–146.

Carlon, D.B. 1999. The evolution of mating systems in tropical reef corals. Tree, 14: 491–495.

Carr, M.E., Friendrichs, M.A.M., Schmeltz, M. and Aita, M.N. 2006. A comparison of global estimates of marine primary production from ocean color. Deep-Sea Res II, 53: 741–770.

Carson, H.S. and Hentschel, B. 2006. Estimating the dispersal potential of polychaete species in the Southern California Bight: implications for designing marine reserves. Mar Ecol Prog Ser, 316: 105–113 (with an appendix of 28 pages).

Carvalho, A.B. 2003. The advantages of recombination. Nat Genet, 34: 128–129.

Carvalho, R.A., Martins-Santos, I.C. and Dias, A.L. 2008. B-chromosomes: an update about their occurrence in freshwater neotropical fishes (Teleostei). J Fish Biol, 72: 1907–1932.

Cassani, J.R. 1990. A new method for early evaluation of grass carp larvae. Prog Fish Cult, 52: 207–210.

Central Marine Fisheries Research Institute (CMFRI). 2013. *Handbook of Marine Prawns of India*. Central Marine Fisheries Research Institute, Kochi, p 144.

Charlesworth, B. and Charlesworth, D. 2000. The degeneration of Y chromosomes. Phil Trans R Soc London, 355B: 1563–1572.

Chia, F.-S., Atwood, D. and Crawford, B. 1975. Comparative morphology of echinoderm sperm and possible phylogenetic implications. Am Zool, 15: 533–565.

Chowdaiah, B.N. 1965. Cytological studies of some Indian Diplopoda. Cytologia, 31: 294–301.

Christensen, B. 1961. Study on cyto-taxonomy and reproduction in the Enchytraeida with notes on parthenogenesis and polyploidy in animal kingdom. Heredity, 47: 437–450.

Christoffersen, M.L. and de Assis, J.E. 2015. Pentastomida: Revisita. IDE, 986: 1–10.

Ciplak, B., Sirin, D., Taylan, M.S. and Kaya, S. 2008. Altitudinal size clines, species richness and population density: case studies in Orthoptera. J Orthop Res, 17: 157–163.

Clark, K.B. and Jensen, K.R. 1981. A comparison of egg size, capsule size and development patterns in the order Ascoglossa (Sacoglossa) (Mollusca: Ophisthobranchia). Int J Invert Reprod, 3: 57–64.

Clewing, C., Bossneck, U., Oheimb, P.V.v. and Albrecht, C. 2013. Molecular phylogeny and biogeography of a high mountain bivalve fauna: the Sphaeriidae of the Tibetan Plateau. Malacologia, 56: 231–252.

Clewing, C., Oheimb, P.V.V., Vinaiski, M. et al. 2014. Freshwater molluscs diversity at the roof of the world: Phylogenetic and biogeographical affinities of Tibetan plateau *Valvata*. J Moll Stud, 1–4, doi: 10.1093/mollus/eyu016.

Cloney, R.A. 1990. Larval tunic and the function of the test cells in ascidians. Acta Zool, 71: 151–159.

Collins, J.J. 2017. Platyhelminthes. Curr Biol, 27: 252–256.

Collins, M.A., Yea, C., Allcock, L. and Thurston, M.H. 2001. Distribution of deep-water benthic and benthic-pelagic cephalopods from the north-east Atlantic. J Mar Biol Ass UK, 81: 105–117.

Costello, M.J. and Chaudhary, C. 2017. Marine biodiversity, biogeography, deep-sea gradients and conservation. Curr Biol, 27: 511–527.

Crews, D. and Bull, J.J. 2008. Sex determination, some like it hot (and some do not). Nature, 7178: 527–528.

Crowe-Riddell, J.M., Snelling, E.P., Watson, A.P. et al. 2016. The evolution of scale sensilla in the transition from land to sea in elapid snakes. Open Biol, 6: 160054, *http://dx.doi.org/10.1098/rsob.160054*.

Crump, R.G. and Barker, M.F. 1985. Sexual and asexual reproduction in geographically separated populations of the fissiparous asteroid *Coscinasterias calamaria* (Gray). J Exp Mar Biol Ecol, 88: 109–127.

Cumberlidge, N., Hobbs, H.H. and Lodge, D.M. 2015. Class Malacostraca, Order Decapoda. In: *Ecology and General Biology: Thorp and Covich's Freshwater Invertebrates.* (eds) Thorp, J. and Rogers, D.C., Academic Press, San Diego, USA, pp 797–847.

Cuomo, M.C. 1985. Sulphide as a larval settlement cue for *Capitella* sp. Biogeochemistry, 1: 169–181.

Cunha, A., Azevedo, R.-B.R., Emmons, S.W. and Leroi, A.M. 1999. Variable cell number in nematodes. Nature, 402: 253.

Cutter, A.D. 2004. Sperm-limited fecundity in nematodes: How many sperm are enough? Evolution, 58: 551–555.

Dade, H.A. 1977. Anatomy and dissection of the honey bee. Int Bee Res Assoc. p 281.

Daly, J.M. 1978. Growth and fecundity in a Northamberland population of *Spiorbis spiorbis* (Polychaete: Serpulidae). J Mar Biol Ass UK, 58: 177–190.

Damare, S., Raghukumar, C. and Raghukumar, S. 2006. Fungi in deep-sea sediment of the Central Indian Basin. Deep Sea Res, 53: 14–27.

Damken, C., Perry, G. and Beggs, J.R. 2012. Complex habitat changes alone elevational gradients interact with resource requirement of insect specialist herbivores. Ecosphere, 3(12), Doi: 10.1890/ES12-00216.1.

Darnell, R.M. 2015. *The American Sea: A Natural History of the Gulf of Mexico.* Texas A & M University Press, p 768.

Darwin, C. 1859. *On the Origin of Species by Areas of Natural Selection.* Murray, London, p 247.

Dash, M.C. 1987. The other annelids. In: *Animal Energetics: Protozoa through Insecta.* (eds) Pandian, T.J. and Vernberg, F.J., Academic Press, 1: 261–301.

David, A.A. and Williams, J.D. 2016. The influence of hypoosmotic stress on the regenerative capacity of the invasive polychaete *Marenzelleria viridis* (Annelida: Spionidae) from its native range. Mar Ecol, 37: 821–830.

David, C.J. 2004. *Experimental Sperm Preservation and Genetic Studies in Selected Fish.* Ph.D. Thesis, Madurai Kamaraj University, Madurai, India.

David, C.J. and Pandian, T.J. 2006. Cadaveric sperm induces intergeneric androgenesis in the fish *Hemigrammus caudovitatus.* Theriogenology, 65: 1048–1078.

David, C.J. and Pandian, T.J. 2008. Dispermic induction of interspecific androgenesis in the fish Buenos Aires tetra using surrogate widow tetra. Curr Sci, 95: 63–74.

Davison, A. 2006. The ovotestis: an underdeveloped organ of evolution. BioEssays, 28: 642–650.

Dawydoof, C. 1959. Classe des Echiurans. In: *Traite de Zoolgoie.* (ed) Grasse, P.P., Masson Cie, Paris, 6: 855–907.

Debortoli, N., Li, X., Eyres, I. et al. 2016. Genetic exchange among bdelloid rotifers is more likely due to horizontal gene transfer than meiotic sex. Curr Biol, 26: 1–10.

Dehorter, O. and Guillemain, M. 2008. Global diversity of freshwater birds (Aves). Hydrobiologia, 595: 619–626.

Delay, B. 1992. Aphally versus euphally in self-fertile hermaphrodite snails from the species of *Bulinus truncatus* (Gastropoda: Planorbidae). Am Nat, 139: 424–434.

Delvi, M.R. and Pandian, T.J. 1979. Ecological energetic of the grasshopper *Poecilocerus pictus* in Bangalore fields. Proc Ind Acad Sci, 88B: 241–256.

Devi, N.P. and Jauhari, R.K. 2004. Reappraisal on anopheline mosquitoes of Garhwal region, Uttarakhand, India. J Vec Borne Dis, 45: 112–123.

Diaz-Cosin, D.J., Nova, M. and Fernandez, R. 2011. Reproduction in earthworms: Sexual selection and parthenogenesis. In: *Biology of Earthworms.* (ed) Karaca, A., Springer Verlag, Berlin, pp 69–86.

Diotel, N., Do Rogo, J.-L., Anglada, I. et al. 2011. Activity and expression of steroidogenic enzymes in the brain of adult zebrafish. Eur J Neurosci, doi: 10.111/j.1460-9568.

Dronen, N.O. Jr. 1978. Host-parasite population dynamics of *Haematoloechus coloradensis* Cort, 1915 (Digenea: Plagiorchiidae). Am Midland Natl, 99: 330–340.

Duggal, C.L. 1978. Copulatory behaviour of male *Panagrellus redivivus.* Nematologia, 24: 257–268.

Dumont, H.J. and Negrea, S. 2002. *Introduction to the Class Branchiopoda: Guides to the Identification of the Microinvertebrates of the Continental Waters of the World.* Backhuys Publishers, Leiden, 19: 1–397.

Dunlap-Pianka, H.L., Boggs, C. and Gilbert, L.E. 1977. Ovarian dynamics in *Heliconius* butterflies: correlation among daily oviposition rates, egg weights and quantitative aspects of oogenesis. J Insect Physiol, 25: 741–749.

Eckelbarger, K.J. 1983. Evolutionary radiation in polychaete ovaries and vitellogenic mechanisms and their role in life history patterns. Can J Zool, 61: 487–504.

Eckelbarger, K.J. 1989. Ultrastructure and development of dimorphic sperm in the abyssal echinoid *Phrissocystis multispina* (Echinodermata: Echinoidea) in Bermuda. Bull Mar Sci, 82: 381–403.

Eckelbarger, K.J. 2005. Oogenesis and oocytes. Hydrobiologia, 535/536: 179–198.

Edward, J.K.P., Mathews, G., Raj, K.D. et al. 2012. Coral reefs of Gulf of Mannar, India–signs of resilience. Proc 12th Int Coral Reef Symp, Cairns, Australia, July 2012.

Emig, C.C. 1989. Phoronida. In: *Reproductive Biology of Invertebrates.* (eds) Adiyodi, K.G. and Adiyodi, R.G., Oxford IBH Publishing, New Delhi, 4B: 165–184.

Equardo, J. and Marian, A.R. 2012. A model to explain spermatophore implantation in cephalopods (Mollusca) and a discussion on its evolutionary origins and significance. Biol J Linn Soc, 105: 711–726.

Ereskovsky, A.V. 2018. Reproductive overview of phylogeny. In: *Encyclopedia of Reproduction.* Elsevier, Vol 6, doi.org./10.1016/B978-0-12.809633-8.20596-7.

Escobar, J.S., Auld, J.R., Correa, A.C. et al. 2011. Patterns of mating system evolution in hermaphroditic animals: Correlation among selfing rate, inbreeding depression, and the timing of reproduction. Evolution, 65: 1233–1253.

Fageria, N.K., Moreira, A., Ferreira, E.P.B. and Knupp, A.M. 2013. Pottasium-use efficiency in upland rice genotypes. Commu Soil Sci Plant Anal, 44: 2656–2665.

Fautin, D.G., Spaulding, J.G. and Chia, Fu-S. 1989. Cnidaria. In: *Reproductive Biology of Invertebrates.* (eds) Adiyodi, K.G. and Adiyodi, R.G., Oxford & IBH Publishing, New Delhi, 4A: 43–62.

Fell, P.E. 1989. Porifera. In: *Reproductive Biology of Invertebrates.* (eds) Adiyodi, K.G. and Adiyodi, R.G., Oxford & IBH Publishing, New Delhi, 4A: 1–41.

Fell, P.E. 1993. Porifera. In: *Reproductive Biology of Invertebrates: Asexual Propagation and Reproductive Strategies.* (eds) Adiyodi, K.G. and Adiyodi, R.G., Oxford & IBH Publishing, New Delhi, 6A: 1–44.

Feng, Z.F., Zhang, Z.F., Shao, M.Y. and Zhu, W. 2011. Developmental expression pattern of the F_c-*Vasa*-like gene, gonadogenesis and development of germ cells in Chinese shrimp *Fenneropenaeus chinensis*. Aquaculture, 314: 202–209.

Ferguson, J.C. 1971. Uptake and release of free aminoacids by star fishes. Biol Bull, 141: 22–29.

Ferguson, J.C. 1972. A comparative study of the net metabolic benefits derived from the uptake and release of free amino acids by marine invertebrates. Biol Bull, 162: 1–7.

Ferraguti, M. 1984. Cytological aspects of oligochaete spermiogenesis. Hydrobiologia, 115: 59–64.

Fiala-Medioni, A. 1987. Lower chordates. In: *Animal Energetics: Bivalvia through Reptilia.* (eds) Pandian, T.J. and Vernberg, F.J., Academic Press, 2: 323–357.

Field, C.B., Behrenfeld, M.J., Randerson, J.T and Falkowski, P. 1998. Primary production of the biosphere: integrating terrestrial and oceanic components. Science, 281: 237–240.

Finch, C.E. 1990. *Longevity, Senescence and the Genome.* University of Chicago Press, p 938.

Fink, P., Pflitsch, C. and Marin, K. 2011. Dietary essential amino acids affect the reproduction of the keystone herbivore *Daphnia pulex*. PLoS One, 6: e28498.

Finn, J.K. 2009. Systematics and biology of the argonauts or 'paper nautiluses (Cephalopoda: Argonautidae). Ph.D. Thesis, La Trobe University, Bundoora, Australia.

Finn, J.K. and Norman, M.D. 2010. The Argonaut shell: gas mediated buoyancy control in a pelagic octopus. Proc R Soc, DOI: 10.1098/rspb.2010.0155.

Fischer, D.O., Dickman, C.R., Jones, M.E. and Blomber, S.P. 2013. Sperm competition drives the evolution of suicidal reproduction in mammals. Proc Natl Acad Sci USA, 110: 17910–17914.

Fong, P.P. and Pearse, J.S. 1992. Photoperiodic regulation of parturition in the self fertilizing viviparous polychaete *Neanthes limnicola* from central California. Mar Biol, 112: 81–89.

Fox, S.W. 1965. *The Origins of Prebiological Systems and of Their Molecular Matrices*. Academic Press, p 504.

Frank, U., Plickert, G. and Muller, W.A. 2009. Cnidarian interstitial cells: The dawn of stem cell research. In: *Stem Cells in Marine Organisms*. (eds) Rinkevich, B. and Matranga, V., Springer, Dordrecht, p 33–61.

Fritz, R.S., Stamp, N.E. and Halverson, T.G. 1982. Iteroparity and semelparity in insects. Am Nat, 120: 264–268.

Frost, T.H. 1987. Porifera. In: *Animal Energetics*. (eds) Pandian, T.J. and Vernberg, F., Academic Press, 1: 27–54.

Fryer, G. 1997. A defene of arthropod polyphyly. In: *Arthropod Relationships, Systematics, Association*. (eds) Fortey, R.A. and Thomas, R.H., Chapman and Hall, London, pp 23–33.

Fukumoto, M. 1996. Ascidian fertilization—Its morphological aspects. Annu Rev (Inst Nat Sci Nagoya City Univ), 1: 9–30.

Furuya, H. and Tsunaki, K. 2003. Biology of dicyemid myxozoans. Zool Sci, 20: 519–532.

Gadgil, M. 1972. Male dimorphism as a consequence of sexual selection. Am Nat, 106: 574–580.

Gaither, M.R., Violi, B., Gray, H.W.I. et al. 2016. Depth as a driver of evolution in the deep sea: insights from grenadiers (Gadiformes: Macrouridae) of the genus *Coryphaenoides*. Mol Phylogen Evol, 104: 73–82.

Galtsoff, P.S. 1964. *The American Oyster Crassostrea virginica Gmelin*. Fish Bull Fish Wildlife Ser, 64: 1–480.

Gamo, T. and Shitashima, K. 2018. Chemical characteristics of hadal waters in the Izu-Ogasawara Trench of the western Pacific Ocean. Proc Japan Acad Ser, 94B: 45–55.

Gasparini, F., Manni, L., Cima, F. et al. 2015. Sexual and asexual reproduction in the colonial ascidian *Botryllus schlosseri*. Genesis, 53: 105–120.

Gates, C.E. 1971. On reversions to former ancestral conditions in megadrile oligochaetes. Evolution, 25: 245–248.

George, T. and Pandian, T.J. 1995. Production of ZZ females in the female heterogametic molly *Poecilia sphenops* by endocrine sex reversal and progeny testing. Aquaculture, 136: 81–90.

George, T. and Pandian, T.J. 1997. Interspecific hybdirization in poeciliids. Ind J Exp Biol, 35: 628–637.

Gibbons, M.J., Janson, L.A., Ismail, A. and Samaai, T. 2010. Life cycle strategy, species richness and distribution in marine hydrozoa (Cnidaria, Medusozoa). J Biogeogr Biol, 37: 441–448.

Gierer, A., Berking, S., Bode, H. et al. 1972. Regeneration in *Hydra* from reaggregated cells. Nature New Biol, 239: 98–101.

Gilbert, J.J., Simpson, T.L. and de Nagy, G.S. 1975. Field experiments on egg production in the freshwater sponge *Spongilla lacustris*. Hydrobiologia, 46: 17–27.

Gilbert, J.J. 1993. Rotifera. In: *Reproductive Biology of Invertebrates: Asexual Propagation and Reproductive Strategies*. (eds) Adiyodi, K.G. and Adiyodi, R.G., Oxford & IBH Publishing, New Delhi, 6A: 231–264.

Gilbert, T.J. 1993. Rotifera. In: *Reproductive Biology of Invertebrates*. (eds) Adiyodi, K.G. and Adiyodi, R.G., Oxford IBH Publishing, New Delhi, 5A: 231–263.

Gill, H.K., Goyal, G. and Chahil, G. 2017. Insect diapause: A review. J Agri Sci Tech, 7: 454–473.

Goransson, U., Jacobsson, E., Strand, M. and Andersson, H.S. 2019. The toxins of nemertean worms. Toxins, 11: 120. DOI: 10.3390/toxins.11020120.

Gordon, M. 1952. Sex determination in *Xiphophorus* (*Platypoecilis*) *maculatus*. 3. Differentiation of gonads in platyfish from broods having sex ratio of the three female to one male. Zoologica, 37: 91–100.

Gould, M.C., Stephano, J.L., Ortiz-Barron, B.J. and Perez-Quezada, I. 2001. Maturation and fertilization in *Lottia gigantea* oocytes: intracellular pH, Ca2+, and electrophysiology. J Exp Zool, 290: 411–420.

Govedich, F.R., Bain, B.A., Moser, W.E. et al. 2010. Annelida (Clitellata): Oligochaeta, Branchiobdellida, Hirudinida and Acanthobdellia. In: *Ecology and Classification of North*

American Freshwater Invertebrates. (eds) Covich, A.P. and Thorp, J.H., Academic Press, New York, pp 385–436.

Graves, J.A. 2013. How to evolve new vertebrate sex determining genes. Dev Dynam, 242: 354–359.

Gray, J. 1928. The growth of fish. III. The effect of temperature on the development of the eggs of *Salmo fario*. J Exp Biol, 6: 125–130.

Griffiths, C.L. and Griffths, R.J. 1987. Bivalvia: In: *Animal Energetics: Bivalvia through Reptilia*. (eds) Pandian, T.J. and Vernberg, F.J., Academic Press, 2: 2–89.

Gross, V., Theffkorn, S. and Mayer, G. 2015. Tardigrada. In: *Evolutionary Developmental Biology of Invertebrates: Ecdysozoa*. (ed) Wanninger, A., Springer Verlag, Wien, 3: 35–52.

Grove, D.J., Loizides, L.G. and Nott, Z. 1978. Satiation amount, frequency of feeding and stomach emptying in *Salmo gaidneri*. J Fish Biol, 12: 507–516.

Gruhl, A. 2015. Myxozoa. In: *Evolutionary Developmental Biology of Invertebrates*. (ed) Wanninger, A., Springer Verlag, Wien, 1: 165–178.

Guerrero, P.C., Rosas, M., Arroyo, M.T.K. and Wiens, J.J. 2013. Evolutionary lag times and recent origin of the biota of an ancient desert (Atacama-Sechura). Proc Natl Acad Sci USA, 110: 11469–11474.

Guisande, C. and Gliwicz, Z.M. 1992. Egg size and clutch size in two *Daphnia* species grown at different food levels. J Plankton Res, 14: 997–1007.

Gusamao, L.F.M. and McKinnon, A.D. 2009. Sex ratios, intersexuality and sex change in copepods. J Plank Res, 31: 1101–1117.

Haag, R. and Staton, J.L. 2003. Variation in fecundity and other reproductive traits in freshwater mussels. Freshwat Biol, 48: 2118–2130.

Hadfield, M.G., Nedved, B.T., Wilbur S. and Koehl, M.A.R. 2014. Biofilm cue for larval settlement in *Hydrodoides elegans* (Polychaeta): is contact necessary? Mar Biol, 161: 2577–2587.

Hahn, D.A. and Denlinger, D.L. 2010. Energetics of insect diapause. Annu Rev Entomol, 56: 103–121.

Haniffa, M.A. and Pandian, T.J. 1978. Morphometry, primary productivity and energy flow in a tropical pond. Hydrobiologia, 59: 23–48.

Haniffa, M.A., Thomas Punitham, M.T. and Arunachalam, S. 1988. Effect of larval nutrition on survival, growth and reproduction in the silkworm *Bombyx mori*. Sericologia, 28: 563–575.

Harding, J.A., Almany, G.R., Houck, L.D. and Hixon, M.A. 2003. Experimental analysis of monogamy in the Caribbean cleaner goby, *Gobiosoma evelynae*. Ani Behav, 65: 865–874.

Hart, M.W., Bryne, M. and Smith, M.J. 1997. Molecular phylogenetic analysis of life-history evolution in asterinid starfish. Evolution, 51: 1848–1861.

Hartman, W.D. and Reiswig, H.M. 1973. The 'individuality of sponges'. In: *Animal Colonies: Development and Function Through Time*. (ed) Boardman, R.S., Van Nostrand Reinhold, pp 567–584.

Harzsch, S., Muller, C.H.G. and Perez, Y. 2015. Chaetognatha. In: *Evolutionary Development Biology of Invertebrates* (ed) Wanninger, A., Springer Verlag, Wein, 1: 215–240.

Hayakawa, Y., Munehara, H. and Komaru, A. 2002. Obstructive role of the dimorphic sperm in a non-copulatory marine sculpin *Hemilepidotus gilberti* to prevent other male's sperm from fertilization. Env Biol Fish, 64: 419–427.

Heip, C.M., Smol, V.N. and Vranken, G. 1982. The systematic and ecology of free-living marine nematodes. Helminthol Abstract, Plant Nematol, 51B: 1–31.

Hejnol, A. 2015. Cycloneuralia. In: *Evolutionary Developmental of Invertebrates: Ecdysozoa I: Non-Tetraconata*. (ed) Wanninger, A., Springer Verlag, Wein, 3: 1–13.

Heller, J. 1993. Hermaphroditism in molluscs. Biol J Linn Soc, 48: 19–42.

Hembry, D.H., Yoder, J.B. and Goodman, K.R. 2014. Coevolution and the diversification of life. Am Nat, 184: 425–438.

Hentig, R.V. 1971. Einfluss von Salzgehalt und Temperatur auf Entwicklung, Wachstum, Fortzflansung von *Artemia salina*. Mar Biol, 45: 255–260.

Hepburn, H.R., Radloff, S.E. and Oghiakhe, S. 2000. Mountain honeybees of Africa. Apidologie, 31: 205–221.

Hess, R.A. and Franca, L.R. 2007. Spermatogenesis and cycle of the seminiferous epithelium. In: *Molecular Mechanism of Spermatogenesis.* (ed) Cheng, C.Y., Landes Bioscience, Austin, Texas, USA, pp 1–5.

Hickmann, C.S. and Porter, S. 2007. Nocturnal swimming, aggregation at light traps and mass spawning of scissurellid gastropods (Mollusca: Vestigastropoda). Invert Biol, 126: 10–17.

Higaki, S., Eto, Y., Kawakami, Y. et al. 2010. Production of fertile zebrafish (*Denio rerio*) possessing germ cells (gametes) originated from primordial germ cells recovered from vitrified embryo. Reproduction, 139: 733–740.

Higgins, R.P. 1974. Kinorhyncha. In: *Reproduction of Marine Invertebrates.* (eds) Giese, A.C. and Pearse, J.S., Academic Press, New York, 1: 507–518.

Hill, S.D. 1970. Origin of regeneration blastema in polychaete annelids. Am Zool, 10: 101–112.

Hillis, D.M. and Green, D.M. 1990. Evolutionary changes of heterogametic sex in the phylogenetic history of amphibians. J Evol Biol, 3: 49–64.

Hirayama, K. Takagi, K. and Kimura, H. 1979. Nutritional effect of eight species of marine phytoplankton on population growth of the rotifer *Brachionusplicatilis.* Bull Jap Soc Sci Fish, 45: 11–16.

Hoberg, E.P., Brooks, D.R. and Siegel-Causly, D. 1997. Host-parasitic cospeciation: history, principles and prospects. In: *Host-Parasite Evolution: General Principles and Avian Models.* (eds) Clayton, D.H. and Moore, J., Oxford University Press, Oxford, pp 212–235.

Hocking, B. 1968. *Six-Legged Science.* Schenkman Publishing Company, p 199.

Hodgkin, J. 1990. Sex determination compared in *Drosophila* and *Caenorhabditis.* Nature, 344: 35–47.

Holter, P. 2016. Herbivore dung as food for dung beetles: elementary coprology for entomologists. Evol Entomol, 41: 367–377.

Hotzi, B., Kosztelnik, M., Hargitai, B. et al. 2017. Sex-specific regulation of aging in *Caenorhabditis elegans.* Aging Cell, 17: e12724.

Hu, S., Dilcher, D.L., Jarzen, D.M. and Taylor, D.W. 2008. Early steps of angiosperm–pollinator coevolution. Proc Natl Acad Sci USA, 105: 1–13.

Hughes, R.N., Wright, P.J., Carvalho, G.R. and Hutchinson, W.F. 2009. Patterns of self compatibility, inbreeding depression, outcrossing and sex allocation in a marine bryozoan suggest the predominating influence of sperm competition. Biol J Linn Soc, 98: 519–531.

Hughes, V., Benfey, T.J. and Martin-Robichand, D.J. 2008. Effects of temperature on sex ratio in juvenile Atlantic halibut, *Hippoglossus hippoglossus.* Env Biol Fish, 81: 415–419.

Hyman, E.H. 1940. *The Invertebrates: Protozoa through Ctenophora.* McGrew-Hill Book Co, New York, Vol 1, p 726.

Hyman, L.H. 1951a. *The Invertebrates: Platyhelminthes and Rhynchocoela.* McGraw-Hill Book, New York, Vol 2, p 550.

Hyman, L.H. 1951b. *The Invertebrates: Acanthocephala, Aschelminthes, and Entoprocta.* McGraw-Hill Book, New York, Vol 3, p 572.

Hyman, L.H. 1955. *The Invertebrates: Echinodermata.* McGraw-Hill Book, New York, Vol 4, p 763.

Hyman, L.H. 1959. *The Invertebrates: Smaller Coelomate Groups.* McGraw-Hill Book, New York, Vol 5, p 783.

Hyman, L.H. 1967. *The Invertebrates: Mollusca.* McGrew Hill Book, New York, p 702.

IPCC (International Panel on Climate Change). 2013. Climate changes 2013. *The Physical Basis Contribution of Working Group 1 to the Fifth Assessment Report of Intergovernmental Panel on Climate Change.* (eds) Stocker, T.F., Qiu, D., Plahter, G.K. et al., Cambridge University Press, Cambridge.

Iyer, R.G., Rogers, D.V., Levine, M. et al. 2019. Reproductive differences among species and between individuals and cohorts in the leech genus *Helobdella* (Lophotrochozoa: Annelida; Clitellata; Hirudinida, Glossiphoniidae) with implications for reproductive resource allocation to hermaphrodite. PLoS ONE, 14: e0214581.

Jacobsen, D., Schultz, R. and Encalada, A. 1997. Structure and diversity of stream invertebrate assemblages: the influence of temperature with altitude and latitude. Freshwat Biol, 38: 247–261.

Jacobsen, D. 2003. Altitudinal changes in diversity of macroinvertebrates from small streams in the Euadorian Andes. Arch Hydrobiol, 158: 145–167.

Jacobsen, D. and Dangles, O. 2017. *Ecology of High Altitude Waters.* Oxford University Press, p 313.

Jamieson, B.G.M. and Rouse, G.W. 1989. The spermatozoa of the Polychaeta (Annelida): An ultrastructural review. Biol Rev, 64: 93–157.

Jangoux, M. 1990. Disease of echinoderms. In: *Diseases of Marine Animals.* (ed) Kinne, O. Biologische Anstalt Helgoland, Hamburg, Vol 3, pp 439–567.

Jankowski, T., Collins, A.G. and Campbell, R. 2008. Global diversity of inland water cnidarians. Hydrobiologia, 595: 35–40.

Jarne, P. and Charlesworth, D. 1993. The evolution of the selfing rate in functionally hermaphrodite plants and animals. Annu Rev Ecol Syst, 24: 441–466.

Jarne, P., Perdien, M.A., Pernof, A.F. et al. 2000. The influence of self-fertilization and grouping on fitness attributes in the freshwater snail *Physaacuta*: population and individual inbreeding. J Evol Biol, 13: 645–655.

Jarne, P. and Auld, J.R. 2006. Animals mix it up too: the distribution of self-fertilization among hermaphrodite animals. Evolution, 60: 1816–1824.

Jaroll, E.L. Jr. 1980. Population dynamics of *Bothriocephalus rarus* (Cestoda) in *Notophthalmus viridiscens*. Am Midland Natl, 103: 360–366.

Jennings, J.B. 1997. Nutritional and respiratory to parasitism exemplified in the Turbellaria. Int J Parasitol, 27: 679–691.

Jimenez, J.J. and Decaens, T. 2000. Vertical distribution of earthworms in grassland soils of the Colombian Liano. Biol Fert Soil, 32: 463–473.

Johnson, S.D. and Anderson, B. 2010. Coevolution between food-rewarding flowers and their pollinators. Evol Edu Outreach, 3: 32–39.

Johnson, S.C., Lively, C.M. and Schrag, S.J. 1995. Evolution and ecological correlates of uniparental reproduction in freshwater snails. Experientia, 51: 498–509.

Johnson, W.S., Stevens, M. and Walting, L. 2001. Reproduction and development of marine peracaridans. Adv Mar Biol, 39: 105–260.

Kabata, Z. 1984. Diseases caused by Metazoans: Crustaceans. In: *Diseases of Marine Animals.* (ed) Kinne, O., Biologische Anstalt Helgoland, Hamburg, 4A: 321–399.

Kallman, K.D. 1984. The new book at sex determination in poeciliid fishes. In: *Evolutionary Genetics of Fishes.* (ed) Turner, B.J., Plenum Publishers, New York, pp 95–171.

Kammerer, C.F., Nesbitt, S.J., Flynn, J.J. et al. 2020. A tiny ornithodiran archosaur from the Triassic of Madagascar and the role of miniaturization in dinosaur and pterosaur ancestry. Proc Natl Acad Sci, USA, 117: 17932–17936.

Kavumpurath, S. and Pandian, T.J. 1992. Hybridization and gynogenesis in two species of the genus *Brachydanio*. Aquaculture, 105: 107–116.

Kavumpurath, S. and Pandian, T.J. 1993a. Masculinisation of *Poecilia reticulata* by dietary administration of synthetic or natural androgens to gravid females. Aquaculture, 116: 83–89.

Kavumpurath, S. and Pandian, T.J. 1993b. Production of YY female gubby *Poecilia reticulata* by endocrine sex reversal and progeny testing. Aquaculture, 116: 183–189.

Kavumpurath, S. and Pandian, T.J. 1994. Masculinizing of fighting fish *Betta splendens* using synthetic or natural androgens. Aquacult Fish Mgmt, 25: 389–393.

Keats, D.W., Steele, D.H. and South, G.R. 1984. Depth-dependent reproductive output of the green sea urchin *Strongylocentrotus droebachiensis* (O.F. Muller) in relation to the nature and availability of food. J Exp Mar Biol Ecol, 80: 77–91.

Keller, L. and Genoud, M. 1997. Extraordinary lifespans in ants: A test of evolutionary theories of ageing. Nature, 389: 958–960.

Keller, R.P., Drake, J.M. and Lodge, D.M. 2007. Fecundity as a basis for risk assessment of nonindigenous freshwater molluscs. Conservation Biol, 21: 191–200.

Kenchington, E., MacDonald, B., Coe, B. et al. 2002. Genetics of mother-dependent sex ratio in the blue mussel (*Mytilus* spp) and implication for doubly uniparental inheritance of mitochondrial DNA. Genetics, 161: 1579–1588.

Kiflawi, M., Mazeroll, A.I. and Goulet, D. 1998. Does mass spawning enhance fertilization in coral reef fish? A case study of the brown surgeon fish. Mar Ecol Prog Ser, 172: 107–114.

Kinne, O. 1983a. *Diseases of Marine Animals: Bivalvia to Schaphopoda*. Biologische Anstalt Helgoland, Hamburg, Vol 2, Part 2, p 1038.

Kinne, O. 1983b. *Diseases of Marine Animals: Cephalopoda–Urochordata*. Biologische Anstalt Helgoland, Hamburg, Vol 3, p 696.

Kinne, O. 1984. *Diseases of Marine Animals: Pisces*. Biologische Anstalt Helgoland, Hamburg, Vol 4, Part 1, p 540.

Kinne, O. 1985. *Diseases of Marine Animals: Reptilia, Aves, Mammalia*. Biologische Anstalt Helgoland, Hamburg, Vol 4, Part 2, p 543–884.

Kiorboe, T. and Sabitini, M. 1995. Scaling of fecundity, growth and development in marine planktonic copepods. Mar Ecol Prog Ser, 120: 285–298.

Kirankumar, S. 2003. *Induction of Intraspecific and Interspecific Androgenesis in Fish*. Ph.D. Thesis, Madurai Kamaraj University, Madurai, India.

Kirankumar, S. and Pandian, T.J. 2003. Production of androgenetic tiger barb *Pantinus tetrazona*. Aquaculture, 228: 37–51.

Kirankumar, S. and Pandian, T.J. 2004. Interspeicific androgenic restoration of rosy barb using cadaveric sperm. Genome, 47: 66–73.

Kitazawa, C. and Amemiya, S. 2011. Regulation potential in development of a direct developing echinoid *Peronella japonica*. Dev Growth Differ, 43: 73–82.

Klass, M.R. 1977. Ageing in the nematode *Caenorahdbitis elegans*: major biological and environmental factors influencing life span. Mech Age Dev, 6: 413–423.

Kleiman, D.G. 1977. Monogamy in mammals. Q Rev Biol, 52: 39–69.

Klug, H. 2018. Why monogamy? A review on potential ultimate drivers. Front Ecol Evol, 6: 30, doi: 10.3389/evo.2018.00030.

Kobayashi, T., Takeuchi, Y., Yashizaki, G. and Takeuchi, T. 2003. Cryopreservation of trout primordial germ cells. Fish Physiol Biochem, 28: 479–480.

Kobayashi, T., Takeuchi, Y., Takeuchi, T. and Yoshizaki, G. 2007. Generation of viable fish from cryopreserved primordial germ cells. Mol Reprod Dev, 74: 207–213.

Kocot, K.M., Christiane, T., Nina, M.T. and Kenneth, H.M. 2019. Phylogenomics of Aplacophora (Mollusca, Aculifera) and a solenogaster without a foot. Proc R Soc, 286B: 20190115.

Kohler, S.L. 2008. The ecology of host–parasite interactions in aquatic insects. In: *Aquatic Insects*. (eds) Lancaster, J. and Briers, R.A., CABI International, London, pp 55–80.

Korner, C. 2012. *Alpine Treelines: Functional Ecology of the Global High Elevation Tree Limits*. Springer, p 220.

Kot, M. 2001. *Elements of Mathematical Ecology*. Cambridge University Press, p 453.

Koteeswaran, R. and Pandian, T.J. 2002. Live sperm from post-mortem preserved Indian catfish. Curr Sci, 82: 447–450.

Kramer, B.H. and Schaible, R. 2013. Colony size explains the lifespan differences between queens and workers in eusocial Hymenoptera. Biol J Linn Soc, 109: 710–724.

Kratochwil, A. and Schwabe, A. 2001. Evolution, coevolution and biodiversity. In: *The Living World: Discovery and Spoliation of the Biosphere*. (eds) Box, E. and Pigantti, S., Academic Press, San Diego, pp 395–419.

Kristensen, R.M. 2002. An introduction to Loricifera, Cycliophora and Micrognathozoa. Integ Comp Biol, 42: 641–651.

Kristiansen, T.S., Jorstad, K.E., Ottera, H. et al. 1997. Estimates of larval survival of cod by release of genetically marked yolk-sac larvae. J Fish Biol, 51A: 264–283.

Krug, P.J. 1998. Poecilogony in an estuarine opisthobranch planktotrophy, lecithotrophy and mixed clutches in a population of the ascoglossan *Alderia modesta*. Mar Biol, 132: 483–494.

Kumari, S. and Pandian, T.J. 1991. Interaction of ration and unilateral eyestalk ablation on energetic of female *Macrobrachium nobilii*. Asian Fish Soc, 4: 227–244.

Kupriyanova, E.K. 2006. Fertilization success in *Galeolaria caespitosa* (Polychaeta, Serpulidae): gamete characteristics, role of sperm dilution, gamete age and contact time. Scient Mari, 70: 309–317.

Kuwamura, T. 1986. Parental care and mating systems of cichilid fishes in Lake Tanganyika: A preliminary survey. J Ethol, 4: 129–146.

Kuwamura, T. 1997. The evolution of parental care and mating systems among Tanganyikancichilids. In: *Parental Care and Mating Systems*. (eds) Kawanabe, H., Hori, M. and Nagoshi, M., Chukyo University Press, Yagota, Nagoya, pp 59–86.

Lackenby, J.A., Chambers, C.B., Ernst, I. and Whittington, I.A. 2007. Effect of water temperature on reproductive development of *Benedenia seriolae* (Monogenea: Capsalidae) from *Seriola lalandi* in Australia. Dis Aquat Org, 74: 235–242.

Lashmanova, E., Zemskaya, N., Proshkina, E. et al. 2017. The evaluation of geroprotective effects of selected flavonoids in *Drosophila melanogaster* and *Caenorhabditis elegans*. Front Pharmacol, 8: 884.

Laumer, C.E. and Giribet, G. 2014. Inclusive taxon sampling suggests a single stepwise origin of ectolecithality in Platyhelminthes. Biol J Linn Soc, 111: 570–588.

Laws, A.N. 2009. Density dependent reductions in grasshopper fecundity in response to nematode parasitism. Can Ent, 141: 415–421.

Le Page, Y., diotel, N., Vaillant, C. et al. 2010. Aromatase, brain sexualization and plasticity: the fish paradigm. Eur J Neurosc, 32: 2105–2115.

Lee, K.Y. Huang, H., Ju, B. et al. 2002. Cloned zebrafish by nuclear transplantation from long-term cultured cells. Natl Biotech, 20: 795–799.

Legendre, M.G., Teugels, G., Cantry, C. and Jalabert, B. 1992. A comparative study on morphology, growth rate and reproduction of *Clarias gariepinus* (Burchell, 1822), *Heterobranchus longifilis* Valanciennes, 1840 and reciprocal hybrids (Pisces: Clariidae). J Fish Biol, 40: 59–79.

Lemen, C.A. and Voris, H.K. 1981, A comparison of reproductive strategies among marine snakes. J Anim Ecol, 50: 89–101.

Levitan, D.R. 1991. Influence of body size and population density on fertilization success and reproductive output in a free spawning invertebrate. Biol Bull, 181: 261–268.

Levitan, D.R., Sewell, M.A. and Chia, F.S. 1992. How distribution and abundance influence fertilization success in the sea urchin *Strongylocentrotus franciscanus*. Ecology, 73: 248–254.

Littleton, J. 2005. Fifty years of chimpanzee demography at Taronga Park Zoo. Am J Primatol, 67: 281–298.

Littlewood, D.T.J., Bray, R.A. and Waeschenbach, A. 2015. Phylogenic patterns of diversity in cestodes and trematodes. In: *Parasitic Diversity and Diversification: Evolutionary Ecology Meets Phylogenetics*. (eds) Morand, S., Korasnov, B.R. and Littlewood, D.T.J., Cambridge University Press, pp 304–319.

Liu, S.J., Liu, Y., Zhou, G.J. et al. 2001. The formation of tetraploid stocks of red crucian carp X common carp hybrids as an effect of interspecific hybridization. Aquaculture, 192: 172–186.

Liu, S.J., Sun, B.D., Zhang, C. et al. 2004. Production of gynogenetic progeny from allotetraploid hybrids red crucian carp x common carp. Aquaculture, 236: 193–200.

Lobo, J.M., Chehlarov, E. and Gueorguiev, B. 2007. Variation in dung beelte (Coleoptera: Scarabaeoidea) assemblages with altitude in the Bulgarian Rhodopes Mountains: A comparison. Eur J Ent, 104: 489–495.

Love, R.M. 1980. *The Chemical Biology of Fishes: Advances 1968–1977*. Academic Press, New York.

Lowe, C.N. and Butt, K.R. 2005. Culture techniques for small soil dwelling earthworms: A review. Pedobiologia, 49: 401–419.

Lukas, D. and Clulton-Brock, T.H. 2013. The evolution of social monogamy in mammals. Science, 341: 526–530.

Lutzen, J. 1968. Unisexuality in the parasitic family Entoconchidae (Gastropoda, Prosobranchia). Malacologia, 7: 7–15.

Ma, T., Kjesbu, O.S. and Jorgensen, T. 1998. Effects of ration on the maturation and fecundity in captive Atlantic herring (*Clupea harengus*). Can J Fish Aquat Sci, 55: 900–908.

MacGinitie, G.E. 1934. The egg-laying activities of the sea hare, *Tethys californicus* (Cooper). Biol Bull, 67: 300–303.

Mackiewicz, J.S. 1988. Cestode transmission pattern. J Parasitol, 74: 60–71.

Madsen, T. and Shine, R. 1992. Determinants of reproductive success in female adders, *Vipera berus*. Oecologia, 92: 40–47.

Mallon, D.P. 2011. Global hotspots in the Arabian Peninsula. ZoolMid East, 54: 13–20.

Mangold, K. 1987. Reproduction. In: *Cephalopod Life Cycles: Comparative Reviews.* (ed) Boyle, P.R., Academic Press, London, 2: 157–200.

Mann, T. 1984. Arachinida. In: *Spermatophores: Development, Structure, Biochemical Attributes and Role in the Transfer of Spermatozoa.* (ed) Mann, T., Springer, Berlin, pp 147–162.

Manni, L. and Burighel, P. 2006. Common and divergent pathways in alternative developmental processes of ascidians. BioEssays, 28: 902–912.

Manriquez, P.H., Hughes, R.N. and Bishop, J.D.D. 2001. Age-dependent loss of fertility in water borne sperm of the bryozoan *Celleporella hyalina*. Mar Ecol Prog Ser, 224: 87–92.

Maptsone, G.M. 2014. Global diversity and review of Siphonophora (Cnidaria, Hydrozoa). PLoS ONE, 9: e87737.

Marcogliese, D.J. 1995. The role of zooplankton in the transmission of helminth parasites to fish. Rev Fish Biol Fisher, 5: 336–371.

Marconato, A., Tessari, V. and Marin, G. 1995. The mating system of *Xyrichthys novacula*: sperm economy and fertilization success. J Fish Biol, 47: 292–301.

Marian, M.P., Pandian, T.J. and Muthukrishnan, J. 1982. Energy balance in *Speliphron violaceum* (Hymenoptera) and use of meconium weight as an index of bioenergetics components. Oecologia, 55: 264–267.

Marian, M.P., Christopher, M.S.M., Selvaraj, A.M. and Pandian, T.J. 1983. Studies on predation of the mosquito *Culex fatigans* by *Rana tigrina* tadpoles. Hydrobiologia, 106: 59–63.

Martin, P., Martens, K. and Goddeeris, B. 1999. Oligochaeta from the abyssal zone of Lake Baikal (Siberia, Russia). Hydrobiologia, 406: 165–174.

Mathavan, S. and Pandian, T.J. 1974. Use of fecal weight as an indicator of food consumption in some lepidopterans. Oecologia, 15: 177–185.

Mathavan, S. 1976. Satiation time and predatory behavior of dragon fly nymph *Mesogomphus lineatus*. Hydrobiology, 50: 55–64.

Mathavan, S. and Pandian, T.J. 1977. Patterns of emergence, import of egg energy and energy export via emerging dragonfly populations in a tropical pond. Hydrobiologia, 54: 257–272.

Mathavan, S. 1990. Effect of temperature on the bio-energetics of the larvae of *Bachythemis contaminata* (Fabricius) and *Orthetrum sabina* (Drury) (Anisoptera: Libellulidae). Odonatolgica, 19: 153–165.

Matsuda, M., Matsuda, C., Hamaguchi, S. and Sakaizumi, M. 1998. Identification of the sex chromosomes of the medaka *Oryzias latipes* by fluorescence *in situ* hybridization. Cytogent Cell Genet, 82: 257–262.

Matsuda, M., Nagahama, Y., Shinonuya, A. et al. 2002. *Dmy* is a Y specific Dm-domain gene required for male development in the medaka fish. Nature, 417: 555–563.

Matsuda, M., Sato, T., Tyozaki, Y. et al. 2003. *Oryzias curvinotus* has *Dmy*, a gene that is required for male development in the medaka *O. latipes*. Zool Sci, 20: 159–161.

Matta, S., Vilela, D.A., Godinho, H.P. and Franca, L.R. 2002. The goitgogen-6-n-propyl-2-thiouracil (PTV) given during testis development increases Sertoli and germ cell number per cyst in fish the Tilapia (*Oreochromis niloticus*) model. Endocrinology, 143: 970–978.

Mayer, G., Franke, F.A., Treffkorn, S. et al. 2015. Onychophora. In: *Evolutionary Developmental Biology of Invertebrates: Ecdysozoa 1: Non-Tetraconata.* (ed) Wanninger, A., Springer Verlag, Wien, pp 53–98.

McCain, C.M. and Gytnes, J.A. 2010. Elevational gradients in species richness. In: *Encyclopedia of Life Sciences.* John Wiley, Chichester, doi: 10.1002/9780470015902.10022548.

McLachlan, A.J. and Yonow, T. 1989. Reproductive strategies in rain-pool dwellers and the model freshwater insect. Hydrobiologia, 171: 223–230.

McMillan, M. and Wilcove, D. 1994. Gone but not forgotten: why have species protected by the endangered species act become extinct? Endangered Species Update, 11: 5–6.

Meadow, P.S., Reichelt, A.C., Meadow, A. and Waterworth, J.S. 1994. Microbial and meiofaunal abundance, redox potential, pH and shear strength profiles in deep sea Pacific sediments. J Geo Sa, 151: 377–390.

Menge, B.A. 1975. Brood or broadcast? The adaptive significance of different reproductive strategies in the two intertidal sea stars *Leptasterias hexactis* and *Pisaster ochraceus*. Mar Biol, 31: 87–100.

Menge-Najera, J. 1994. Reproductive trends, habitat types and body characteristics in velvet worms (Onychophora). Rev Biol Trop, 42L3: 611–622.

Metzler, S., Heinze, J. and Schrempf, A. 2016. Mating and longevity in ant males. Ecol Evol, 6: 8903–8906.

Micaletto, G., Gambi, M.C. and Cantone, G. 2002. A new record of the endosymbiont polychaete *Veneriserva* (Dorvilleidae) with description of a new subspecies and relationships with its host *Laetmonice producta* (Polychaeta: Aphroditidae) in Southern ocean waters (Antarctica). Mar Biol, 141: 691–698.

Michiels, N. and Newman, L. 1998. Sex and violence in hermaphrodites. Nature, 391: 647.

Miller, S.L. 1953. A production of amino acids under possible primitive earth conditions. Science, 117: 528–529.

Mire, J.P. and Millet, L. 1994. Size of mother does not determine size of egg or fry in the Owens Pupfish *Cyprinodon radiosus*. Copeia, 1994: 100–107.

Misra, K.G., Singh, V., Yadava, A.K. and Misra, S. 2020. Treeline migration and settlement recorded by Himalayan pencil cedar tree-rings in the highest alpine zone of western Himalya, India. Curr Sci, 118: 192–195.

Miura, I. 2017. Sex determination and sex chromosomes in Amphibia. Sex Dev, 11: 298–306.

Monroy, F., Aira, M., Velando, A. and Dominguez. 2005. Size assortive matings in the earthworm *Eisenia fetida* (Oligochaeta: Lumbricidae). J Ethol, 23: 69–70.

Moorthy, V.N. 1938. Freshwater nematodes from the intestine of fish. Proc Helm Soc Wash, 5: 24–26.

Mora, C., Tittensor, D.P., Adl, S. et al. 2011. How many species are there on earth and in the ocean? PLoS Biol, 9: e1001127.

Moreira-Filho, O., Bertollo, L.A.C. and Galetti, P.M. Jr. 1993. Distribution of sex chromosome mechanisms in Neotropical fish and description of ZZ/ZW system in *Parodon hilarii*. Caryologia, 46: 115–125.

Morgan, T.H. 1889. Experimental studies of the regeneration in *Planaria lugubris*. Arch Entwicklungsmech, 13: 364–397.

Morgan, T.H. 1901. *Regeneration*. MacMillan, New York, p 316.

Morgan, T.H. and Bridges, C.B. 1916. *Sex-linked Inheritance in Drosophila*. Carnegie Institution, Washington.

Mukherjee, S. 2016. *The Gene: An Intimate History*. Penguin Books, p 593.

Muller, C. 1908. Regenerationverscuche an *Lumbriculus variegatus* und *Tubifex rivulorum*. Heredity, 6: 55–76.

Muller, H.J. 1928. The problem of genetic modification. Proc 5th Int Cong, 1: 234–260.

Munuswamy, N. and Subramoniam, T. 1985. Oogenesis and shell gland activity in a freshwater fairy shrimp *Streptocephalus dichotomus* Baird (Crustacea: Anostraca). Cytobios, 44: 137–147.

Murphy, M.T. 1996. Survivorship, breeding dispersal and mate fidelity in eastern kingbirds. The Condor, 98: 82–92.

Murugavel, P. and Pandian, T.J. 2000. Effect of altitude on hydrobiology, productivity and species richness in Kodayar—a tropical peninsular Indian aquatic system. Hydrobiologia, 430: 35–57.

Murugesan, P., Khan, S.A. and Ajithkumar, T.T. 2007. Temporal changes in the benthic community structure of the marine zone of Vellar estuary, southeast coast of India. J Mar Biol Ass India, 49: 154–158.

Murugesan, P., Balasubramanian, T. and Pandian, T.J. 2010. Does hemocoelom exclude embryonic stem cells and asexual reproduction in invertebrates? Curr Sci, 98: 768–770.

Muthukrishnan, J. and Pandian, T.J. 1987. Insecta: In: *Animal Energetics: Protozoa through Insecta*. (eds) Pandian, T.J. and Vernberg, F.J., Academic Press, 1: 374–512.

Muthukrishnan, J. and Pandian, T.J. 1987. Relation between feeding and egg production in some insects. Proc Indian Acad Sci, 96: 171–179.

Muthukrishnan, J. and Senthamizhselvan, M. 1987. Oviposition energetics of some wasps: An analysis of behaviors (unpublished).

Muthukrishnan, J. 1994. Arthropoda–Insecta. In: *Reproductive Biology of Invertebrates*. (eds) Adiyodi, K.G. and Adiyodi, R.G., Oxford & IBH Publishing, New Delhi, 6B: 166–292.

NaNakorn, U., Rangsin, W. and Witchasunscul, S. 1993. Suitable conditions for induction of gynogenesis in the catfish *Clarias macrocephalus* using sperm of *Pangasius sutchi*. Aquaculture, 118: 53–63.

Nanda, I., Kondo, M., Honung, U. et al. 2002. A duplicated copy of *DMRT1* in the sex determination region of the Y chromosome of the medaka *Oryzias latipes*. Proc Natl Acad Sci, USA, 99: 11778–11783.

Nanda, I., Schlupp, I., Lamatsch, D.K. et al. 2007. Stable inheritance of host species derived microchoromosomes in the gynogenetic fish *Poecilia formosa*. Genetics, 177: 917–926.

Nandini, S., Sarma, S.S.S. and Rao, T.R. 1998. Effect of co-existence of the population growth of rotifers and cladocerans. Russ J Aquat Ecol, 8: 1–10.

Nandini, S., Ortiz, A.R.N. and Sarma, S.S.S. 2011. *Elaphoidella grandidieri* (Harpecticoida: Copepoda): Demographic characteristics and possible use as live prey in aquaculture. J Env Biol, 32: 505–511.

Narendran, T.C. 2002. Parasitic Hymenoptera and biological control. In: *Biocontrol Potential and its Exploitation in Sustainable Agriculture*. (eds) Upadhyay, R.K. et al., Kluwer Academic/ Plenum Publishers, New York, pp 1–12.

Nelson, D.R., Guidetti, R. and Rebecchi, L. 2010. Tardigrada. In: *Ecology and Classification of North American Freshwater Invertebrates*. (eds) Thorp, J.H. and Covich, A.P., Elsevier, Amsterdam, pp 455–484.

Neusser, T.P., Hess, M. and Schrodl, M. 2009. Tiny but complex–interactive 3D visuzalization of the interstitial acochlidian gastropod *Pseudunela cornuta* (Challis, 1970). Front Zool, 6: 20, doi: 10.1186/1742-9994-6-20.

Neves, R.C. and Reichert, H. 2015. Microanatomy and development of the dwarf male of *Symbion pandora* (Phylum Cycliophora): New insights from ultrastructural investigation based on serial section electron microscopy. PLoS ONE, 10: e0122364.

Newton, S.F. and Newton, A.V. 1997. The effect of rainfall and habitat on adundance and diversity of birds in a fenced protected area in the central Saudi Arabian desert. J Arid Environ, 35: 715–735.

Nickol, B.B. 2006. Phylum Acanthocephala. In: *Fish Diseases and Disorders*. (ed) Wao, P.J.K., CABI International, UK, 1: 444–452.

Norena, C., Damborenea, C. and Brusa, F. 2015. Phylum Platyhelminthes. In: *Freshwater Invertebrates: Ecology and General Biology*. (eds) Thorp, J.H. and Rogers, D.C., Elsevier, Amsterdam, pp 181–204.

Oheimb, P.V.v., Albrecht, C., Riedel, F. et al. 2013. Testing the role of the Himalaya Mountains as a dispersal barier in freshwater gastropods (*Gyraulus* spp). Biol J Linn Soc, 109: 526–534.

Oka, H. and Watanabe, H. 1957. Colony-specificity in compound ascidians. Bull Mar Biol Stn, Asamushi, 10: 153–155.

Okazaki, K. and Dan, K. 1954. The metamorphosis of partial larvae of *Peronella japonica* Mortensen, a sand dollar. Biol Bull, 106: 83–99.

Okutsu, T., Suzuki, K., Takeuchi, Y. and Yoshizaki, G. 2006. Testicular germ cells can colonize sexually undifferentiated embryonic gonad and produce functional egg in fish. Proc Natl Acad Sci USA, 103: 2725–2729.

Okutsu, T., Takeuchi, Y., Yashozaki, G. et al. 2007. Production of trout offspring from triploid salmon parents. Science, 317: 1517.

Olejarz, J.W., Allen, B., Veller, C. and Nowak, M.A. 2015. The evolution of non reproductive workers in insect colonies with haplodiploid genetics. eLife, 2015: e08918.

Olesen, J. 2018. Crustacean life cycles–developmental strategies and environmental adaptations. In: *The National History of the Crustacea: Life Histories*. Oxford University Press, London, pp 1–34.

Oliveira, R.F. 2006. Neuroendocrine mechanism of alternative reproductive tactics in fish. In: *Fish Physiology, Behaviour and Fish Physiology.* (eds) Sloman, K., Batshine, S. and Wilson, R., Elsevier, Amsterdam, 24: 297–357.

Oliver, J.H. Jr. 1989. Biology and systematic of ticks (Acari: Ixodida). Annu Rev Ecol Syst, 20: 297–430.

Oparin, A.I. 1938. *The Origin of Life.* 1st Edition, MacMillan, New York, p 270.

Ospina-Alwarez, N. and Piferrer, F. 2008. Temperature-dependent sex determination of fish revisited. Prevalence, a single sex ratio response pattern and possible effect of climate change. PLoS ONE, 3: e2837.

Ostrovsky, A.N. 2013. *Evolution of Sexual Reproduction in Marine Invertebrates: Examples of Gymnolamate Bryozoans.* Springer Verlag, Wien, p 387.

Ostrovsky, A.N., Lidgard, S., Gordon, D.P. et al. 2016. Matrotrophy and placentation in invertebrates: a new paradigm. Biol Rev, 91: 673–711.

Otani, S., Maegawa, S., Inoue, K. et al. 2002. The germ cells lineage identified by *vas*-mRNA during embryogenesis in goldfish. Zool Sci, 19: 519–526.

Pai, S. 1927. Lebenzyklus der *Anguillulaaceti.* EhrbgZoolAnz, 74: 257–270.

Pandian, T.J. 1967. Changes in chemical composition and caloric content of developing eggs of the shrimp *Crangon crangon.* HelgolanderWissenMeeresunters, 16: 216–224.

Pandian, T.J. and Schumann, K.-H. 1967. Chemical composition and caloric content of egg and zoea of the hermit crab *Eupagurus bernhardus.* Helgolander Wissen Meeresunters, 16: 225–230.

Pandian, T.J. and Fluchter, J. 1968. Rate and efficiency of yolk utilization in the sole *Solea solea.* Helgolander Wissen Meeresunters, 18: 53–65.

Pandian, T.J. 1969. Yolk utilization in the gastropod *Crepidula fornicata.* Mar Biol, 3: 117–121.

Pandian, T.J. 1970a. Ecophysiological studies on the developing eggs and embryos of the European lobster *Homarus gammarus.* Mar Biol, 5: 154–167.

Pandian, T.J. 1970b. Yolk utilization and hatching time in the Canadian lobster *Homarus americanus.* Mar Biol, 7: 249–254.

Pandian, T.J. 1972. Egg incubation and yolk utilization in the isopod *Ligia oceanica.* Proc Ind Natl Sci Acad, 38: 430–441.

Pandian, T.J. 1975. Mechanism of Heterotrophy. In: *Marine Ecology.* (ed) Kinne, O., John Wiley, London, 3A: 61–249.

Pandian, T.J. and Vivekanandan, E. 1976. Effects of feeding and starvation on growth and swimming activity in an obligatory air-breathing fish. Hydrobiologia, 49: 33–40.

Pandian, T.J., Mathavan, S. and Jayagopal, C.P. 1979. Influence of temperature and body weight on mosquito predation by the dragonfly nymph *Mesogomphus lineatus.* Hydrobiologia, 62: 99–104.

Pandian, T.J. 1980. Impact of dam building on marine life. Helgolander Wissens Meeresunters, 33: 415–421.

Pandian, T.J. and Balasundaram, C. 1980. Contribution to the reproductive biology and aquaculture of *Macrobrachium nobilii.* Proc Symp Invert Reprod, Madras University, Madras, 1: 183–193.

Pandian, T.J. 1985. Behavioural energetic of some insects. Proc Indian Acad Sci, 94(Anim Sci): 219–224.

Pandian, T.J. and Vivekanandan, E. 1985. Energetics of feeding and digestion. In: *Fish Energetics: New Perspectives.* (eds) Tyler, P. and Calow, P., Croom Helm, London, pp 99–124.

Pandian, T.J. and Marian, M.P. 1985a. Nitrogen content of food as an index of absorption efficiency in fishes. Mar Biol, 85: 301–311.

Pandian, T.J. and Marian, M.P. 1985b. A method for the estimation of assimilation efficiency of reptiles. Proc Acad Sci Lett, 7: 351–354.

Pandian, T.J. and Marian, M.P. 1985c. Prediction of absoption efficiency from food nitrogen in amphibians. Proc Indian Acad Sci, 95: 387–395.

Pandian, T.J. and Marian, M.P. 1985d. Time and energy costs of metamorphosis in the Indian bullfrog *Rana tigrina.* Copeia, 1985: 653–662.

Pandian, T.J. and Marian, M.P. 1985e. Physiological correlates of surfacing behavior: Effect of aquarium depth on growth and metamorphosis in *Rana tigrina*. Physiol Behav, 35: 867–872.

Pandian, T.J. and Marian, M.P. 1986a. Nitrogen content of food as an index of absorption efficiency in fishes. Mar Biol, 85: 301–311.

Pandian, T.J. and Marian, M.P. 1986b. Estimation of absorption efficiency in polychaetes using nitrogen content of food. J Exp Mar Biol Ecol, 90: 289–295.

Pandian, T.J. and Marian, M.P. 1986c. Prediction of assimilation efficiency in lepidopterans. Proc Indian Acad Sci, 95: 641–665.

Pandian, T.J. and Marian, M.P. 1986d. An indirect procedure for the estimation of assimilation efficiency of aquatic insects. Freshwater Biol, 16: 93–98.

Pandian, T.J. and Vernberg, F.J. 1987a. *Animal Energetics: Protozoa through Insecta*. Academic Press, San Diego, Vol 1, p 523.

Pandian, T.J. and Vernberg, F.J. 1987b. *Animal Energetics: Bivalvia through Reptilia*. Academic Press, San Diego, Vol 2, p 631.

Pandian, T.J. 1987c. Fish. In: *Animal Energetics: Bivalvia through Reptilia*. (eds) Pandian, T.J. and Vernberg, F.J., Academic Press, 2: 358–467.

Pandian, T.J. 1994. Crustacea. In: *Reproductive Biology of Invertebrates*. (eds) Adiyodi, K.G. and Adiyodi, R.G., Oxford & IBH Publishing, New Delhi, 6A: 39–166.

Pandian, T.J. and Sheela, S.G. 1995. Hormonal induction of sex reversal in fish. Aquaculture, 138: 1–22.

Pandian, T.J. and Koteeswaran, R. 1999. Natural occurrence of monoploid and polyploids in the Indian catfish *Heteropneustes fossilis*. Curr Sci, 76: 1134–1137.

Pandian, T.J. 2000. Hydrobiologia. Special Issue, 430: 1–205.

Pandian, T.J. 2002. Biodiversity: Status and endeavors of India. ANJAC J Sci, 1: 21–32.

Pandian, T.J. and Kirankumar, S. 2003. Androgenesis and conservation of fishes. Curr Sci, 85: 917–931.

Pandian, T.J. 2011a. *Sexuality in Fishes*. Science Publishers/CRC Press, USA, p 208.

Pandian, T.J. 2011b. *Sex Determination in Fish*. Science Publishers/CRC Press, USA, p 270.

Pandian, T.J. 2012. *Genetic Sex Differentiation in Fish*. CRC Press, USA, p 214

Pandian, T.J. 2013. *Endocrine Sex Differentiation in Fish*. CRC Press, USA, p 303.

Pandian, T.J. 2015. *Environmental Sex Determination in Fish*. CRC Press, USA, p 299.

Pandian, T.J. 2016. *Reproduction and Development in Crustacea*. CRC Press, USA, p 301.

Pandian, T.J. 2017. *Reproduction and Development in Mollusca*. CRC Press, USA, p 299.

Pandian, T.J. 2018. *Reproduction and Development in Echinodermata and Prochordata*. CRC Press, USA, p 270.

Pandian, T.J. 2019. *Reproduction and Development in Annelida*. CRC Press, USA, p 276.

Pandian, T.J. 2020. *Reproduction and Development in Platyhelminthes*. CRC Press, USA, p 303.

Pandian, T.J. 2021. *Reproduction and Development in Minor Phyla*. CRC Press, USA, p 299.

Pandian, T.J. 2022. *Evolution and Speciation in Plants*. CRC Press, USA (in preparation).

Parker, G.A. 1978. Searching for mates. In: *Behavioral Ecology: an Evolutionary Approach*. (eds) Krebs, J.R. and Davies, N.B., Blackwells, Oxford, pp 214–244.

Parmesan, C. and Yohe, G. 2003. A globally coherent fingerprint of climate change impacts across natural systems. Nature, 421: 37–42.

Paterson, A.M., Gray, R.D. and Wallis, G.P. 1993. Parasites, petrels and penguins. Does louse presence reflect seabird phylogeny? Int Parasitol, 23: 515–526.

Pemberton, A.J., Hughes, R.N., Manriquez, P.H. and Bishop, J.D.D. 2003. Efficient utilization of very dilute aquatic sperm: sperm competition may be most likely than sperm limitation, when eggs are retained. Proc R Soc, 270B: 223–226.

Pemberton, A.J., Hansson, L.J., Craig, S.F. et al. 2007. Microscale genetic differentiation in a sessile invertebrate with cloned larvae: investigating the role of polyembryony. Mar Biol, 153: 71–82.

Pennington, J.T. 1985. The ecology of fertilization of echinoid eggs: the consequences of sperm dilution adult aggregation and synchronous spawning. Biol Bull, 169: 417–430.

Pennuto, C.M. and Stewart, T.J. 2001. Oviposition site preference and factors influencing egg mass characteristics of the saw-combed fishfly (Megaloptera: Corydalidae) in Southern Maine. J Freshwat Ecol, 16: 209–217.

Phalee, A., Wongawad, C., Rojanapaibul, A. and Chai, J.Y. 2015. Experimental life history and biological characteristics of *Fasciola gigantea* (Digenea: Fasciolidae). Korean J Parasitol, 53: 59–64.

Phillipson, R.F. 1969. Reproduction of *Nippostrongylus brasiliensis* in rat intestine. Parasitology, 59: 961–967.

Phillipson, R.F. 1970. Experiments on the reproduction of *Nippostrrongylus brasiliensis* in the rat intestine. Parasitology, 61: 317–322.

Pilger, J.F. 1987. Reproductive biology and development of *Themiste lageniformis*. A parthenogenic sipunculan. Bull Mar Sci, 41: 59–67.

Pincheira-Donoso, D., Bauer, A.M., Meiri, S. and Uetz, P. 2013. Global taxonomic diversity of living reptiles. PLoS ONE, 8: 359741.

Poddubnaya, T.L. 1984. Parthenogenesis in Tubificidae. Hydrobiologia, 115: 97–99.

Podolsky, R.D. 2002. Fertilization ecology of the egg coats: physical versus chemical contributions to fertilization success of free spawned eggs. J Exp Biol, 205: 1657–1668.

Pompini, M., Buser, A.M., Thali, M.R. et al. 2013. Temperature-induced sex reversal is not responsible for sex ratio distortions in gralying *Thymallus thymallus* or brount trout *Salmo trutta*. J Fish Biol, 83: 404–411.

Ponniah, A.G. and Pandian, T.J. 1981. Surfacing frequency as a predictor variable of bioenergetics components in air-breathing fishes. Hydrobiologia, 83: 491–497.

Potts, F.A. 1910. Notes on free-living nematodes. J Cell Sci, 55: 433–484.

Poulin, R. 1992. Determinants of host specificity in parasites of freshwater fishes. Int J Parasitol, 22: 753–758.

Poulin, R. and Morand, S. 2000. The diversity of parasites. Q Rev Biol, 75: 277–293.

Premoli, M.C. and Sella, G. 1995. Sex economy in benthic polychaetes. Ethol Ecol Evol, 7: 27–48.

Prevedelli, D. and Vandini, R.Z. 1999. Survival, fecundity and sex ratio of *Dinophilus gyrociliatus* (Polychaeta: Dinophillidae) under different dietary conditions. Mar Biol, 132: 163–170.

Price, P.W. 1975. Reproductive strategies of parasitoids. In: *Evolutionary Strategies of Parasitic Insects and Mites*. (ed) Price, P.W., Plenum Press, New York, pp 88–111.

Priede, I.G., Froese, R., Bailey, D.M. et al. 2006. The absence of sharks from abyssal regions of the world's oceans. Proc R Soc, 273B: 1435–1441.

Pudasaini, R. 2015. Effect of climate change on insect pollinator: A review. New York Sci J, 8: 39–42.

Putter, A. 1909. *Die Einbrung der Wassertiere und der Stoffhaushalt der Gewasser*. G Fischer, Jena.

Quinn, T.P. and Myers, K.W. 2004. Anadromy and the marine migrations of Pacific salmon and trout: Rounsenfell revisited. Rev Fish Biol Fisher, 14: 421–442.

Raghukumar, C., Raghukumar, S., Sheelu, G. et al. 2004. Buried in time: culturable fungi in a deep-sea sediment core from Chagos Trench, Indian Ocean. Deep Sea Res, Oceanogr Res, 51: 1759–1768.

Raghukumar, C., Damare, S. and Singh, P. 2010. A review on deep-sea fungi: occurrence, diversity and adaptations. Bot Mar, 53: 479–492.

Raguso, R.A. 2020. Coevolution as an engine of biodiversity and a cornucopia of ecosystem services. Plants People Planet, 2020: 1–13.

Ramalho, M. 2004. Stingless bees and mass flowering trees in the canopy of Atlantic forests: a tight relationship. Acta Bot Bras, 18: 37–47.

Ramanov, M.N., Farre, M., Lithgow, P.E. et al. 2014. Reconstruction of gross avian genome structure, organization and evolution suggests that the chicken lineage most closely resembles the dinosaur avian ancestor. BMC Genomics, 14: 1060.

Randolph, H. 1891. The regeneration of the tail in *Lumbriculus*. ZoolAnz, 14: 154–156.

Rangel, J. and Fisher II, A. 2019. Factors affecting the reproductive health of honey bee (*Apis mellifera*) drones–a review. Apidologie, 50: 759–778.

Rasmussen, E. 1973. Systematics and ecology of the Isefjord marine fauna (Denmark). Ophelia, 11: 1–495.

Ravaglia, M.A. and Maggese, M.C. 2002. Oogenesis in the swamp eel *Synbranchus marmoratus* (Bloch 1975) (Teleostie: Synbranchidae). Ovarian anatomy stages of oocytes development and micropyle structure. BioCell, 26: 325–337.

Rawlinson, K.A. 2014. The diversity, development and evolution of polyclad flatworm larvae. Evol Dev, 5: 9.

Rawson, P.D. and Hilbish, T.J. 1995. Evolutionary relationships among male and female mitochondrial DNA lineages in the *Mytilus edulis* species complex. Mar Biol Evol, 12: 892–901.

Reddy, S.R. 1973. *Mosquito Control through Larvivorous Predators*. Ph.D. Thesis, Bangalore University, India.

Reeve, M.R. 1963. The filter-feeding in *Artemia*. 1. In pure cultures of plant cells. J Exp Biol, 40: 195–205.

Reynolds, J.W. 1974. Are oligochaetes really hermaphroditic amphimictic organism? Biologists, 56: 98–99.

Reznick, D., Bryant, M. and Holmes, D. 2006. Evolution of senescence and post-reproductive life span in guppies (*Poecilia reticulata*). PLoS Biol, 4, doi: 10.1371/journal.pbio.0010007.

Rice, M.E. 1970. Asexual reproduction in a sipunculan worm. Science, 167: 1618–1620.

Rice, M.E. and Pilger, J.F. 1993. Sipuncula. In: *Reproductive Biology of Invertebrates: Asexual Propagation and Reproductive Strategies*. (eds) Adiyodi, K.G. and Adiyodi, R.G., Oxford & IBH Publishering, New Delhi, 6A: 279–296.

Rice, S.A. 1978. Spermatogenesis and sperm transfer in spionid polychaete (Annelida). Trans Am Microsc Soc, 97: 160–170.

Rice, S.A., Karl, S. and Rice, K.A. 2008. The *Polydora cornuta* complex (Annelida: Polychaeta) contains population that are reproductively isolated and genetically distinct. Am Microsc Soc, 127: 45–64.

Richman, S. 1958. The transmission of energy by *Daphnia pulex*. Ecol Monogr, 28: 273–291.

Ricklefs, R.E. 2010. Evolutionary diversification, coevolution between populations and their antagonists and the filling of niche space. Proc Natl Acad Sci USA, 107: 1265–1272.

Riffell, J.A., Krug, P.J. and Zimmer, R.K. 2002. The effects of sperm density and gamete contact time on fertilization success of blacklip (*Haliotis rubra*, Leach 1814) and greenlip (*H. laeviigata*, Donovan 1808) abalone. J Exp Biol, 205: 1439–1450.

Riffell, J.A. Krug, P.J. and Zimmer, R.K. 2004. The ecological and evolutionary consequences of sperm chemo-attraction. Proc Natl Acad Sci, USA, 101: 4501–4506.

Rinkevich, B. 2009. Stem Cells: Autonomy interactors that emerge as causal agents and legitimate units of selection. In: *Stem Cells in Marine Organisms*. (eds) Rinkevich, B. and Matranga, V., Springer, Dordrecht, pp 1–20.

Rinkevich, Y., Matranga, V. and Rinkevich, B. 2009. Stem cells in aquatic invertebrates: common premises and emerging unique themes. In: *Stem Cells in Marine Organisms*. (eds) Rinkevich, B. and Matranga, V., Springer Verlag, Dordrecht, pp 61–103.

Riser, N.W. 1974. Nemertea. In: *Reproduction of Marine Invertebrates: Acoelomate and Pseudocoelomate Metazoans*. (eds) Giese, A.C. and Pearse, J.S., Academic Press, New York, 1: 359–390.

Rivero-Wendt, C.L.G., Borges, A.C., Oliveira-Filho, E.C. et al. 2014. Effects of 17-methyltestosterone on the reproduction of the freshwater snail *Biomphalaria glabrata*. Gene Mol Res, 13: 605–615.

Robertson, J.G.M. 1990. Female choice increases fertilization success in the Australian frog, *Uperoleia laevigata*. Anim Behav, 39: 639–645.

Romanov, M.N., Farre, M., Lithgow, P.E. et al. 2014. Reconstruction of gross avian genome structure, organization and evolution suggests that the chicken lineage most closely resembles the dinosaur avian ancestor. BMC Genomics, 15: 1060.

Rosa, R., Caetano-Filho, M., Shilbalta, O.A. and Margarido, V.P. 2009. Cytotaxomy in distinct populations of *Hoplias* aff. *malabaricus* (Characiformes, Erythrinidae) from lower Paranapanema River basin. J Fish Biol, 75: 2682–2694.

Rougeot, C., Krim, A., Mandiki, S.N.M. et al. 2007. Sex steroid dynamics during embryogenesis and sexual differentiation in European perch *Perca fluviatilis*. Theriogenology, 67: 1046–1052.

Royer, M. 1975. Hermaphroditism in insects. Studies on *Icerya purchari*. In: *Intersexuality in the Animal Kingdom*. (ed) Rheinboth, R. Springer Verlag, Heidelberg, pp 135–145.

Rozhnov, S.V. 2002. Morphogenesis and evolution of crinoids and other Pelmatozoan echinoderms in the early Paleozoic. Paleontol J, 36S: 525–674.

Runham, N.W. 1993. Mollusca. In: *Reproductive Biology of Invertebrates*. (eds) Adiyodi, K.G. and Adiyodi, R.G., Oxford & IBH Publishers, New Delhi, 6: 311–383.

Rusin, L.Yu. and Malakhov, V.V. 1998. Free-living marine nematodes possess no eutely. Dok Biol Sci, 361: 331–333.

Russel, F.S. 1976. *The Eggs and Planktonic Stages of British Marine Fishes*. Academic Press, New York, p 524.

Rychel, A.L. and Swalla, B. 2009. Regeneration in hemichordates and echinoderms. In: *Stem Cells in Marine Organisms*. (eds) Rinkevich, B. and Matranga, V., Springer Verlag, Dordrecht, pp 245–266.

Saavedra, C., Ryero, M.-G. and Zourus, E. 1997. Male-dependent doubly uniparental inheritance of mitochondrial DNA and female-dependent sex ratio in the mussel *Mytilus galloprovincialis*. Genetics, 145: 1073–1082.

Sabatini, F.M., Borja, J.-A., Sabina, B. et al. 2017. Beta-diversity of Central European forests decreases along an elevational gradient due to the variation in local community assembly processes. Ecography, doi: 10.111/ecog.02809.

Safarik, K.M., Redden, A.M. and Schneider, M.J. 2006. Density dependent growth of the polychaete *Diopatra aciculata*. Scient Mari, 70: 337–341.

Sallar, U. 1990. Formation and construction of asexual buds of the freshwater sponge *Radiospongilla cerebellata* (Porifera, Spongillidae). Zoomorphology, 109: 295–301.

Sample, B.E., Lowe, J., Seeley, P. et al. 2014. Depth of the biologically active zone in upland habitats at the Hanford site Washington: Implications for remediation and ecological risk management. Integ Environ Assess Mgmt, 11: 150–160.

Sankaraperumal, G. and Pandian, T.J. 1991. Effect of temperature and *Chlorella* density on growth and metamorphosis of *Chironomus circumdatus* (Kieffur) (Diptera). Int J Freshwat Entomol, 13: 167–177.

Santagata, S. 2015a. Phoronida. In: *Evolutionary Developmental Biology of Invertebrates*. (ed) Wanninger, A., Springer Verlag, Wein, 2: 231–246.

Santagata, S. 2015b. Ectoprocta. In: *Evolutionary Developmental Biology of Invertebrates: Lophotrochozoa (Spiralia)*. (ed) Wanninger, A., Springer, Wein, 2: 247–262.

Santagata, S. 2015c. Brachiopoda. In: *Evolutionary Developmental Biology of Invertebrates*. (ed) Wanninger, A., Springer Verlag, Wein, 2: 263–278.

Sara, M. 1974. Sexuality in the Porifera. Boll Zool, 41: 327–374.

Saragih, H.T., Muhamad, A.A.K., Alfianto, A. et al. 2019. Effects of *Spirogyra jaoensis* as a dietary supplement on growth, pectoralis muscle performance, and small intestine morphology of broiler chickens. Vet world, 12: 1233–1239.

Sasal, P., Trouve, S., Muller-Graft, C. and Morand, S. 1999. Specificity and host predictability: a comparative analysis among monogenean parasites of fish. J Anim Ecol, 68: 437–444.

Sasson, D.A. and Ryan, J.E. 2016. The sex lives of ctenophores: the influence of light, body size, and self-fertilization on the reproductive output of the sea walnut, *Mnemiopsis leidyi*. Peer J, 4: e1846.

Schaffer, W.M., Jenson, J.B., Hobbs, D.E. et al. 1979. Competition, foraging energetics and the cost of sociality in three species of bees. Ecology, 60: 976–987.

Scharer, L. and Wedekind, C. 1999. Lifetime reproductive output in a hermaphrodite cestode, when reproducing alone or in pairs. A time cost of pairing. Evol Ecol, 13: 381–384.

Schierwater, B. 2005. My favorite animal, *Trichoplax adhaerens*. BioEssays, 27: 1294–1302.

Schmelz, R.M., Niva, C.C., Rombke, J. and Collado, R. 2013. Diversity of terrestrial enchytraeidae (Oligochaeta) in Latin America: Current Knowledge and future research potential. Appl Soil Ecol, 69: 13–20.

Schmid, M. 1983. Evolution of sex chromosomes and heterogametic systems in Amphibia. Differntiation, 23S: 13–22.

Schmidt, G.A. 1932. Dimorphismeembryonaire de *Lineus guserensis* Ruber de la cote Mourmanne et de Roscof et ses relations avec less forms adultes. Ann Inst Oceanogr (Paris), 12: 65–103.

Schmidt-Rhaesa, A. 2014. Phylum Gnathostomulida. In: *Handbook of Zoology*. (eds) Sterrer, W. and Sorenssen, M.A., DeGruyter, Berlin, 3: 135–196.

Schneider, J. and Fromhage, L. 2010. Monogamous mating strategies in spiders. In: *Animal Behaviour: Evolution and Mechanisms*. (ed) Kappeler, P., Springer, Berlin, pp 441–464.

Schockaert, E.R., Hooge, M. Sluys, R. et al. 2008. Global diversity of free-living flatworms (Platyhelminhtes, Turbellaria) in freshwater. Hydrobiologia, 595: 41–48.

Schoener, A. 1972. Fecundity and possible mode of development of some deep sea ophiuroids. Limnol Oceanogr, 17: 193–199.

Schroder, P.C. 1989. Annelida–Polychaeta. In: *Reproductive Biology of Invertebrates*. (eds) Adiyodi, K.G. and Adiyodi, R.G., Oxford IBH Publishing, New Delhi, 4A: 383–442.

Schroder, T., Howard, S., Arroyo, M.L. and Walsh, E.J. 2007. Sexual reproduction and diapauses of *Hexarthra* sp (Rotifera) short-lived in the Chihuahuan Desert. Freshwat Biol, 52: 1033–1042.

Schulz, R.W. and Miura, T. 2002. Spermatogenesis and its endocrine regulation. Fish Physiol Biochem, 26: 43–56.

Schulz, R.W., DeFranca, L.R., Lareyre, J.J. et al. 2010. Spermatogenesis in fish. Gen Comp Endocrinol, 165: 390–411.

Scribner, T., Page, K.S. and Bartron, M.L. 2001. Hybridization in freshwater fishes: A review of case studies and cytonuclear methods of biological inference. Rev Fish Biol Fisher, 10: 293–323.

Seale, D.B. 1987. Amphibia. In: *Animal Energetics: Bivalvia through Reptilia*. (eds) Pandian, T.J. and Vernberg, F.J., Academic Press, 2: 468–552.

Searcy, W.A. and Yasukawa, K. 1995. *Polygyny and Sexual Selection in Red-Winged Blackbirds*. Princeton University Press, Princeton, p 331.

Sebens, K.P. 1987. Coelenterata. In: *Animal Energetics*. (eds) Pandian, T.J. and Vernberg, F.J., Academic Press, 1: 55–120.

Seeley, T.D. 1985. *Honeybee Ecology*. Princeton University Press, Princeton, USA, p 213.

Segers, S. 2008. Global diversity of rotifers (Rotifera) in freshwater. Hydrobiologia, 595: 49–59.

Seigel, R.A. 1983. Natural survival of eggs and tadpoles of the wood frog, *Rana sylvatica*. Copeia, 1983: 1096–1098.

Seito, D., Morinaga, C., Aoki, T. et al. 2007. Proliferation of germ cells during gonadal sex differentiation in medaka. Insights from germ cell-depleted mutant *zenzai*. Dev Biol Bull, 280–290.

Ser, J.R., Roberts, R.B. and Kocher, T.D. 2009. Multiple interacting loci control sex determination in Lake Malawi cichlid fish. Evolution, doi: 10.111/j.1558.5646.2009.00871.x.

Serezli, R., Guzel, S. and Kocabas, M. 2010. Fecundity and egg size of three salmonid species (*Oncorhynchus mykiss, Salmo labrax, Salvelinus fontinalis*) cultured at the same farm condition in North-Eastern, Turkey. J Anim Vet Adv, 9: 576–580.

Serra, M., Snell, T.W. and Wallace, R.L. 2018. Reproduction, Overview by phylogeny: Rotifera. In: *Reference Module in Life Sciences*. (ed) Roitberg, B.D., Elsevier, DOI: 10.1016/B978-0-12-809633-8.20646-8.

Seth, R.K. and Sharma, V.P. 2002. Growth, development, reproductive competence and adult behavior of *Spodoptera litura* (Lepidoptera: Noctuidae) reared on different diets. Proc IAEA Symp, 15–22.

Sewell, J. 2016. Chinese mitten crab, *Eriocheir sinensis*. http://www.nonnativespecies.org/factsheet/downloadFactssheet.cfm?speciesid=1379.

Sewell, M.A. and Young, C.M. 1997. Are echinoderm egg size distribution bimodal? Biol Bull, 193: 297–305.

Shapiro, A.M. 1970. The role of sexual behavior in density related dispersal of pierid butterflies. Am Nat, 104: 367–372.

Sheer, J.F. and Kovacs, K.M. 1997. Allometry of diving capacity in air-breathing vertebrates. Can J Zool, 75: 339–358.

Sheltema, R.S. 1966. Trans-Atlantic dispersal of veliger larvae from shoal water benthic Mollusca. Sec Int Oceanogr Cong (Moscow). Abstract No 375, p 320.

Shepherd, S.A. 1986. Studies on southern Australian abalone (genus *Haliotis*). 7. Aggregative behavior of *H. laevigata* in relation to spawning. Mar Biol, 90: 231–236.

Shields, J.D. and Segonzac, M. 2007. New nemertean worms (Carcinonemertidae) on bythograeid crabs (Decapoda: Brachyura) from Pacific hydrothermal vent sites. J Crust Biol, 27: 681–692.

Shikina, S. and Chang, C.-F. 2018. Cnidaria. In: *Encyclopedia of Reproduction*. Vol 6, doi. org/10.1016/B978-0-12.809633-8.20597-9.

Shillaker, R.O. and Moore, P.G. 1987. The biology of brooding in the amphipods, *Lembos websteri* Bate and *Corophium bonnelli* Milne Edwards. J Exp Mar Biol Ecol, 110: 113–132.

Shirai, Y. 1995. Longevity, flight ability and reproductive performance of the diamondback moth, *Plutella xylostella* (L.) (Lepidoptera: Yponomeutidae), related to adult body size. Res Popul Ecol, 37: 269–277.

Shostak, S. 1993. Cnidaria. In: *Reproductive Biology of Invertebrates: Asexual Propagation and Reproductive Strategies*. (eds) Adiyodi, K.G. and Adiyodi, R.G., Oxford & IBH Publishing, New Delhi, 6A: 45–106.

Shuster, S.M. 1991. Changes in female anatomy associated with the reproductive moult in *Paracerceis sculpta*, a semelparous isopod crustacean. J Zool Lond, 225: 365–379.

Silva, L., Meireless, L., Davila, S. et al. 2013. Life history of *Bulimulus tenuissimus* (O'Orbingny, 1835) (Gastropoda: Pulmonata: Bulimullidae). Effect of isolation on reproductive strategy and in resource allocation over their lifetime. Moll Res, 33: 75–79.

Simonini, R. and Prevedelli, D. 2003. Effects of temperature on two Mediterranean populations of *Dinophilus gyrociliatus* (Polychaeta: Dinophillidae). 1. Effects on life history and sex ratio. J Exp Mar Biol Ecol, 291: 79–93.

Smith, G. and Grenfell, B.T. 1984. The influence of water temperature and pH on the survival of *Fasciola hepatica* miracidia. Parasitology, 88: 97–104.

Smith, J.J.B. 1979. Effect of diet viscosity on the operation of the pharyngeal pump in the blood-feeding bug *Rhodnius prolixus*. J Exp Biol, 82: 93–104.

Sommer, R.J. 2015. Nematoda. In: *Evolutionary Developmental Biology of Invertebrates: Non-Tetracrondata*. (ed) Wanninger, A., Springer Verlag, Wien, 3: 15–34.

Southward, A.J. and Southward, E.C. 1987. Pogonophora. In: *Animal Energetics*. (eds) Pandian, T.J. and Vernberg, F., Academic Press, 2: 201–228.

Southwick, E.E. and Pimental, D. 1981. Energy efficiency of honey production by bees. Bioscience, 31: 730–732.

Spakulova, M. and Casanova, J.C. 2004. Current knowledge on B chromosomes in natural populations of helminth parasites: a review. Cytogent Genome Res, 106: 222–229.

Starkweather, P.L. 1987. Rotifera. In: *Animal Energetics: Protozoa through Insecta*. (eds) Pandian, T.J. and Vernberg, F.J., Academic Press, 1: 159–184.

Steiner, G. 1993. Spawning behavior of *Pulsellum lofotensis* (M. Sars) and *Cadulus subfusiformis* (M. Sars) (Scaphopoda, Mollusca). Sarsia, 78: 31–33.

Stocker, L.J. and Underwood, A.J. 1991. The relationship between the presence of neighbours and rates of sexual and asexual reproduction in a colonial invertebrate. J Exp Mar Biol Ecol, 149: 191–205.

Stohr, S., O'Hara, T.D. and Thuy, B. 2012. Global diversity of brittle stars (Echinodermata: Ophiuroidea). PLoS ONE, 7: e31940.

Strathmann, R.R. and Strathmann, M.F. 1982. The relationship between adult size and brooding in marine invertebrates. Am Nat, 119: 91–101.

Subramoniam, T. 1993. Spermatophores and sperm transfer in marine crustaceans. Adv Mar Biol, 29: 129–214.

Subramoniam, T. 2017. *Sexual Biology and Reproduction in Crustaceans*. Academic Press, p 526.

Sugiyamo, N., Iseto, T., Hirose, M. and Hirose, E. 2010. Reproduction and population dynamics of the solitary entoproct *Loxosomella plakorticola* inhabiting a desmosponge *Plakortis* sp. Mar Ecol Prog Ser, 415: 73–82.

Sulston, J.E. and Horvitz, H.R. 1977. Post-embryonic cell lineages of the nematode, *Caenorhabditis elegans*. Dev Biol, 56: 110–156.

Sun, Y.D., Lui, S.J., Zhang, C. et al. 2003. Chromosome number and gonadal structure of F_0–F_{11} allotetraploid crucian carp. China J Genet, 230: 37–41.

Suomalainen, E. 1962. Significance of parthenogenesis in the evolution of insects. Annu Rev Ent, 7: 349–366.

Sutcliff, D.W. 2010. Reproduction in *Gammarus* (Crustacea: Amphipoda): basic processes. Freshwat Forum, 2: 102–128.

Taborsky, M. 2001. The evolution of bourgeois parasitic and cooperative reproductive behaviours in fishes. J Hered, 92: 100–102.

Tagami, T., Kagami, H., Mutsubara, Y. et al. 2007. Differentiation of female primordial germ cells in the male testes of chicken (*Gallus gallus domesticus*). Mol Reprod Dev, 74: 68–78.

Takegaki, T. and Nakazono, A. 1999. Reproductive behavior and mate-fidelity in the monogamous goby *Valenciennea longipinnis*. Ichthyol Res, 46: 115–123.

Takeuchi, Y., Yoshizaki, G., Kobayashi, T. and Takeuchi, T. 2003. Generation of live fry from intraperitoneally transplanted primordial germ cells in rainbow trout. Biol Reprod, 69: 1142–1149.

Takeuchi, Y., Yoshizaki, G., Kobayashi, T. and Takeuchi, T. 2004. Surrogate-broodstock produces salmonids. Nature, 430: 629–630.

Temereva, E.N. and Malakhov, V.V. 2016. Viviparity of larvae, a new type of development in phoronids (Lophophorata: Phoronida). Doklady Biol Sci, 467: 72–74.

Tews, J., Brose, U., Grimm, V. et al. 2004. Animal species diversity driven by habitat heterogeneity/diversity: the importance of keystone structures. J Biogeogr, 31: 79–92.

Thompson, R.J. 1979. Fecundity and reproductive effort in the blue mussel (*Mytilus edulis*), the sea urchin (*Strongylocentrotus droebachiensis*) and the snow crab (*Chironectes epilio*) from the populations in Nova Scotia and New Foundland. J Fish Res Bd Can, 36: 955–964.

Thorson, G. 1950. Reproductive and larval ecology of marine bottom invertebrates. Q Rev Biol. 25: 1–45.

Thyagarajan, S.P., Jayaram, S., Gopalakrishnan, V. et al. 2002. Herbal medicines for liver diseases in India. J Gastroenterol Hepatol, 17S: 370–376.

Tocque, K. and Tinsley, R.C. 1994. The relation between *Pseudodiplorchus americanus* (Monogenea) density and host resources under controlled conditions. Parasitology, 83: 181–193.

Toledo, A., Crusz, C., Fragoso, G. et al. 1997. *In vitro* culture of *Taenia crassipes* larval cells and cyst regeneration after injection into mice. J Parasitol, 83: 181–193.

Touaylia, S., Garrido, J. and Boumaiza, M. 2013. Abundance and diversity of the aquatic beetles in a Mediterranean stream system (Northern Tunisia). Ann Soc Ent France, 49: 172–180.

Traut, W. 1969a. Zur Sexualitat von *Dinophilus gracialiatus* (Archiannelidae). I. Der einfluss von Aussenbedingungen und genetischenFaktoren auf des Geschlechtsverhaltnis. Biol Zentralbl, 88: 467–695.

Traut, W. 1969b. Zur Sexualitat von *Dinophilus gracialibitus* (Archiannelidae). II. Der Aufbau des Ovars und die Oogenese. Biol Zentrabl, 89: 137–161.

Traut, W. 1970. Zur Sexualitat con *Dinophilus graciliatus* (Archiannelidae). III. Die Geschlectsbetimmung. Biol Zentralbl, 9: 695–714.

Trouve, S., Sasal, P., Jourdane, J. et al. 1998. The evolution of life history traits in parasitic and free-living platyhelminthes: a new perspective. Oeclologia, 115: 370–378.

Tsurusaki, N. and Cokendolpher, J.C. 1990. Chromosomes of sixteen species of harvestmen (Arachinida, Opiliones, Caddidae and Phalangiidae). J Arachnol, 18: 151–166.

Tucker, V.A. 1973. Aerial and terrestrial locomotion: A comparison of energetic. In: *Comparative Physiology: Locomotion, Respiration, Transport and Blood*. (eds) Bolis, L., Schmidt-Nielsen, K. and Maddrell, S.H.P., North-Holland Publishers, Amsterdam, pp 61–76.

Turner, B.J., Eder, J.F. Jr., Laughlin, T.F. et al. 1992. Extreme clonal diversity and divergence in a population of selfing hermaphrodite fish. Fish Sci, 63: 147–148.

Tutman, P., Sifner, S.K., Dulcic, J. et al. 2008. A note on the distribution and biology of *Ocysthoe tuberculata* (Cephalopoda, Ocythoidae) in the Adriatic Sea. Vie et Milieu, 58: 215–221.

Utinomi, H. 1961. A revision of the nomenclature of the family Nephtheidae (Octocorallia: Alcyonacea). 2. The boreal genera *Gersemia*, *Duva*, *Drifa* and *Pseudodrifa*. (n.g.). PublSeto Mar Biol Lab, 9: 229–246.

Vacquier, V.D. 1998. Evolution of gametic recognition proteins. Science, 281: 1995–1998.

Vagelli, A. 1999. The reproductive biology and early ontogeny of the mouth-brooding Banggai cardinalfish *Pterapogon kauderni* (Perciformes: Apogonidae). Environ Biol Fish, 56: 79–82.

Van de Vyver, G. 1970a. La non confluence intraspecifique chez les spongiares et la notion d'individu. Ann Embryol Morph Fr, 3: 251–262.

Van de Vyver, G. 1970b. La non confluence intraspecifique chez les spongiares et la notion d'individu. J Elisha Mitchell Sci Soc, 23: 161–174.

Van Soest, R.W.M., Boury-Esnault, N., Vacelet, J. et al. 2012. Global diversity of sponges (Porifera). PLoS ONE, 7: e35105.

Varadaraj, K. and Pandian, T.J. 1989. Induction of allotriploidy in the hybrid *Oreochromis mossambicus* female X red tilapia male. Proc Ind Acad Sci, 98: 351–358.

Vargas, R.I. and Chang, H.B. 1991. Evolution of oviposition similarities for mass production of melon fly, oriental fruit fly and Mediterranean fruit fly (Diptera: Tephritidae). J Econ Ent, 84: 1695–1698.

Vences, M. and Kohler, J. 2008. Global diversity of amphibians (Amphibia) in freshwater. Hydrobiolgia, 595: 569–580.

Veron, G., Patterson, B.D. and Reeves, R. 2008. Global diversity of mammals (Mammalia) in freshwater. Hydrobiologia, 595: 607–617.

Vianey-Liaud, M. 1995. Bias in the production of heterozygous pigmented embryos from successively mated *Biomphalaria glabrata* (Gastropoda: Pulmonata). Malacol Rev, 28: 97–106.

Vidal, M.A., Soto, E.R.M. and Veloso, A. 2009. Biogeograpy of Chilean herpetofauna: Distributional patterns of species richness and endemism. Amphibia-Reptilia, 30: 151–171.

Vidal-Martinez, V.M., Torres-Irineo, E. and Aguirre-Macedo, M.L. 2016. A century (1914–2014) of studies on marine fish parasites published in The Journal of Parasitology. In: *A Century of Parasitology: Discoveries, Ideas and Lesson Learned by Scientists*. (eds) Jonovy, J. Jr. and Esch, G.W., John Wiley, pp 57–74.

Vilela, D.A.R., Silva, G.B., Pexito, M.T.D. et al. 2003. Spermatogenesis in teleosts: insights from Nile tilapia (*Oreochromisnilotica*) model. Fish Physiol Biochem, 28: 187–150.

Villalobos, F.B. 2005. Reproduction and larval biology of North Atlantic asteroids related to the invasion of the deep sea. Ph.D. Thesis, University of Southampton.

Vincent, M. and Thomas, J. 2011. *Kryptoglanisshajii*, an enigmatic subterranean-spring catfish (Siluriformes, *Incertaesedis*) from Kerala, India. Ichthyol Res, 58: 161–165.

Vitturi, R. and Catalano, E. 1988. The male XO sex-determining mechanism in *Theodoxusmeridionalis* (Linnaeus, 1758) (Mollusca: Prosobranchia). J Hered, 89: 538–543.

Vivekanandan, E., Ali, H., Jasper, B. and Rajagopalan, M. 2009. Vulnerability of corals to warming of Indian Seas: A projection for the 21st Century. Curr Sci, 97: 1654–1658.

Vivekanandan, E. 2011. *Climate Change and Indian Marine Fisheries*. Central Marine Fisheries Research Institute, Kochi, Spl Pub, 105: 1–97.

Vivekanandan, E. and Jeyabaskaran, R. 2012. *Marine Mammal Species of India*. Central Marine Fisheries Research Institute, Kochi, p 228.

Volff, J.N., Nanda, I., Schmid, M. and Schartl, M. 2007. Governing sex determination in fish: Regulatory putches and ephemeral dictators. Sex Dev, 1: 85–99.

Vreys, C. and Michiels, N.K. 1998. Sperm trading by volume in a hermaphroditic flatworm with mutual penis intromission. Anim Behav, 56: 777–785.

Vrijenhoek, R.C., Dawley, R.M., Cole, C.J. and Bogart, J.P. 1989. A list of known unisexual vertebrates. In: *Evolution and Ecology of Unisexual Vertebrates*. (eds) Dawley, R. and Bogart, J.P., New York State Mus Bull, 466: 19–23.

Wakayama, T. and Yanagimachi, R. 1998. Development of normal mice from oocytes injected with freeze-dried spermatozoa. Nat Biotech, 16: 639–641.

Waldschmidt, S.R., Jones, S.M. and Porter, W.P. 1987. Reptilia. In: *Animal Energetics: Bivalvia through Reptilia*. (eds) Pandian, T.J. and Vernberg, F.J., Academic Press, pp 553–620.

Walker, C.W., Unuma, T., McGinn, N.A. et al. 2001. Reproduction in sea urchins. In: *Edible Sea Urchins: Biology and Ecology*. (ed) Lawrence, J.M., Elsevier, Amsterdam, pp 5–26.

Wallace, J.B., Webster, J.R. and Woodall, W.R. 1977. The role of filter feeders in flowing waters. Arch Hydrobiol, 79: 506–532.

Wallace, R.A. 1991. *Biology: The World of Life*. Harper Collin Publishers, USA, p 695.

Wallace, R.L. and Snell, T.W. 2010. Rotifera. In: *Ecology and Classification of North American Freshwater Invertebrates*. (eds) Thorp, J.H. and Covich, A.P., Elsevier, pp 173–235.

Walne, P.R. 1964. Observation on the fertility of the oyster (*Ostrea edulis*). J Mar Biol Ass UK, 44: 293–310.

Wang, H., Matzke-Karasz, R., Horne, D.J. et al. 2020. Exceptional preservation of reproductive organs and giant sperm in Cretaceous ostracods. Proc R Soc, 287B: 2020661.

Wang, J., Vanga, S.K. and Saxena, R. et al. 2018. Effect of climate change on the yield of cereal crops: A review. Climate, 6: 41, doi: 10.3390/cli6020041.

Wang, K., Shen, Y., Gan, X. et al. 2019. Morphology and genome of a snailfish from the Mariana Trench provide insights into deep-sea adaptation. Nat Ecol Evol, 3: 823–833.

Wang, M., Hour, J. and Lei, Y. 2014. Classification of Tibetan lakes based on variations in seasonal lake water temperature. Chi Sci Bull, 49: 4847–4855.

Wanninger, A. 2015. Entoprocta. In: *Evolutionary Developmental Biology of Invertebrates*. (ed) Wanninger, A., Springer Verlag, Wein, 2: 89–101.

Warburg, M.R. and Cohen, N. 1991. Reproductive pattern, allocation, and potential in a semelparous isopod from the Mediterranean region of Israel. J Crust Biol, 11: 368–374.

Warburton, F.E. 1958. Reproduction of fused larvae in the boring sponge *Cliona celata* Grant. Nature, 181: 493–494.

Warburton, F.E. 1961. Inclusion of parental somatic cells in sponge larvae. Nature, 191: 1317.

Warburton, F.E. 1966. The behavior of sponge larvae. Ecology, 47: 672–674.

Ward, D. 2009. *Biology of Deserts*. Oxford University Press, p 352.

Wats, S. 2003. *Lupinus arcticus*: Growing and using native plans in the northern interior of B.C. Symbios Research & Restoration.

Watson, J.D. and Crick, F.H.C. 1953. The structure of DNA. Cold Spring Harb Symp Quant Biol, 18: 123–131.

Watts, H.E. and Holekamp, K.E. 2009. Ecological determinants of survival and reproduction in the spotter hyena. J Mammol, 90: 461–471.

Weeks, S.C., Sanderson, T.J., Zofkova, M. and Knott, B. 2008. Breeding systems in the clam shrimp family Limnadiidae (Branchiopoda, Spinicaudata). Invert Biol, 127: 336–349.

Wennberg, S. 2008. Aspects of priapulid development. Uppsala University Dissertation, Faculty of Science and Technology, p 451.

West, H.H., Harrington, F. and Pierce, S.K. 1984. Hybridization of two populations of marine opisthobranchs with different developmental patterns. Am Malacol Bull, 9: 9–12.

Westheide, W. 1969. Spermatodes meneineinfacher spermatophorentypbeiinterstitiellen Polychaeter *Hesinida gohari* Hartmann-Schroeder (hesionidae). Natuwissenschaffen, 12: 641–642.

Weygoldt, P. 1990. Arthropoda–Chelicerata. In: *Reproductive Biology of Invertebrates*. (eds) Adiyodi, K.G. and Adiyodi, R.G., Oxford IBH Publishing, New Delhi, 4B: 77–119.

Whittington, I.D. and Horton, M.A. 1996. A revision of *Neobenedenia* Yamaguti, 1963 (Monogenea: Capsalidae) including a redescription of *N. melleni* (MacCallum, 1927) Yamaguti 1963. J Nat Hist, 30: 1113–1156.

Williams, J.D. and Boyko, C.B. 2012. The global diversity of parasitic isopods associated with crustacean hosts (Isopoda: Bopyroidea and Cryptoniscoidea). PLOS ONE, 7: e35350.

Wilson, G.D.F. 2008. Global diversity of isopod crustaceans (Crustacea: Isopoda) in freshwater. In: *Freshwater Animal Diversity Assessment*. (eds) Balian, E.V., Leveque, C., Segers, H. and Martens, K., Springer, Netherlands, pp 231–240.

Wilson, H.V. 1907. On some phenomena of coalescence and regeneration in sponges. J Exp Zool, 5: 245–258.

Wilson, W.H. 1991. Sexual reproductive modes in polychaetes: Classification and diversity. Bull Mar Sci, 48: 500–516.

Windsor, D.A. 1998. Most of the species on earth are parasites. Int J Parasitol, 28: 1939–1942.

Wong, J.W.Y., Meunier, J. and Kolliker, M. 2013. The evolution of parental care in insects: the roles of ecology, life history and the social environment. Ecol Ent, 38: 123–137.

Wong, T.-T., Saito, T., Crodian, J. et al. 2010. Zebrafish germline chimeras produced by transplantation of ovarian germ cells sterile host larva. Biol Reprod, 84: 1190–1197.

Wootton, R.J. 1973. The effect of size of food ration on egg production in the female three-spined stickleback *Gasterosteus aculeatus*. J Fish Biol, 5: 89–96.

Wright, J.C. 2012. Myriapoda (Including Centipedes and Millipedes). In: *Evolution & Diversity of Life*. John Wiley, Chichester, DOI: 10.1002/9780470015902.a0001607.pub3.

Wulff, J.L. 1990. Patterns and processes of size change in Caribbean desmosponges of branching morphology. In: *New Perspectives in Sponge Biology*. (ed) Rutzler, K., Smithsonian Inst. Press, Washington, D.C., pp 425–435.

Wulff, J.L. 1991. Asexual fragmentation, genotype success, and population dynamics of erect branching sponges. J Exp Mar Biol Ecol, 149: 227–247.

Ya'cob, Z., Takaoka, H., Pramual, P. et al. 2016. Breeding habitat preference of preimaginal black flies (Diptera: Simuliidae) in Peninsular Malaysia. Acta Tropica, 153: 57–63.

Yamaha, E., Murakami, M., Hada, K. et al. 2003. Recovery of fertility in male hybrids of cross between goldfish and common carp by transplantation of PGC (Primordial Germ Cell) containing graft. Genetica, 119: 121–131.

Yamaki, M., Kawakami, K., Taniura, K. and Arai, K. 1999. Live haploid mosaic charr *Salvelinus leucomaeus*. Fish Sci, 65: 736–741.

Yasui, G.S., Fujimato, T., Sakao, S. et al. 2011. Production of loach (*Misgurnus anguillicaudatus*) germline chimera using transplantation of primordial germ cells isolated from cyropreserved blastomeres. J Anim Sci, 89: 2380–2388.

Yokouchi, K., Fukuda, N., Miller, M.J. et al. 2012. Influence of early habitat use on the migratory plasticity and demography of Japanese eels in central Japan. Estuar Coast Shelf Sci, 107: 132–140.

Yoon, C., Kawakami, K. and Hopkins, N. 1997. Zebrafish *vasa* homologue RNA is localized to the cleavage planes of 2- and 4-cell stage embryos and expressed in the primordial germ cells. Development, 124: 3157–3166.

Yoshizaki, G., Takeuchi, G., Sakatani, S. and Takeuchi, T. 2000. Germline specific expression of green fluorescent protein in transgenic rainbow trout under control of the rainbow trout *vasa*-like promoter. Int J Dev Biol, 44: 323–326.

Yusa, Y., Yamato, S. and Marumura, M. 2001. Ecology of parasitic bannacle *Koleolepas avis*; relationships to the hosts distribution left-right symmetry and reproduction. J Mar Biol, Ass UK, 81: 781–788.

Zattara, E.E. and Bely, a.E. 2016. Phylogenetic distribution of regeneration and sexual reproduction in annelids: regeneration is ancestral and fission involves regenerative clades. Invert Biol, 135: 409–414.

Zervos, S. 1988. Population dynamics of thelastomatid nematode of cockroaches. Parasitology, 96: 353–358.

Zhang, Z.-Q. 2011. Animal biodiversity: An introduction to higher-level classification and taxonomic richness. Zootaxa, 3148: 7–12.

Zhang, Z.-Q. 2013. Animal biodiversity: An update of classification and diversity in 2013. Zootaxa, 3703: 5–11.

Zhao, S., Dai, Q. and Fu, J. 2009. Do rivers function as genetic barriers for the plateau wood frog at high elevations? J Zool, 279: 270–276.

Zigler, K.S. and Lessios, H.A. 2003. 250 million years of bindin evolution. Biol Bull, 205: 8–15.

Zobell, C.E. 1954. Some effects of high hydrostatic pressure on apparatus observed on the Danish Galathea Deep-sea Expedition. Deep-Sea Res, 2: 24–32.

Zouros, E., Ball, A.O., Saavedra, C. and Freeman, K.R. 1994a. Mitochondrial DNA inheritance. Nature, 368: 818.

Zouros, E., Oherhanser, B.A., Saavedra, C. and Freeman, K.R. 1994b. An unusual type of mitochondrial DNA inheritance in the blue mussel *Mytilus*. Proc Natl Acad Sci USA, 91: 7463–7467.

Zupo, V. 2000. Effect of microalgal food on the sex reversal of *Hippolyte inermis* (Crustacea, Decapoda). Mar Ecol Prog Ser, 201: 251–259.

Zupo, V. 2001. Influence of diet on sex differentiation of *Hippolyte inermis* (Crustacea, Natantia) in the field. Hydrobiologia, 449: 131–140.

Author Index

Species Index

A

Abies squamata, 37
Abyssidrilus stilus, 23-24
Abyssothyris wyvillee, 24
Acanthina punctulata, 46
Acanthinucella spirata, 127
Acanthinula, 75
Acanthochromis polyacanthus, 214, 279
Acanthocyclops vernalis, 138
Acanthopagrus schlegelii, 244-245
Acanthurus, 231
A. nigrofuscus, 181, 184
Acartia centura, 138
Acartia clause, 269
A. erythreae, 269
A. keralensis, 138
A. negligens, 138
A. plumosa, 138
A. spinicauda, 138
A. steueri, 269
A. tonsa, 184, 269
Acartiella gravelyi, 138
Acaulus, 100
Acrocalanus gracilis, 246
Acropora tenuis, 79
A. valida, 79
Acrosternum graminea, 215
Adalaria proxima, 206
Aedes aegypti, 185
Aeolosoma viridae, 96
Aiolopus thallassimus, 215
Albatrossaster richardi, 24
Alcyanidium duplex, 71
Alderia willowi, 219
Allomuraytrema robustum, 213, 220
Allostichaster insignis, 106
Alona pulchella, 268
Alosa pseudoharengus, 165
A. sapidissima, 165
Alpheus angulosus, 152
A. armatus, 152

A. heterochaelis, 152
A. roquensis, 152
Alsophila pometaria, 162-163
Alutera, 231
Amanses, 231
Amnicola limosa, 189
Amphipholis squamata, 82, 229
Amphiporus lactifloreus, 242
Amphiprion, 159, 187
A. alkalopsis, 159
Amphiprion bicinctus, 159
Amphiura filiformis, 172
A. monorina, 70
Anabas scandens, 57
Ancylus fluviatilis, 81
Angraecum sesquipedale, 41
Anguilla anguilla, 239
Anilocra apogyne, 229
A. fontinalis, 76
Anisogammarus anandalei, 138
Anodonta, 76, 238
Anopheles, 58
Antechinus, 162
Antedon bifida, 172
Anthonomus grandis, 159
Anthopleura elongantissima, 100
A. stellula, 100
Aphrocallistes, 102
Apis, 185
Apis mellifera, 45, 160
Aplysia californica, 81, 90, 157
Aplysina cavernicola, 126
Apogon lineatus, 200
Apogonichthys menesmus, 214
A. waikiki, 214
Aporrectodea rosea, 137
Arachnopusia monoceros, 24
Aracia sinaloae, 82
Arbacia punctulata, 217-218
Arctopsyche irrorata, 52
Arenaria bryophylla, 37
Arenicola marina, 181, 184

Subject Index

Author's Biography

Recipient of the S.S. Bhatnagar Prize, the highest Indian award for scientists, one of the ten National Professorships, T.J. Pandian has served as editor/ member of editorial boards of many international journals. His books on Animal Energetics (Academic Press) identify him as a prolific but precise writer. His five volumes on Sexuality, Sex Determination and Differentiation in Fishes, published by CRC Press, are ranked with five stars. He has authored a multi-volume series on Reproduction and Development of Aquatic 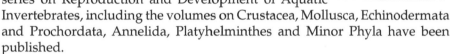 Invertebrates, including the volumes on Crustacea, Mollusca, Echinodermata and Prochordata, Annelida, Platyhelminthes and Minor Phyla have been published.

Milton Keynes UK
Ingram Content Group UK Ltd.
UKHW031143141024
449569UK00024B/1120